Food Protection and Security

Food Protection and Security

Preventing and Mitigating Contamination during Food Processing and Production

Edited by

S. Kennedy

ELSEVIER

AMSTERDAM • BOSTON • HEIDELBERG • LONDON
NEW YORK • OXFORD • PARIS • SAN DIEGO
SAN FRANCISCO • SINGAPORE • SYDNEY • TOKYO
Woodhead Publishing is an imprint of Elsevier

WP
WOODHEAD PUBLISHING

Woodhead Publishing is an imprint of Elsevier
The Officers' Mess Business Centre, Royston Road, Duxford, CB22 4QH, United Kingdom
50 Hampshire Street, 5th Floor, Cambridge, MA 02139, United States
The Boulevard, Langford Lane, Kidlington, OX5 1GB, United Kingdom

Notice
Knowledge and best practice in this field are constantly changing. As new research and experience broaden our understanding, changes in research methods, professional practices, or medical treatment may become necessary.

Practitioners and researchers must always rely on their own experience and knowledge in evaluating and using any information, methods, compounds, or experiments described herein. In using such information or methods they should be mindful of their own safety and the safety of others, including parties for whom they have a professional responsibility.

To the fullest extent of the law, neither the Publisher nor the authors, contributors, or editors, assume any liability for any injury and/or damage to persons or property as a matter of products liability, negligence or otherwise, or from any use or operation of any methods, products, instructions, or ideas contained in the material herein.

British Library Cataloguing-in-Publication Data
A catalogue record for this book is available from the British Library

Library of Congress Cataloging-in-Publication Data
A catalog record for this book is available from the Library of Congress

ISBN: 978-1-78242-251-8 (print)
ISBN: 978-1-78242-261-7 (online)

Publisher: Nikki Levy
Acquisition Editor: Rob Sykes
Editorial Project Manager: Karen Miller
Production Project Manager: Kirsty Halterman and Karen East
Designer: Victoria Pearson

Contents

3. Creating a Food Defense/Response Plan in Food Processing Facilities

C.L. Lorenzen and C.N. Cutter

4. Creating a Food Defense and Response Plan in Complex Food Production Systems

M. Seeger, T. Sellnow and E.L. Petrun

8. Personal Hygiene and Good Maintenance Practices for the Servicing of Food Processing Equipment

F. Moerman

List of Contributors

J. Barros-Velázquez University of Santiago de Compostela, Lugo, Spain

K. Böhme Agrifood Technological Centre of Lugo (CETAL), Lugo, Spain

P. Calo-Mata University of Santiago de Compostela, Lugo, Spain

C.N. Cutter Pennsylvania State University, University Park, PA, United States

K. Everstine University of Minnesota, Minneapolis, MN, United States

N.P. Greis University of North Carolina at Chapel Hill, Chapel Hill, NC, United States

C.L. Lorenzen University of Missouri, Columbia, MO, United States

K. Lorenzen European Hygienic Engineering & Design Group, Frankfurt, Germany

F. Moerman Catholic University of Leuven – KU Leuven, Leuven, Belgium

M.L. Nogueira University of North Carolina at Chapel Hill, Chapel Hill, NC, United States

E.L. Petrun University of Maryland, College Park, MD, United States

M. Seeger Wayne State University, Detroit, MI, United States

T. Sellnow University of Central Florida, Orlando, FL, United States

Chapter 1

Supply Chain Complexity and Economically Motivated Adulteration

K. Everstine
University of Minnesota, Minneapolis, MN, United States

Globalization of our food supply increases many types of risk, not the least of which is the risk of economically motivated adulteration (EMA) or food fraud (the intentional adulteration or misrepresentation of food for economic gain). Increasing complexity reduces the ability of both regulators and industry to effectively oversee food supply chains. A brief description of each of the main themes included in this chapter is given here.

- *Regulatory and supply chain control challenges and globalization:* this section will provide a brief background and examples of some of the challenges in overseeing increasingly globalized food supply chains.
- *Economically motivated adulteration and food fraud: definitions and scope:* this section will define EMA and food fraud, discuss what is currently known about the scope of the problem given the available data, and describe various methods of perpetrating EMA.
- *Drivers of EMA opportunity and incentive:* this section will discuss the factors that drive the opportunity for EMA and the incentive behind EMA.
- *Assessing the vulnerability of foods and ingredients to EMA:* EMA risk cannot be assessed and mitigated using traditional food safety control frameworks. This section will present the general framework for evaluating EMA vulnerability in foods and ingredients, and briefly discuss one guidance document created for use by industry in conducting food fraud vulnerability assessments.
- *Future trends: legislation and EMA mitigation efforts:* the development of risk mitigation methods for industry and government, as well as new regulations for EMA control, will continue to evolve over the coming years around the world. This section will focus on recent developments in US-based legislation and one UK government-commissioned report to highlight future trends.

Food Protection and Security. DOI: http://dx.doi.org/10.1016/B978-1-78242-251-8.00001-1

1.1 REGULATORY AND SUPPLY CHAIN CONTROL CHALLENGES AND GLOBALIZATION

Our food supply is becoming increasingly globalized. Imports of food and agricultural commodities in many developed countries are rising. The US Food and Drug Administration (FDA) reported increases of 10% per year in shipments of FDA-regulated foods between 2002 and 2009 (United States Food and Drug Administration, 2011). The percentage of volume of US food consumption attributed to imports rose from 11% in 1989 to almost 17% in 2009 (http://www.ers.usda.gov/topics/international-markets-trade/us-agricultural-trade/import-share-of-consumption.aspx). A 2010 FDA report projected that future growth in imports of regulated products would exceed growth of domestic products. The European Union (EU) is the top agricultural importer, by value (European Commission, 2013). Total agricultural imports into the EU increased an estimated 24% between 2000 and 2008 (von Witzke and Noleppa, 2010). Globalization of the food supply facilitates market growth and consolidation, gives populations in many countries a year-round supply of food products that cannot be grown domestically, and can help drive down production costs. It can also result in long, interconnected, multinode and complex supply chains, which can be challenging to oversee and regulate.

The United States has less direct regulatory oversight for imported food products than those that are domestically produced. Foreign facility inspections are more expensive than domestic inspections, and the FDA performs inspections of foreign food facilities at a far lower rate than domestic facilities (United States Food and Drug Administration, 2011). Globalization of the food supply has dramatically increased the distances that food products and ingredients travel, as well as the number of intermediate parties between primary production and the ultimate consumer ("farm to fork"). In 2005, a large recall was initiated in the United Kingdom as a result of contamination of chili powder with the industrial dye Sudan 1 (http://www.theguardian.com/society/2005/feb/23/food.foodanddrink1). The chili powder supply chain involved transactions among at least six different companies in India and the United Kingdom over a time period of more than two years (http://www.telegraph.co.uk/news/uknews/1484427/Tracking-down-the-rogue-powder.html). The chili powder was subsequently used in the production of Worcestershire sauce, which was then sold as an ingredient to at least 60 manufacturers and suppliers and incorporated into more than 600 finished food products. Fig. 1.1 shows a visual representation of the complexity and breadth of the reported supply chain for the recalled chili powder. In this example, various factors contributed to the loss of oversight of the supply chain and hindered trace-back and trace-forward investigations. These factors included the number of intermediate parties, the lack of transparency throughout the supply chain, the physical nature of the product (it was sold

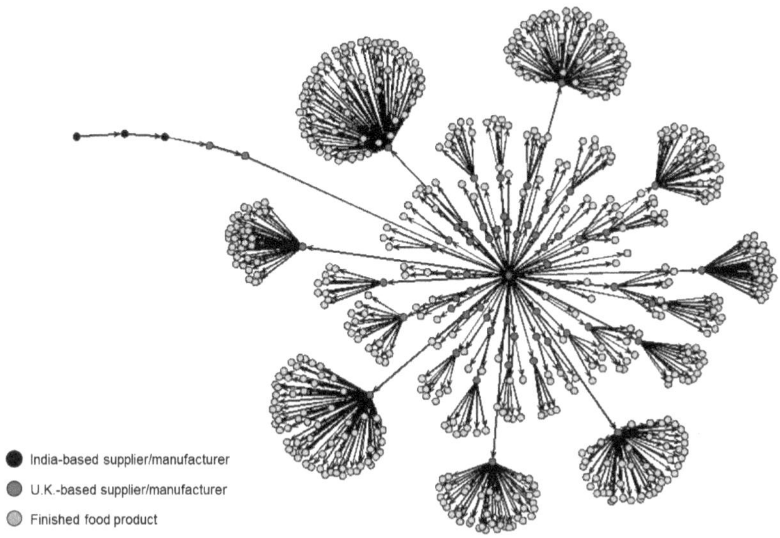

FIGURE 1.1 Visual representation of the reported supply chain for recalled chili powder. *Data sources: Food Standards Agency of the U.K. National Archives and* The Guardian.

in powdered form), and the amount of time that elapsed between production and ultimate retail sale. The recall of more than 600 finished food products cost the United Kingdom an estimated £100 million. In a separate example, the tragic 2008 incident of melamine contamination of dairy supplies in China, the resulting illnesses in hundreds of thousands of infants were confined to China. However, public health, laboratory, regulatory, and other government resources throughout the world were needed to respond to the incident, conduct product testing and recalls, and determine the risk of human exposure to melamine (http://www.ncbi.nlm.nih.gov/pmc/articles/PMC2799451/). Ultimately, food product recalls occurred in at least 47 countries.

Arguably, the increasingly complex nature of supply chains for food products increases the risk of contamination, both unintentional and intentional. It also increases the burden of ensuring authenticity along the supply chain. Routine laboratory testing of ingredients for quality assurance can be costly, and it is impractical to test food ingredients for a wide range of adulterants during each transaction along the supply chain. Therefore, one of the risks that increase when supply chain oversight and visibility are reduced is economically motivated adulteration or food fraud (Everstine et al., 2013). In a 2010 report, the FDA cited EMA as "perhaps the most serious challenge on the horizon" for the foods, drugs, and medical devices that the agency regulates (United States Food and Drug Administration, 2011). The 2013 horse meat adulteration incident in Europe prompted a UK review of the systems that assure food integrity and outlined a number of recommendations to improve the deterrence, detection, and prevention of food fraud and food crime (HM Government, 2014). As noted in the report, "much less attention has been focused on food authenticity, food fraud and food crime" than on food safety, and there is a need to "protect consumers and honest businesses through an effective regulatory framework."

1.2 ECONOMICALLY MOTIVATED ADULTERATION AND FOOD FRAUD: DEFINITIONS AND SCOPE

FDA adopted the term "economically motivated adulteration" to refer to what is more commonly known as food fraud. The FDA proposed the following working definition of EMA: "fraudulent, intentional substitution or addition of a substance in a product for the purpose of increasing the apparent value of the product or reducing the cost of its production, i.e., for economic gain" (Lutter, 2009). This appears to limit the definition of EMA to incidents involving the addition or substitution of a substance. However, FDA's general definition of "adulteration" is more comprehensive and includes additional forms of misrepresentation of food. Per the US Federal Food Drug and Cosmetic Act, a food is deemed "adulterated" if it is contaminated with a potentially poisonous substance or otherwise may be "injurious to health"; if any valuable constituent has been omitted, substituted, added to increase bulk or weight, or if "damage or inferiority has been concealed in any manner"; and if it was previously denied admission into the United States (21 U.S.C.). Furthermore, food is deemed "misbranded" if it bears a false or misleading label or container, is offered for sale under another name, and does not conform to the standard of identity that it represents. Using the example of the sale of a product labeled as olive oil that actually consists of soybean oil, this form of fraud could either be viewed as adulteration of olive oil with soybean oil, or misbranding of soybean oil as olive oil. A 1966 review of legal cases of "economic adulteration" noted the overlap between the two definitions, and stated that the "statutory provisions [related to economic adulteration] in the act are general, vague, complex, and abstruse." The author reviewed multiple cases in which the courts issued "diverse and conflicting opinions" and concluded that there was a "patent and immediate need for a revised economic adulteration statute" (Forte, 1966).

The consequence of EMA and/or food fraud is that the purchaser does not have accurate information about the true identity of the product. Although the FDA definition of EMA appears more limited than the definition of food fraud, the agency ultimately has regulatory status over all forms of food misrepresentation. Furthermore, in the event of an EMA or food fraud incident, a determination of the specifics of the infraction and the type of adulteration or misbranding is typically decided by the courts. Therefore, a clear distinction between the terms "food fraud" and "EMA" is both challenging and impractical. The Food Standards Agency of the United Kingdom and the Grocery Manufacturers Association in the United States each define food fraud as "deliberately placing food on the market, for financial gain, with the intention of deceiving the consumer" (HM Government, 2014; Grocery Manufacturers Association and Kearney, 2010). The "consumer" may be a food company purchasing from a supplier, another intermediary party along a food supply chain, or the ultimate purchaser of the product at retail. Herein, the terms "EMA" and "food fraud" will be used interchangeably to refer to the intentional adulteration or misrepresentation of foods or food ingredients for economic gain.

EMA perpetrators do not intend to cause illnesses or deaths in consumers. Health effects in consumers may result in detection of the adulteration and subsequent investigation, while the intent of the act is deception and financial gain. However, errors and misjudgments by EMA perpetrators have occurred. Concerns about fraudulent food have been increasing over the past decade, particularly in response to three incidents with broad health and economic consequences. In 2007, pet foods were recalled in the United States and other countries following melamine contamination of Chinese-produced wheat gluten used as an ingredient (http://www.ncbi.nlm.nih.gov/pubmed/18689873). Although no human illnesses were known to occur, thousands of pets in the United States became ill or died and melamine-contaminated animal feed entered the human food supply chain. Following that incident, in 2008, melamine contamination again caused widespread recalls, this time resulting from contamination of dairy supplies in China. This incident resulted in hundreds of thousands of illnesses and the deaths of at least six infants who consumed formula produced with adulterated milk (http://www.sciencemag.org/content/322/5906/1310). Most recently, in 2013, horse meat contamination of ground beef in Europe resulted in the recall of 50,000 tons of meat and affected more than 20 brands. Although no human illnesses are known to have resulted from consumption of horse meat, government health authorities expended substantial resources conducting a risk assessment for the presence of drug residues in horse meat and the potential risk for consumers.

EMA happens in a variety of ways. Research at the Food Protection and Defense Institute at the University of Minnesota defined the following methods of EMA (http://www.foodfraudresources.com/ema-incidents):

Substitution: complete replacement of a food product/ingredient with an alternate food product/ingredient. One example of substitution is the intentional misrepresentation of fish fillets as an alternate and more expensive species.

Dilution: partial replacement of a food product/ingredient with an alternate food product/ingredient. This includes the addition of an alternate ingredient to increase the overall weight or volume. Examples include the dilution of honey with other sugar syrups and the dilution of extra virgin olive oil with lower quality or alternate oils.

Artificial Enhancement: the addition of an unapproved chemical additive to artificially enhance the quality of a product. These types of additives can include industrial dyes, fungicides, artificial ripening agents, etc. One example is the addition of Sudan dyes to chili powder to enhance the bright red color of the spice.

Mislabeling: intentional misrepresentation with respect to harvesting or processing techniques or other quality attributes. Examples include false labeling of organic and/or cage-free eggs and misrepresentation of halal or kosher processing of meats.

Transshipment/Origin Masking: misrepresentation of the geographic origin of a product. This can happen through false declaration of customs documents or mislabeling at retail. The shipment of Chinese-origin honey through intermediate countries and subsequent false labeling upon entry into the United States is one example of transshipment; this allows perpetrators to avoid antidumping duties placed on Chinese honey by the US International Trade Commission.

Counterfeit: fraudulent labeling of a product by an unauthorized party as a brand-name product. Examples include the fraudulent production, labeling and sale of brand-name infant formula.

Theft and Resale: theft of a food product and resale into commerce through unapproved channels. Theft of products can occur at retail or prior to retail ("cargo theft"). One example is the theft of infant formula from grocery stores and subsequent resale to small retail establishments or customers.

Intentional Distribution of Contaminated Product: the intentional sale of a product despite knowledge of foodborne contamination. One example is the falsification of documents and subsequent sale of *Salmonella*-contaminated peanut products that resulted in hundreds of illnesses in the United States in 2008.

EMA is a concern for many reasons. From a public health and food protection perspective, past incidents have illustrated that unintended health consequences can result and they can be catastrophic. EMA results in a loss of supply chain transparency and control, and therefore may hinder traceability efforts. Food safety programs such as Hazard Analysis and Critical Control Points (HACCP) are based on knowledge of the true identity of a food product. Therefore, EMA compromises food safety efforts and regulatory oversight. EMA can also have a negative effect on markets through evasion of antidumping duties and creation of unfair competition. Finally, EMA reduces consumers' ability to make informed food choices.

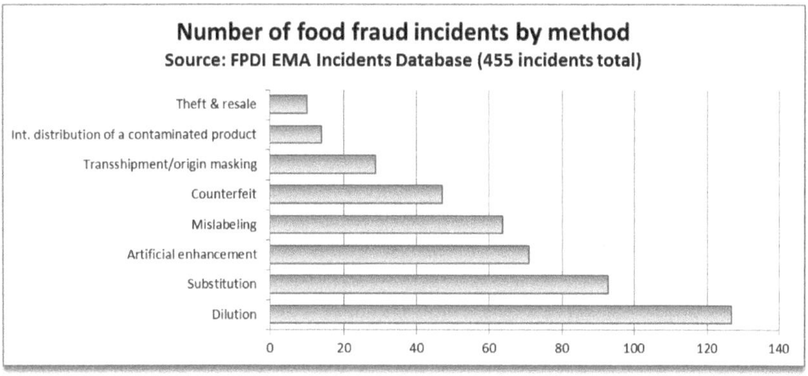

FIGURE 1.2 The number of EMA incidents by method of adulteration. *Source: FPDI EMA Incidents Database, November 2015 (http://www.foodshield.org).*

We cannot reliably estimate the true scope of EMA in foods since most incidents are likely undetected or unreported. A 2014 report based on research conducted in the United Kingdom estimated that only between 3% and 4% of fraud is detected (Gee et al., 2014). There are various databases that catalogue records of food fraud, including the FPDI EMA Incidents Database (https://www.foodshield.org/index.cfm/discover-tools-links/tools/), the United States Pharmacopeia (USP) Food Fraud Database (http://www.foodfraud.org/), the Rapid Alert System for Food and Feed (RASFF) portal (http://ec.europa.eu/food/safety/rasff/index_en.htm), and the Food Standards Agency Food Fraud Database (https://www.food.gov.uk/enforcement/enforcework/foodfraud/foodfrauddatabase). Information from the first three databases is currently either publicly available or available upon request. Each database uses different criteria for compiling records, and each provides a unique perspective on the true incidence of EMA. There is general agreement about the food product categories that appear, from available data, to be the most prone to fraud. These categories include oils, spices, milk, fruit juices, honey, seafood, alcoholic beverages, and grains (Johnson, 2014).

The FPDI EMA Incidents Database, a US-based repository, contains more than 400 records of publicly documented EMA incidents. The incidents catalogued in the database are compiled through manual searches of media sources, scientific journals, and publicly available data from regulatory agencies. The database is intended to provide contextual information about the EMA incidents, such as food product category, location of production and distribution, the number of illnesses and deaths, and the method of adulteration. Fig. 1.2 shows the number of incidents in the database attributed to each EMA method. Dilution and substitution account for about half of the total incidents and artificial enhancement accounts for about 16% of incidents.

1.3 DRIVERS OF EMA OPPORTUNITY AND INCENTIVE

There are strong economic incentives behind food fraud. In the 1980s, executives of Beech-Nut Corporation were found guilty of violating US federal laws by selling adulterated and misbranded apple juice (Buder, 1988). The company purchased fraudulent apple juice concentrate at 20–25% the price of competing products, enabling the company to maintain profitability. The addition of melamine to milk in China was an efficient and inexpensive means of increasing the apparent protein content of the milk. Melamine contains 67% nitrogen while protein contains about 16% nitrogen, and milk contains about 3.5% protein. Therefore, the addition of small quantities of inexpensive scrap melamine could substantially increase the apparent protein content of milk, and potentially allow dilution with water. This enabled producers of substandard milk that did not meet protein content requirements to continue to sell their supplies. Finally, using estimated 2013 costs of €4.00/kg for beef and €1.00/kg for horse meat, substitution of only 5% horse meat into ground beef would have saved €15,000.00 per 100,000 kg of beef.

Although the underlying motivation is economic, there are many factors that drive the incentive and the opportunity for EMA (http://www.ncbi.nlm.nih.gov/pubmed/23575142). The incentive may be driven by business or market pressures, price fluctuations, increases or decreases in rates of duty, and changes in supply or demand of ingredients. The opportunity to perpetrate EMA can be driven by many factors, including

- supply chain characteristics and oversight (e.g., the degree of vertical integration or decentralization, the number of intermediate parties involved, and the number and type of controls in place along the supply chain);
- the availability of effective analytical methods for testing food ingredients and the cost of these methods;
- the existence of federal standards of identity or other industry-wide standards;
- the prevalence of use of third-party or shared auditing programs; and
- the existence and degree of active involvement of industry trade organizations.

Given the intentional nature of EMA and the various factors that drive the incentive and opportunity for fraud, risk assessment strategies tailored toward unintentional foodborne contamination will not lead to effective risk management programs for EMA. The general consensus in recent years is that regulatory agencies and food companies should conduct EMA vulnerability assessments for food products, in order to prioritize resource allocation to those products most vulnerable to fraud.

1.4 ASSESSING THE VULNERABILITY OF FOODS AND INGREDIENTS TO EMA

EMA presents a different set of challenges from those involved in preventing either unintentional foodborne contamination or terrorism through intentional contamination of food supplies. With both unintentional foodborne contamination and terrorism, the identification of consumer illnesses or deaths prompts an outbreak investigation which leads to the identification of the contaminated food product. Recovery efforts and implementation of future preventive controls efforts may be put in place following the event. The food safety model, in particular, is based on detection of an expected set of contaminants. HACCP and other food safety programs, as well as public health disease surveillance systems, are typically built around well-characterized risks and robust sampling data.

The risk of intentional adulteration of food supplies for ideological reasons (terrorism, political protest, etc.) is usually addressed through facility-level vulnerability assessments, food sector vulnerability assessments, and the development of agent detection methods. Risk assessment for intentional adulteration usually includes an analysis of the most likely threat agents. The risk of intentional adulteration of the food supply is considered to be very low, but could cause very serious consequences.

In contrast to unintentional foodborne contamination and ideologically motivated intentional contamination, EMA involves perpetrators who are intent on evading detection (and, therefore, avoiding immediate health effects). They also usually have knowledge of the existing food safety structure. The risk of EMA in the food supply is much higher than that of intentional adulteration (based on the available data we have on each), but there are usually no immediate health consequences. Due to the intentional nature of EMA, the challenges with predicting human behavior, the use of a variety of adulterants, the potential for the introduction of unexpected adulterants, and a lack of available data on the true scope of the problem, the development of traditional risk assessment methods for EMA is not feasible. However, using what is known about the incentive and opportunity drivers of EMA, as well as additional factors, we can build methods for evaluating the vulnerability of food products and ingredients to EMA. This will allow regulatory agencies and food companies to include a consideration of fraud vulnerability into their existing food protection strategies, and devote additional resources to those products and ingredients determined to be the most vulnerable.

The first publicly available guidance for evaluating food fraud vulnerabilities was published in the *Third Supplement to the Food Chemicals Codex 9* by the United States Pharmacopeial Convention (USP) (United States Pharmacopeia, 2014). This guidance document describes a holistic strategy for qualitatively evaluating the vulnerability of food ingredients to fraud. It is intended for use by the food industry in evaluating their ingredient supply chains. The USP guidance document advocates an evaluation of both "controllable" factors and "uncontrollable" factors that contribute to food fraud vulnerability. Controllable factors include supply chain, audit strategy, supplier relationship, history of supplier quality and safety issues, testing frequency, and susceptibility of quality assurance methods and specifications. Uncontrollable factors include fraud history and economic anomalies. "Geopolitical considerations" is the final contributing factor, which may be controllable if there are options for choosing the geographic source of an ingredient.

Application of this general framework on a broader scale using tailored sources of information and data would allow the development of regulatory agency-level EMA vulnerability assessments for the products they oversee. The results of these vulnerability assessments could then be incorporated into existing agency-level frameworks for evaluating risk and allocating regulatory resources. Academic and nongovernmental institutions are ideally positioned to bring together multidisciplinary teams of researchers to synthesize data available from a variety of sources and develop quantitative methods for evaluating each of the contributing factors to EMA vulnerability.

1.5 FUTURE TRENDS: LEGISLATION AND EMA MITIGATION EFFORTS

As a result of multiple large-scale incidents of EMA over the past decade, countries around the world have promised to increase EMA protections in the food supply. The development of risk mitigation tools and methods for both industry and government will continue to evolve over the coming years. This section will highlight, in particular, recent developments in legislation in the United States and recent recommendations to the government in the United Kingdom. These two examples demonstrate the reluctance by government agencies to impose strict EMA regulations on industry, but also the acknowledged need for innovative approaches to risk assessment and information sharing.

The US FDA Food Safety Modernization Act (FSMA) was signed into law in 2011 and was intended to be the most substantial improvement to food safety laws in the United States in 70 years (http://www.fda.gov/Food/GuidanceRegulation/FSMA/). Since passage, the FDA has released two rules that discuss or address EMA in foods for human consumption. Although the risk of intentional adulteration has typically been addressed through vulnerability assessments and corresponding mitigation strategies, the agency concluded that intentional adulteration for economic gain would be better addressed as part of food safety plans (hazard analysis and preventive controls). As stated in the proposed rule "Focused Mitigation Strategies to

Protect Food Against Intentional Adulteration" (http://www.fda.gov/Food/GuidanceRegulation/FSMA/ucm378628.htm), "[t]he nature of economically motivated adulteration makes it difficult to identify all relevant factors to be considered in a vulnerability assessment to predict when novel events of economic adulteration are expected to occur." In September 2015, the FDA released the final rule "Current Good Manufacturing Practice, Hazard Analysis, and Risk-Based Preventive Controls for Human Food" (hereafter, the "FSMA PC rule") (http://www.fda.gov/Food/GuidanceRegulation/FSMA/ucm334115.htm). The agency included EMA mitigation efforts as part of a facility's food safety plan, indicating that EMA-associated adulterants should be included in hazard analysis, preventive controls, and supply chain programs. In addition, the agency indicated that the focus should be on those adulterants with "the potential to cause illness or injury." EMA that only affects the quality of a product, but is not a food safety hazard, was cited as outside the scope of the rule.

Following the horsemeat adulteration incident in early 2013 in Europe, the UK government requested a review of weaknesses in food supply networks and recommendations for improving the integrity of the food supply. The final report, "Elliott Review into the Integrity and Assurance of Food Supply Networks," was released in July 2014 and contained recommendations for both industry and government. Among others, these recommendations included:

- "Work with industry to ensure that opportunities for food fraud, food crime, and active mitigation are included in company risk registers"
- "Encourage industry to conduct sampling, testing and supervision of food supplies at all stages of the food supply chain"
- "Work with the industry to help it establish its own 'safe haven' to collect, collate, analyze and disseminate information and intelligence"
- "Facilitate work to standardize the approaches used by the laboratory community testing for food authenticity"
- "Work in partnership with Public Health England and local authorities with their own laboratories to consider appropriate options for an integrated shared scientific service around food standards"
- "Encourage third party accreditation bodies undertaking food sampling to incorporate surveillance sampling in unannounced audits to a sampling regime set by the standard holder" and
- "Work with industry and regulators to introduce anti-fraud auditing measures."

Most notably, the report cited the importance of the partnership among government, regulators, and industry for addressing the problem of food fraud.

The solution to the problem of EMA and food fraud must be largely industry-driven, since the food industry has primary responsibility for supply chain control and oversight. The Elliott Review recommended limited additional food fraud regulations be placed on the food industry; similarly, the EMA provisions of the FSMA PC rule are fairly narrow in scope. However, the Elliott Review also laid out clear recommendations for how government and other organizations could best facilitate the prevention of food fraud in collaboration with industry. Recently in the United Kingdom, progress has been made toward collaborative food fraud prevention efforts. Since release of the Elliott Review, a Food Crime Unit was formed in the United Kingdom, and the BRC Global Standard for Food Safety Issue 7 was released, which included updates for minimizing fraud risk. The Elliott Review also noted that one of the most important aspects of collaboration would be the creation of a "safe haven" or protected environment for information sharing between industry and government. Compilation and analysis of anonymized industry and regulatory intelligence could be a powerful means of identifying fraudulent ingredients before they arrive at retail. There would be significant legal and cultural hurdles to overcome to implement this type of information sharing, especially in the United States. However, an integrated food protection system that proactively reduces the risk of EMA will almost certainly require it.

1.6 SOURCES FOR FURTHER INFORMATION

The United States Pharmacopeia (USP) Food Fraud Database (www.foodfraud.org) catalogs thousands of records of food fraud. The FPDI EMA Incidents database (www.foodshield.org) provides contextual information about hundreds of EMA incidents. A guidance document for conducting food fraud vulnerability assessments was published by USP in the *Third Supplement to Food Chemicals Codex 9* (http://www.usp.org/food-ingredients/food-chemicals-codex) and is also available for download at www.foodfraud.org. SSAFE provides a downloadable Excel document for conducting a food fraud vulnerability assessment at http://www.ssafe-food.org/our-projects/. Various private companies and industry groups are also developing commercial products tailored at reducing the risk of food fraud to companies.

The Global Food Safety Initiative (GFSI) released a position paper on mitigating the risk of food fraud (http://www.mygfsi.com/news-resources/news/295-gfsi-position-paper-on-mitigating-the-public-health-risk-of-food-fraud.html). GFSI also recently added new requirements for food organizations to have in place a food fraud vulnerability assessment procedure. More information about the BRC Global Standard for Food Safety Issue 7, a GFSI-approved standard which includes food fraud prevention measures, can be found here: http://www.brcglobalstandards.com/Manufacturers/Food/FoodIssue7.aspx#.Vgrov5dL1Ng. Finally, the Elliott review (available at https://www.gov.uk/government/publications/elliott-review-into-the-integrity-and-assurance-of-food-supply-networks-final-report) describes an ambitious and well-planned national framework for preventing food fraud.

A list of additional selected publications generally addressing EMA and food fraud is given below:

- Everstine K, Spink J, Kennedy S. (2013) Economically motivated adulteration (EMA) of food: common characteristics of EMA incidents. J Food Prot. 2013 Apr; 76(4):723–35. doi: 10.4315/0362-028X.JFP-12-399.
- Moore JC, Spink J, Lipp M. (2012) Development and application of a database of food ingredient fraud and economically motivated adulteration from 1980 to 2010. J Food Sci. 2012 Apr;77(4):R118–26. doi: 10.1111/j.1750-3841.2012.02657.x. Epub 2012 Apr 4.
- Spink J, Moyer DC. (2011) Defining the public health threat of food fraud. J Food Sci. 2011 Nov-Dec;76(9):R157–63. doi: 10.1111/j.1750-3841.2011.02417.x.
- Johnson, Renee. Food Fraud and "Economically Motivated Adulteration" of Food and Food Ingredients. Congressional Research Service. January 2014. Available at: https://www.fas.org/sgp/crs/misc/R43358.pdf.
- D. I. Ellis, V. L. Brewster, W. B. Dunn, J. W. Allwood, A. P. Golovanov and R. Goodacre (2012) Fingerprinting food: current technologies for the detection of food adulteration and contamination. Chem. Soc. Rev., 2012, 41, 5706–5727.
- United States Government Accountability Office Report to Congressional Requesters. (2011) Food and Drug Administration: Better Coordination Could Enhance Efforts to Address Economic Adulteration and Protect the Public Health. GAO-12-46: Published: Oct 24, 2011. Publicly Released: Nov 23, 2011.
- Grocery Manufacturers Association (GMA), GMA Science and Education Foundation, & A.T. Kearney. (2010) Consumer Product Fraud: Deterrence and Detection. Available at: http://www.atkearney.com/consumer-products-retail/consumer-product-fraud-deterrence-and-detection.
- United States Government Accountability Office Report to the Ranking Member, Subcommittee on Oceans, Atmosphere, Fisheries, and Coast Guard, Committee on Commerce, Science, and Transportation, U.S. Senate. (2009) FDA Program Changes and Better Collaboration among Key Federal Agencies Could Improve Detection and Prevention. GAO-09-258: Published: Feb 19, 2009. Publicly Released: Mar 20, 2009.
- Coley, Noel. (2005) "The fight against food adulteration," Royal Society of Chemistry, Education in Chemistry, Issue March 2005.
- Forte, Wesley E. (1966) "The Food and Drug Administration and the Economic Adulteration of Foods," Indiana Law Journal: Vol. 41: Iss. 3, Article 2.

REFERENCES

21 U.S.C., Title 21, Chapter 9, Subchapter IV, Section 342.

Buder, L. 2 Former Executives of Beech-Nut Guilty in Phone Juice Case. The New York Times. 18 February 1988.

European Commission. 2013. Monitoring Agri-trade Policy – Agricultural trade in 2012: A good story to tell in a difficult year?.

Everstine, K., Spink, J., Kennedy, S., 2013. Economically Motivated Adulteration (EMA) of Food: Common Characteristics of EMA Incidents. Journal of Food Protection 76, 723–735.

Forte, W.E., 1966. The Food and Drug Administration and the Economic Adulteration of Foods. Indiana Law Journal vol. 41 (3).

Gee, J., Jack, L., and Button, M., 2014. Minimising fraud and maximizing value in the UK food and drink sector 2014. Available at: <http://www.pkf-littlejohn.com/food-fraud-report-2014.php>.

Grocery Manufacturers Association and A.K. Kearney. Consumer Product Fraud: Deterrence and Detection. 2010. Accessed at: <http://www.gmaonline.org/downloads/research-and-reports/consumerproductfraud.pdf>.

HM Government. 2014. Elliott Review into the Integrity and Assurance of Food Supply Networks – Final Report (A National Food Crime Prevention Framework).

Johnson, R. Food Fraud and 'Economically Motivated Adulteration' of Food and Food Ingredients. Congressional Research Service. January 2014.

Lutter, R. Addressing Challenges of Economically-Motivated Adulteration. Presentation at the FDA Public Meeting on Economically Motivated Adulteration. 1 May 2009. Accessed at <http://www.fda.gov/NewsEvents/MeetingsConferencesWorkshops/ucm163619.htm>.

United States Food and Drug Administration. 2011. A Special Report: Pathway to Global Product Safety and Quality.

United States Pharmacopeia. 2014. Appendix XVII: Guidance on Food Fraud Mitigation. Available at <http://www.foodfraud.org>.

von Witzke, H. and Noleppa, S. 2010. EU agricultural production and trade: Can more efficiency prevent increasing "land-grabbing" outside of Europe?

Chapter 2

Fingerprinting for Detecting Contaminants in Food

K. Böhme[1], J. Barros-Velázquez[2] and P. Calo-Mata[2]
[1]*Agrifood Technological Centre of Lugo (CETAL), Lugo, Spain,* [2]*University of Santiago de Compostela, Lugo, Spain*

2.1 INTRODUCTION

Food quality and safety are increasingly important public health issues. Every year, about 600 million persons fall ill after consuming contaminated food or water and 420,000 of them die. Diseases caused by unsafe food can cause severe diarrhea, life-threatening intoxications, and cancer (WHO, 2015). Consumers need to purchase safe products that do not involve any kind of health risk. The aim of "food safety" is to avoid health hazards for the consumer, such as microbiological and chemical contaminants, as well as adulteration.

Chemical food contaminants are veterinary drugs, feed additives, growth promoters, dioxins, heavy metals, and pesticides that are known or suspected to be carcinogenic, causing cardiovascular disease, kidney and liver dysfunction, etc., when humans are exposed by ingestion of the contaminated food or water. Thus, food products are under stringent laboratory control to assure they comply with the regulatory limits for residues and contaminants. The progress made in analytical chemistry in the last decades toward significantly higher sensitivity and specificity now allows the detection of chemical contaminants in complex food matrices, down to parts per billion. Routine analytical methods for the detection of single or multiple chemical contaminants include rapid screening tests, such as enzyme-linked immunosorbent assays and microbial inhibition tests, as well as complex multitarget instrumental analysis based on chromatography, mass spectrometry, and vibrational and atomic spectroscopy.

The contamination of food products with microorganisms presents a problem of global concern, since the growth and metabolism of microorganisms can cause serious foodborne intoxications. Thus, the safety of a food product depends in great part on the presence and nature of the microorganisms.

Food Protection and Security. DOI: http://dx.doi.org/10.1016/B978-1-78242-251-8.00002-3

Besides molds and yeasts, bacteria are the principal microorganisms responsible for various types of food spoilage and foodborne intoxications. It should be mentioned that a food product naturally contains an indigenous microbiota that can include pathogenic bacterial species; however, most bacterial contamination occurs during processing and manipulation of the food products. The global incidence of foodborne disease is difficult to estimate, but it has been reported that every year 230,000 people die due to diarrheal diseases after consuming contaminated food or water. In industrialized countries, the percentage of the population suffering from foodborne diseases each year has been reported to be up to 30% (WHO, 2015). In order to control and minimize the microbial hazard, pathogenic bacteria need to be identified in a rapid and unequivocal way. Traditionally, bacterial species have been identified by classic tools relying on culturing processes coupled to morphological, physiological, and biochemical characterization. In the last few decades, the field of microbiological identification has turned to more rapid and sensitive methods, including antibody-based assays and DNA-based methods, together with important advances in bioinformatics tools. Thus, some methodologies such as ELISA or PCR have already become classics. Recently, the development of rapid and highly sensitive techniques, such as real-time PCR, DNA microarrays, and biosensors, has provoked the replacement of traditional culturing methods in the field of bacterial identification in clinical diagnostics, as well as in the food sector. Furthermore, Fourier transform infrared spectroscopy (FT-IR) has been described as a new method for rapid and reliable bacterial identification (Sandt et al., 2006). At the same time, proteomic tools such as mass spectrometry have been introduced for the identification of microorganisms (Klaenhammer et al., 2007).

This chapter aims to review the detection of food contaminants by fingerprinting techniques. The term fingerprinting has its origin in forensics, where specific DNA profiles are used to differentiate individuals. DNA fingerprinting is the most common approach for species differentiation and identification, with many applications in the food sector, as much for food authentication purposes as for microbial pathogen identification. The different techniques, such as amplified fragment length polymorphism (AFLP), random amplified polymorphic DNA (RAPD), repetitive sequence-based PCR (rep-PCR), multiple-locus variable number tandem repeat analysis (MLVA), multilocus sequence typing (MLST), and pulsed-field gel electrophoresis (PFGE), create sets of characteristic DNA profiles, specific for every individual. All these techniques have been extensively studied with the aim of determining plant varieties, animal species, and the source and/or geographic origin of a food or food ingredient. Besides food authenticity, DNA fingerprinting is a common tool for bacterial species differentiation at the genus, species, and strain levels (Mandal et al., 2011). On the one hand, identification of the microbiota of a food product allows shelf life to be determined and any necessary measures to be taken to assure quality. On the other hand, the detection of certain microbial contaminants with a foodborne pathogenic character is crucial to avoid a microbial health risk for consumers. Furthermore, DNA typing of bacterial strains at the subspecies level is carried out to classify the strains in relation to their origin, antibiotic resistance and virulence, being crucial for microbial source tracking (MST), and epidemiological studies.

Nowadays, molecular fingerprinting is not restricted to DNA-based methods, but describes a variety of analytical methods that can measure the composition of a sample in a nonselective way, such as by collecting a spectrum. In following sections, molecular fingerprinting techniques, different from DNA fingerprinting, and their applications for detecting food contaminants are described. Analytical methods, such as spectroscopy and spectrometry generate spectral profiles that represent fingerprints of the analyzed target. As with DNA fingerprints, the information obtained may be used to differentiate and identify individuals for food authenticity and microbial identification purposes. This chapter focuses especially on (1) the detection of bacterial contaminants and (2) bacterial discrimination at the species and subspecies levels. Closely related species are sometimes difficult to distinguish by traditional and DNA-based methods, but the pathogenic character may differ significantly. This is even more important in those cases where different strains of the same species exhibit different virulent potentials. That is why identification at the species level is not always sufficient and bacterial typing methods that give information about virulence factors and antibiotic resistance are key aspects of ongoing research. Table 2.1 gives an overview of the main foodborne pathogenic bacterial species studied by molecular fingerprinting techniques and the corresponding references are listed. Spectral approaches are also being applied to the detection of mycotoxins and food contaminants different from those of microbial origin. In this chapter, the use of spectral fingerprinting techniques to detect chemical contaminants is summarized, including detection of antibiotics, drugs, hormones, melamine, pesticides, and some further banned food ingredients.

2.2 MALDI-TOF MASS SPECTROMETRY

Matrix-assisted laser desorption/ionization (MALDI) is a very soft ionization technique in that large molecules, such as proteins, can be analyzed without fragmentation of the molecules. For this the sample is mixed and crystallized with a matrix solution that protects the target molecules when submitted to shots from a laser beam, absorbing the applied energy. Following this, the matrix ions transfer the energy to the sample molecules, resulting in ions of low charge ($+1/+2$). Subsequently, the ions produced are separated in an electric field with relation to their mass/charge (m/z) ratio by a time of flight (TOF) analyzer.

TABLE 2.1 Bacterial Food Contaminants Studied by Fingerprinting Techniques With the Corresponding References

Species	Method	Reference
Acinetobacter baumanii		
	MALDI-TOF MS	Alvarez-Buylla et al. (2012), Sousa et al. (2014)
***Bacillus* spp.**		
	MALDI-TOF MS	Branquinho et al. (2014), Fernández-No et al. (2013)
	SERS	Patel et al. (2008)
***Campylobacter* spp.**		
	MALDI-TOF MS	Bessède et al. (2011), Kolínská et al. (2008), Mandrell et al. (2005), Zautner et al. (2013)
***Clostridium* spp.**		
	MALDI-TOF MS	Grosse-Herrenthey et al. (2008), Reil et al. (2011)
	FT-IR	Kirkwood et al. (2006)
Escherichia coli		
	MALDI-TOF MS	Christner et al. (2014), Clark et al. (2013), Khot and Fisher (2013), Matsumura et al. (2014), Novais et al. (2014), Siegrist et al. (2007)
	FT-IR	Al-Qadiri et al. (2006)
	Raman/SERS	Cho et al. (2015), Yang and Irudayaraj (2003)
	LIBS	Barnett et al. (2011), Diedrich et al. (2007), Marcos-Martinez et al. (2011), Mohaidat et al. (2011), Multari et al. (2013)
***Legionella* spp.**		
	MALDI-TOF MS	Gaia et al. (2011), Moliner et al. (2010), Pennanec et al. (2010)
***Listeria* spp.**		
	MALDI-TOF MS	Barbuddhe et al. (2008), Jadhav et al. (2014)
	FT-IR	Al-Holy et al. (2006), Janbu et al. (2008), Rebuffo et al. (2006)
	SERS	Green et al. (2009)
Salmonella enterica		
	MALDI-TOF MS	Dieckmann et al. (2008), Dieckmann and Malorny (2011), Kuhns et al. (2012), Sparbier et al. (2012)
	FT-IR	Al-Qadiri et al. (2008), Baldauf et al. (2006), Männig et al. (2008)
	SERS	Duan et al. (2015)
	LIBS	Barnett et al. (2011), Marcos-Martinez et al. (2011), Multari et al. (2013)

(*Continued*)

TABLE 2.1 (Continued)

Species	Method	Reference
***Shigella* spp.**		
	MALDI-TOF MS	Khot and Fisher (2013)
Staphylococcus aureus		
	MALDI-TOF MS	Böhme et al. (2012), Carbonnelle et al. (2007), Du et al. (2002), Dubois et al. (2010), Jackson et al. (2005), Josten et al. (2013), Szabados et al. (2010), Wolters et al. (2010)
	FT-IR	Amiali et al. (2011)
	Raman	Harz et al. (2005)
	LIBS	Barnett et al. (2011)
***Vibrio* spp.**		
	MALDI-TOF MS	Dieckmann et al. (2010), Erler et al. (2015), Hazen et al. (2009)
Yersinia enterocolitica		
	MALDI-TOF MS	Ayyadurai et al. (2010), Lasch et al. (2010), Stephan et al. (2011)
	FT-IR	Kuhm et al. (2009)

MALDI-TOF MS emerged as a new tool for bacterial differentiation due to its speed, simplicity, and cost effectiveness. Whole bacterial cells can be analyzed directly without any sample pretreatment and the resulting spectral profiles are highly specific, representing a fingerprint for the corresponding organism (Clark et al., 2013). Furthermore, culturing conditions and growth media have been shown to influence the fingerprints to a lower extent, allowing the comparison and identification of bacterial strains even if different protocols have been applied. With MALDI-TOF MS soluble and low-weight proteins (1,500–20,000 Da) are detected and the majority have been associated with ribosomal proteins and to a lower degree with structural proteins, such as cold-shock proteins and DNA-binding proteins (Ryzhov and Fenselau, 2001). The number of studies aimed at the application of MALDI-TOF MS fingerprinting to microbial identification has increased significantly in the last decades and several detailed reviews give an overview of the methodology, sample preparation and application to bacteria, yeasts, and fungi (Clark et al., 2013; Welker and Moore, 2011). The applicability of MALDI-TOF MS fingerprinting to clinical routine analysis has been demonstrated in a number of studies, achieving 92–98% of correct species identification. This is a significantly better result than that obtained with commonly applied microbial identification tools. MALDI-TOF MS has been compared to conventional phenotypic and molecular methods, highlighting the higher discrimination potential, in addition to the speed and cost effectiveness of the proteomic approach. More recently, MALDI-TOF MS fingerprinting has also been applied to the detection of foodborne pathogens isolated from food products (Böhme et al., 2013).

Bacterial identification by MALDI-TOF MS is carried out by comparing the spectral profile to a previously created library of reference spectra. The whole spectral profile is representative for the corresponding strain and the determination of a number of characteristic peaks allows the classification of the strain at the genus, species, and even subspecies or strain level.

2.2.1 Spectral Databases

Whereas the first work in this area was realized at an intralaboratory level with few reference spectra, some manufacturers of mass spectrometers created commercial spectral databases. The MALDI Biotyper system from Bruker Daltonics (Bremen, Germany) includes an ample database of bacterial strains, mycobacteria, and fungi. A robust, standardized procedure for automated bacterial analysis, including sample preparation and data analysis, has been described (Sauer et al., 2008). The entries in the database consist of representative main spectra that have been created by replicative measurements of the corresponding reference strain. The Biotyper system has been validated in several studies, such as the identification of 1371 clinical isolates, achieving 93.2% of correct identification (Bizzini et al., 2010) and the study of 980 clinical isolates, resulting in 92.2% correct species identification (van Veen et al., 2010). Another microbial identification system based on MALDI-TOF MS, including a spectral archive, is the VITEK MS platform from bioMérieux (Marcy-l'Étoile, France). This database consists of SuperSpectra that are created by the determination of a set of biomarker peaks, representative for the corresponding genera, species, and strain. The VITEK MS platform has been validated for the identification of 1129 clinical isolates, achieving 93% of correct identification (Martiny et al., 2012). When compared to the Biotyper database, similar results were obtained. The Biotyper platform has been approved by the Food and Drug Administration (FDA) for the official identification of 40 bacterial species and the VITEK MS database for 194 species (Deak et al., 2015). Further databases are the Andromas database from Andromas SAS (Paris, France) and the MicrobeLynx bacterial identification system from Waters Corporation (Manchester, UK). Today, the Andromas database is frequently used for clinical routine analysis in Europe and includes reference data for bacteria, mycobacteria, yeasts, and fungi (Dupont et al., 2010). It has been implemented into the clinical microbiology laboratory of the Necker-Enfants Malades Hospital to identify all microorganisms isolated routinely, achieving 93% and 99% of correct species identification after single and two acquisitions, respectively (Bille et al., 2012). Likewise, the MicrobeLynx database has been successfully applied to identify clinical isolates (Rajakaruna et al., 2009).

The spectral libraries mentioned are only available commercially and require high charges for access. In this sense, the main drawback of the MALDI-TOF MS fingerprinting approach is the lack of public spectral libraries. In addition, most studies carried out are aimed at the identification of clinically relevant strains. Although many human diseases are caused by the consumption of contaminated food and the previously described commercial databases also include bacterial species that are of interest in the field of food safety, the application of these libraries to food control is not always indicated due to the lack of bacterial strains isolated from food products. For this reason, the laboratory intern database SpectraBank has been created and made publicly available (www.spectrabank.org) (Böhme, Fernández-No et al., 2012a). The whole process for the identification of an unknown bacterial strain, including sample preparation, MALDI-TOF MS measurement and data analysis by the web tool Speclust (Alm et al., 2006) and comparison to the SpectraBank library, are explained in detail. Fig. 2.1 shows the schematic procedure of the identification process. At present, SpectraBank includes open access spectral information obtained by MALDI-TOF MS for more than 200 bacterial strains of 56 different bacterial species with interest in food safety and quality. Continuous extension of the data is intended, by adding new strains and improving the data processing, analysis', and sharing. Recently, a further in-house spectral database has been created for *Vibrio* spp. (VibrioBase), since the entries in the Biotyper database were not sufficient for correct species identification of *Vibrio* spp. strains (Erler et al., 2015).

2.2.2 MALDI-TOF MS Fingerprinting for the Detection of Bacterial Food Contaminants

Most reviews and studies that have been published about MALDI-TOF MS fingerprinting for bacterial identification have been aimed at application to the clinical sector and the routine identification of human pathogens in clinical isolates. Nevertheless, as mentioned before, many bacterial foodborne pathogens cause human infectious diseases and are therefore included in these studies, such as *Bacillus* spp., *Escherichia coli*, *Listeria* spp., *Klebsiella pneumoniae*, *Pseudomonas aeruginosa*, *Salmonella enterica*, *Staphylococcus aureus*, and *Streptococcus* spp. (Farfour et al., 2012; Hsieh et al., 2008; Smole et al., 2002). In Table 2.1 studies based on MALDI-TOF MS analysis of bacterial species with foodborne pathogenic character are listed. Recently, the fast, accurate, and easy-to-handle MALDI-TOF MS fingerprinting technology has also been applied to bacterial identification in veterinary diagnostics, environmental isolates, and food samples. Mazzeo et al. (2006) and Böhme et al. (2013)

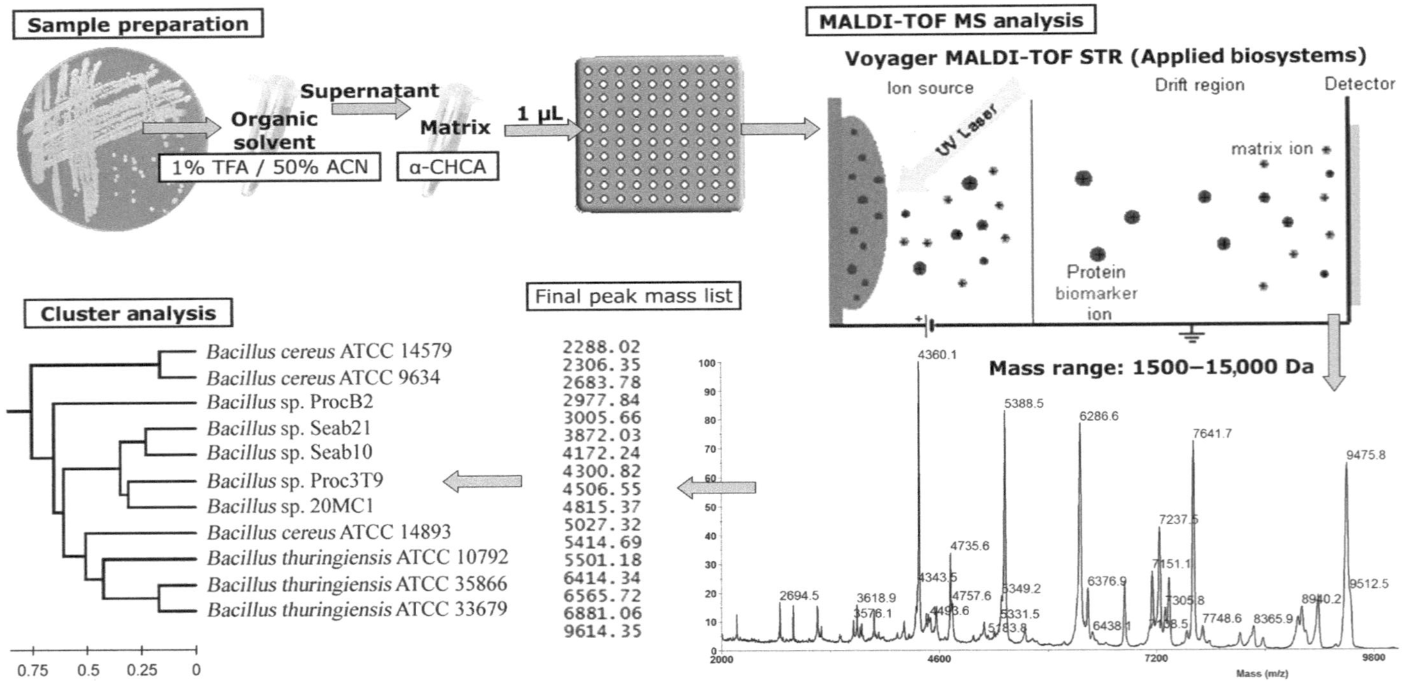

FIGURE 2.1 Scheme of the bacterial identification process by MALDI-TOF MS fingerprinting.

constructed spectral libraries containing spectra of 24 and 58 foodborne bacterial species, respectively. Mazzeo et al. (2006) made the spectral profiles and peak mass lists freely available on the Web (http://bioinformatica.isa.cnr.it/Descr_Bact_Dbase.htm). The library includes the genera *Escherichia*, *Yersinia*, *Proteus*, *Morganella*, *Salmonella*, *Staphylococcus*, *Micrococcus*, *Lactococcus*, *Pseudomonas*, *Leuconostoc*, and *Listeria*. Böhme et al. (2013) created the spectral library SpectraBank (www.spectrabank.org), where other researchers can download spectral information of 58 bacterial species that are of interest for food safety and quality, including the genera *Acinetobacter*, *Aeromonas*, *Bacillus*, *Carnobacterium*, *Clostridium*, *Listeria*, *Photobacterium*, *Pseudomonas*, *Stenotrophomonas*, *Shewanella*, *Staphylococcus*, *Vibrio*, and a number of genera of the Enterobacteriaceae family.

In a further study, histamine-producing bacterial species have successfully been differentiated by MALDI-TOF MS fingerprinting (Fernández-No et al., 2010).

In most studies, MALDI-TOF MS analysis has been applied to bacterial reference strains. Although these strains correspond to bacterial species for which the pathogenic potential is known, it has been shown that real isolates can differ significantly from type strains in their phenotypic and proteotypic properties, due to modifications caused by environmental changes or the food matrices. To perform efficient bacterial species identification of foodborne pathogens that are isolated from food products, it is important to include spectral information of strains isolated from different food matrices into the databases. Therefore, for subtyping of *Yersinia enterocolitica* (Stephan et al., 2011) and *E. coli* (Novais et al., 2014) a few strains isolated from food products have been included. Likewise, four strains isolated from food have been studied for species level differentiation of *Bacillus pumilus* and *Bacillus safensis* (Branquinho et al., 2014). Dubois et al. (2010) studied 92 *Staphylococcus* spp. strains isolated from food and plants in relation to species identification inside the genus *Staphylococcus* (Dubois et al., 2010). In further studies, Dieckmann et al. (2011) analyzed *S. enterica* subsp. *enterica*, *Aeromonas* spp., and *Vibrio* spp. strains, including isolates from food, environment, animals, and humans. *Cronobacter sakazakii* is getting attention as an emerging foodborne pathogen and due to its presence in infant formulas. MALDI-TOF MS has been successfully applied to identify *Cronobacter* spp. strains at the species level with high accuracy. In these studies, besides reference strains, strains have been isolated from infant formulas, milk powder producing plants, and further food samples (Stephan et al., 2010; Zhu et al., 2011). *Vibrio parahaemolyticus* is a main causative agent of pandemic outbreaks of seafood-borne gastroenteritis. MALDI-TOF MS has been applied to distinguish *V. parahaemolyticus* from other *Vibrio* spp. For this, *V. parahaemolyticus* strains have been isolated from illness-related human and food samples of diverse outbreaks and analyzed by

MALDI-TOF MS (Hazen et al., 2009). A new approach for the rapid identification of *Listeria monocytogenes* has been described by Jadhav et al. (2014). Isolates from different food products (UHT milk, cheese, chicken paté, and cantaloupe), spiked with bacterial pathogens, have been submitted directly to analysis in selective enrichment broth, after incubation of 30 h (24 h first enrichment and additional enrichment of 6 h). This methodology represents a fast and simple way to identify *L. monocytogenes* at very low concentrations in food products. MALDI-TOF MS has been successfully applied to the differentiation of the closely related species *Streptococcus uberis* and *Streptococcus parauberis* and to the correct identification of two *S. parauberis* strains isolated from vacuum-packaging refrigerated seafood products (Fernández-No et al., 2012). Some species of the genus *Streptococcus* are of special interest for the dairy industrial sector, since they are important mastitis-causing agents. In this sense, MALDI-TOF MS has been successfully applied to differentiate mastitis-causing *Streptococcus* spp. strains, isolated from mastitis causes, blood and food at the species and subspecies level (Raemy et al., 2013; Schabauer et al., 2014). Finally, in a number of studies of Böhme et al. the ability of MALDI-TOF MS fingerprinting in identification of bacterial strains isolated from seafood products has been tested. For that, a total of 50 bacterial strains have been isolated from different fish and processed seafood products.

In general, MALDI-TOF mass spectral profiles exhibit a high interspecific variability and at the same time a high intraspecific similarity for most bacterial species, allowing identification at the genus and species levels. Exceptions have been reported for some very closely related species, such as *Escherichia coli* and *Shigella* spp., *Streptococcus* spp., and *Listeria* spp. that could not be differentiated by the commonly applied databases at the species level (Farfour et al., 2012; Risch et al., 2010). Nevertheless, MALDI-TOF MS demonstrated a high discriminatory potential at the intraspecies level. The differentiation of subspecies and serotypes is of crucial importance for risk assessment in the food sector, due to the varying pathogenic character. Likewise, determination of clonal lineages is fundamental for epidemiological studies of foodborne disease outbreaks. Clustering of mass spectral data has been successfully applied for chemotaxonomic studies of bacterial strains and compared to phylogenetic trees based on DNA analysis. The high similarity observed is not surprising, since the molecules detected by MALDI-TOF MS are generally attributed to ribosomal proteins that serve as taxonomic markers for the corresponding genus, species, or strain (Welker and Moore, 2011). In addition, in many cases a higher discriminatory potential has been observed for the proteomic approach when compared to conventional bacterial classification tools (Böhme et al., 2013; Risch et al., 2010). Recent reviews pay special attention to MALDI-TOF MS applications as a bacterial typing tool, with the aim of detecting antimicrobial resistance and carrying out epidemiological studies (Clark et al., 2013; Sandrin et al., 2013). In some cases, the available spectral databases do not exhibit sufficient resolution at the species, subspecies or strain levels to perform bacterial typing of serotypes, pathotypes, or clonal lineages. Thus, the use of bioinformatics tools and the determination of subtype-specific biomarker peaks is required (Suarez et al., 2013). In this sense, subtyping of *Y. enterocolitica* (Stephan et al., 2011) and *Yersinia pestis* (Ayyadurai et al., 2010) strains into different serotypes could be achieved. Likewise, strains of the highly infective *Campylobacter jejuni* have been classified by MALDI-TOF MS, resulting in the separation of hyperinvasive strains and strains with an extended amino acid metabolism from the other strains (Mandrell et al., 2005). A critical aspect for *E. coli* identification is the high similarity to *Shigella* spp. The differentiation of *E. coli* and *Shigella* spp. by MALDI-TOF MS has been demonstrated recently by determining 15 biomarker peaks with the ClinProTool from Bruker (Khot and Fisher, 2013). A further challenge is the bacterial typing of *E. coli* isolates and differentiation of clonal groups to realize epidemiological studies. In a number of studies, MALDI-TOF MS has been successfully applied to the differentiation of *E. coli* clones (Christner et al., 2014; Matsumura et al., 2014; Novais et al., 2014). In these studies, clonal groups related to extended-spectrum-β-lactamase (ESBL) producers have been identified at high percentages. Likewise, in the spectral profiles of the Shiga-Toxigenic *E. coli* O104:H4 two peaks have been determined that were not present in the spectra of preoutbreak strains. MALDI-TOF MS has also been successfully applied for discrimination of the five most important serovars of *S. enterica* subsp. *enterica* based on a decision tree and specific biomarkers (Dieckmann and Malorny, 2011). In another study, the differentiation of *S. enterica* typhy from nontyphy was not possible with the Biotyper database, but clear identification could be achieved after identifying serovar-specific biomarkers (Kuhns et al., 2012).

MST aims at the detection of foodborne pathogens in the food chain and determination of the source of contamination and consequent corrective actions to be taken, as well as prevention of foodborne outbreaks. MALDI-TOF MS fingerprinting has been successfully applied to analyze *Enterococcus faecium* and *Enterococcus faecalis* and differentiate the strains in relation to their isolation sources (meat or dairy products) (Quintela-Baluja et al., 2013). Similarly, spectral variability could be observed at the strain level in relation to the geographical origin and moment of isolation of *V. paraheamolyticus* strains (Hazen et al., 2009).

High interest exists in the ability to distinguish methicillin-resistant *S. aureus* strains (MRSA) from methicillin-sensitive *S. aureus* strains (MSSA). MALDI-TOF MS fingerprinting has been successfully applied to this matter and was also able to distinguish clonal types of *S. aureus* (Du et al., 2002; Jackson et al., 2005; Wolters et al., 2010). The analysis of MRSA strains isolated during an outbreak allowed differentiation from

MSSA strains, as well as rapid typing of the outbreak strains and detection of epidemic lineages (Josten et al., 2013). In Fig. 2.2, spectral differences observed for *S. aureus* at the strain level are highlighted. The corresponding study demonstrated the discriminatory potential of MALDI-TOF MS, since the studied strains exhibited 100% identical 16S rRNA gene sequences, but could be classified into subgroups by the spectral profiles (Böhme, Morandi, et al., 2012b). Nevertheless, the subtypes could not be related to the production of toxins, as also confirmed by further studies (Szabados et al., 2010).

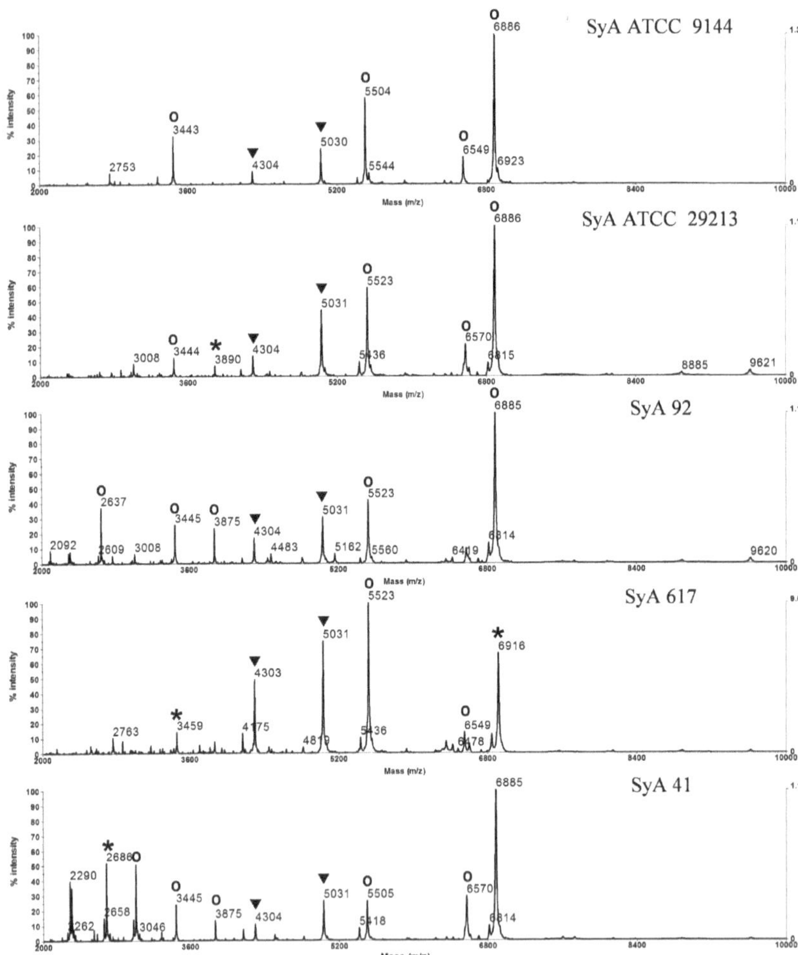

FIGURE 2.2 Spectral profiles of different *S. aureus* strains. Symbols correspond to species-specific (▼), subgroup-specific (*), and further characteristic peaks (○).

The genus *Bacillus* is known to comprise closely related species, with differentiation by DNA-based approaches not always possible. However, the different species corresponding to the *Bacillus cereus/thuringiensis* and the *Bacillus subtilis/amyloliquefaciens* complexes exhibit very different pathogenic and spoilage potential, thus requiring their correct identification. With MALDI-TOF MS and cluster analysis, a clear grouping of the *B. subtilis* strains was achieved (Böhme et al., 2013). In the case of *B. cereus* and *B. thuringiensis*, the 16S rRNA sequences are nearly 100% identical, inhibiting any differentiation at the species level. By phyloproteomic clustering, the strains could not be identified; however, a differentiation of the strains into subgroups was obtained. These findings were confirmed in a further study, where a number of *Bacillus* spp. strains, isolated from fresh and processed food products, were analyzed by MALDI-TOF MS (Fernández-No et al., 2013).

2.3 VIBRATIONAL SPECTROSCOPY

Vibrational spectroscopy consists of two approaches, infrared (IR) absorption and Raman scattering, and provides structural and chemical information about molecules based on their vibrational transitions. Both approaches are fast, low-cost, highly specific for target molecules, robust, and easy to use. Furthermore, no or a minimum sample preparation is required and chemical constituents can be determined qualitatively and quantitatively down to very low concentrations, with these characteristics making vibrational spectroscopy very suitable for food contamination detection. Furthermore, the advantage over other analytical techniques is the fact that the measurements are nondestructive, reagentless, and may be applied directly to food surfaces. In addition, the instruments can be transferred to portable devices, allowing immediate analysis of the food products and onsite evaluations throughout the whole food-processing chain.

In the last few years, the number of studies aimed at the use of vibrational spectroscopy for food safety applications increased markedly and different applications have been reviewed recently for FT-IR (Huang et al., 2008; Jimaré Benito et al., 2008; Woodcock et al., 2008) and Raman spectroscopy (Craig et al., 2013; Yang and Ying, 2011), as well as for different food classes, such as milk (Cattaneo and Holroyd, 2013), fish (Cheng et al., 2013), and water (Lopez-Roldan et al., 2013).

IR spectroscopy has widely been studied for bacterial identification purposes and in lesser extents for the detection of chemical contaminants, such as melamine and pesticides. Raman spectroscopy gives complementary information to IR spectroscopy and has gained increasing attention in the last decades for both foodborne pathogen and chemical contaminants detection. Although IR spectroscopy is cheaper and easier in instrumentation and handling, Raman spectroscopy has several advantages over conventional IR spectroscopy, such as less interference of water and more detectable features over the same wave-number range (Lu et al., 2011). Nevertheless, Raman signals are weak and the spectra are highly interfered with by noise signals and fluorescence background, making it difficult to obtain good results for low concentrations of contaminants. New advances in nanotechnology have made it possible to enhance Raman signals by surface-enhanced Raman spectroscopy (SERS). Raman spectroscopy is combined with nanomaterials, such as gold or silver nanosphere colloids, solid surface gold-coated nanosubstrates, bimetallic nanosubstrates, or spherical magnetic-core gold-shell nanoparticles (Fan et al., 2014).

Fingerprints obtained by vibrational spectroscopy are affected by various factors and other nontargeted food components may significantly interfere with the obtained signals. To overcome these challenges and permit its use in routine analysis of food control laboratories, advanced data preprocessing and statistical analyses are required. The use of chemometrics for the interpretation of high-dimensional spectral fingerprints have made complex analyses possible.

2.3.1 Vibrational Spectroscopy for Bacterial Identification

Vibrational spectroscopy has been extensively applied to the study of bacteria in food stuffs (Lu et al., 2011; Pahlow et al., 2015). The characteristic fingerprints generated contain information about the biochemical constitution of bacterial cells and enable the differentiation of the organisms at the species and even strain level.

Many studies have analyzed bacterial foodborne pathogens and were aimed at either the identification of bacterial species present in food products or the detection and in some cases quantification of a concrete species with pathogenic character. In further studies, various foodborne pathogenic bacterial species have been studied together and could be successfully differentiated and identified by SERS approaches (Fan et al., 2011; Sundaram et al., 2013; Xie et al., 2013). A spectral library of 19 bacterial species of the most important harmful and nonpathogenic bacteria associated with meat and poultry has been created and tested for the identification of spiked meat samples. The test samples were correctly assigned to their genus and in most cases down to the species level by Raman fingerprinting and a three-level classification model by means of support vector machines (Meisel et al., 2014).

Bacterial identification by FT-IR achieved 100% of correct species identification for Gram-positive bacteria and 80% for Gram-negative bacteria (Janbu et al., 2008; Sandt et al., 2006). IR and Raman spectroscopy have also been applied to differentiate strains at the subspecies level with the aim of typing different biotypes and serotypes. In this sense, an FT-Raman procedure was successful in discriminating different *E. coli* strains on whole apples and accurately differentiated the nonpathogens from pathogens, including *E. coli* O157:H7 (Yang and Irudayaraj, 2003). Likewise, strains of *Y. enterocolitica* have been distinguished into the main biotypes and serotypes (Kuhm et al., 2009). In this study, species of the genus *Yersinia* that cannot be differentiated by conventional biochemical methods exhibit specific IR fingerprints, allowing the clear discrimination at the species level. In addition, the presence of the ail gene, one of the main pathogenicity markers, was demonstrated using FT-IR and correct identification of isolates concerning the ail gene was achieved in 98.5%. Fan et al. (2011) successfully applied SERS to identify *E. coli* O157:H7 and *Staphylococcus epidermidis* in a mixed bacterial sample. Nevertheless, analysis of bacterial mixtures of different species remains challenging. Furthermore, bacterial identification generally includes cultivation procedures previous to analysis. This implies that only cultivable bacterial cells can be detected and the risk of the presence of injured but viable cells, spores, and already produced toxins remains. Al-Qadiri et al. (2008) applied FT-IR spectroscopy to detect sublethally heat-injured *S. enterica* var. *typhimurium* and *L. monocytogenes* cells. The studies of injured bacterial cells by vibrational spectroscopy have been reviewed and discussed by Lu et al. (2011). The detection of thermo-resistant spores of *Bacillus* spp. has also been successfully carried out by SERS (Alexander and Le, 2007; He et al., 2013). Further works have reported the application of Raman and FT-IR fingerprinting to detect the contamination of food products with mycotoxins, such as aflatoxins in maize kernels (Lee et al., 2014) and deoxynivalenol in ground wheat and barley (Liu et al., 2009).

To avoid time-consuming culturing steps, different approaches have been described to isolate bacterial strains from food matrices previous to IR or Raman analysis. The most promising practice is the implementation of nanoparticles that are targeted against the bacterial species of interest, concentrated, and then submitted to spectral analysis. In this sense, magnetic nanoparticles have been applied, functionalized with anti-*E. coli* O157:H7 or anti-*Salmonella typhimurium* antibodies (Ravindranath et al., 2009). The pathogens could be detected using a portable mid-IR spectrometer in complex food matrixes with a detection limit of 10(4)–10(5) CFU/mL. In further studies, bacterial cells have been captured by using antibody–antigen interaction with nanoparticles immobilized on SERS active surfaces and detected by direct analysis with spectral fingerprinting (Chae et al., 2013).

Besides bacterial species identification and detection of single pathogens, the monitoring of microbial spoilage of food products by FT-IR spectroscopy has been described for fish (Tito et al., 2012), meat (Ellis et al., 2002), and milk (Nicolaou and Goodacre, 2008), allowing the quantification of the microbial load by direct analysis on the food surface without any culturing or isolation processes.

2.3.2 Vibrational Spectroscopy for Detection of Chemical Contaminants

The capability of vibrational spectroscopy to detect all the biochemical compounds of a cell has led to the emergence of FT-IR and SERS as competent tools for the analysis of trace amounts of chemical hazards in various food products. The applications of vibrational spectroscopy fingerprinting to detect chemical contaminants in various food stuffs, such as dairy products, fish, fruits, vegetables, and condiments, have been reviewed in detail (Fan et al., 2014; Zheng and He, 2014).

Most studies reported are targeted on detecting melamine in milk, milk powder, infant formula, and egg white (Betz et al., 2012; Hu et al., 2015; Wang et al., 2015). Besides detection of melamine at very low concentrations (63 ppb to 2 ppm), quantification was also possible.

The characteristic of vibrational spectroscopy to be applicable directly on food surfaces without destroying the food gives it an enormous potential to detect pesticides on fruit and vegetable surfaces. In this sense, pesticide residues could be detected on apple surfaces and in apple juice samples at concentrations of $0.125\,\mu g/cm^2$ and $3\,\mu g/mL$ (3 ppm), respectively (Wijaya et al., 2014). Pesticides could also be detected in mixtures (Zhai et al., 2015) and on the surfaces of further fruits, such as bananas and citrus fruits (Müller et al., 2014).

Further chemical contaminants that have been investigated by FT-IR and Raman spectroscopy are banned food additives, such as sudan I, rhodamine B, and malachite green (He et al., 2015). Likewise, the contamination of fish with heavy metals could be detected by vibrational spectroscopy (Chen et al., 2013).

2.4 LASER-INDUCED BREAKDOWN SPECTROSCOPY

Laser-induced breakdown spectroscopy (LIBS) is another spectral fingerprinting technology that has emerged in the last few years. It uses a laser pulse as excitation source and emits an atomic spectrum of the analyzed target, representing the whole elemental composition. Since a very small amount of material is required, normally a few picograms, LIBS is also considered a nondestructive technique. Measurements can be carried out directly on surfaces of food or surfaces used for food manipulating and processing, with minimal or no sample preparation. Thus, real-time analysis can be carried out, exhibiting high sensitivity and the ability to detect all elements, including biological and chemical hazards. In addition, LIBS is a cost-effective, fast and robust technique that can be implanted into portable systems for nonspecialist users. All these attributes make LIBS fingerprinting a competent tool for food analysis purposes. Nevertheless, since it is a very recent methodology, only a few reports exist describing the use of LIBS for the detection of food contaminants. More work has been done and reviewed recently for biomedical and military applications (Pathak et al., 2012; Rehse et al., 2012).

A challenge of LIBS fingerprinting is the complex data analysis, since the obtained spectral profiles contain mostly all the same elements. The use of chemometrics for computerized data analysis is necessary and allows a search for similarities and specificities between two spectral fingerprints. The so-called discriminant function analysis (DFA) is a computational discrimination technique that analyzes the entire observed elemental composition of a sample target and reduces the entire information of a spectrum to a quantity known as the discriminant function score. The scores from each spectrum are plotted against each other and the grouping obtained represents the differences and similarities that are immediately visible (Rehse et al., 2012).

2.4.1 LIBS for Bacterial Identification

Most of the reported studies are targeted at the identification of bacterial pathogens, responsible for human diseases. The differences detected by LIBS analysis are based on the chemical compositions of the outer membranes, which vary between different species of bacteria. These membranes contain divalent cations such as Mg^{2+} and Ca^{2+} that are the dominating spectral features in the LIBS profiles. Emission lines from other trace inorganics such as iron, potassium, sodium, manganese, and phosphorus, as well as molecules such as CN, OH, and NH, are also present and create all together a spectral fingerprint of the analyzed target organism.

LIBS fingerprinting has been successfully applied to the discrimination of bacterial foodborne pathogens, including specimens from 13 distinct taxonomic bacterial classes representative of 5 bacterial genera (Putnam et al., 2013). In further studies, the foodborne pathogens *E. coli*, *S. enterica*, and *S. aureus* have been analyzed by LIBS and could be clearly identified from each other due to the obtained specific spectral profiles (Barnett et al., 2011; Multari et al., 2013a; Singh, 2014). The authors also showed the applicability of LIBS to the detection of *S. enterica* and *E. coli* in different food matrices, such as milk, chicken, eggshell, ground beef, bologna, and lettuce. For bacterial identification the spectral fingerprints of an unknown bacterial strain are compared against a previously compiled spectral library. The development of an application of LIBS together with neural networks (NNs) for analysis and comparison of the spectra achieved over 95% correct species identification for bacterial strains of the species *P. aeroginosa*, *E. coli*, and *S. typhimurium* (Marcos-Martinez et al., 2011). LIBS also exhibits great potential to discriminate bacterial strains at the intraspecies level. In this sense, three clonal methicillin-resistant *S. aureus* (MRSA) strains could be distinguished and differentiated from one methicillin-sensitive *S. aureus* (MSSA) strain (Multari et al., 2010). Likewise, an *E. coli* O157:H7 strain could be clearly discriminated from nonpathogenic commonly occurring environmental *E. coli* strains (Diedrich et al., 2007).

When studying different culture conditions, the fingerprints obtained by LIBS apparently have proven to be independent to growth conditions and culture media. Those spectra of the same strain cultured under different conditions have demonstrated that they are more similar than those spectral profiles obtained for different strains grown under the same conditions (Diedrich et al., 2007; Marcos-Martinez et al., 2011; Rehse et al., 2010). In addition, LIBS profiles did not change between viable and killed bacterial cells, nor with time when aging on abiotic surfaces, this being of special interest for the analysis of processed food products and processing surfaces (Mohaidat et al., 2011; Multari, Cremers, Dupre, et al., 2013a).

Quantification of bacterial pathogens and detection of bacterial mixtures by LIBS fingerprinting remains challenging, but has been reported by Rehse et al. (2010). The authors showed that the intensity of the LIBS spectrum is linearly dependent on the cell number, but does not influence the specificity. When analyzing bacteria in mixed samples, the dominant bacterial component could be reliably identified if it comprised 70% or more of the mixture.

2.4.2 LIBS for the Detection of Chemical Contaminants

Advances in LIBS spectral data analysis and chemometrics techniques for data analysis have led to the ability to detect chemical contaminants in complex matrices such as foods. Pesticides and dioxins could be successfully detected in tissue fats and rendering oils. The pesticide concentrations in the samples ranged from 0.005 to 0.1 μg/g (Multari et al., 2013b).

Similarly, spinach and rice have been analyzed by LIBS with the aim of detecting contamination with pesticides. The results demonstrated that the LIBS technique together with a chemometric method (PLS-DA) could be a great tool to distinguish pesticide-contaminated samples from pesticide-free samples in a rapid manner (Kim et al., 2012).

In a further study, heavy metals have been analyzed in different tissues of contaminated fish samples. External calibration has been applied to quantify the contamination in the food products (Wan et al., 2013). Similarly, harmful metals have been detected by LIBS in rice (Hemati Farsani et al., 2014).

2.5 FUTURE TRENDS

The spectral fingerprinting techniques described here all represent accurate and fast methods in comparison to other bacterial identification techniques. However, the challenge of food security and foodborne pathogen detection is that very low doses of a critical strain may cause serious infectious diseases after multiplying rapidly in a few hours or days. Unfortunately, recent identification methods also require multiplication of bacterial cells to reach a detectable concentration. In addition, one disadvantage of most fingerprinting techniques is that isolated bacterial strains are required. That means that culturing steps are still necessary to isolate and concentrate the bacterial strains from the food matrices. In this sense, approaches that apply spectral fingerprinting directly to a sample without or with minimal sample pretreatment are challenging. In clinical routine analysis, direct bacterial identification by MALDI-TOF MS has been achieved for urine and blood samples (Clark et al., 2013). Similar approaches could be applied to liquid food samples, as already demonstrated for contaminated water or the rinse water after washing lettuce and cotton cloth (Holland et al., 2000). Nevertheless, a critical point of food matrices is the presence of a mixture of different bacteria and, until now, the application of MALDI-TOF MS fingerprinting for microbial mixtures has not yet been demonstrated. For that to occur, specific biomarker proteins have to be determined that allow an unequivocal identification of the corresponding strain on the basis of a few peaks. The identification of biomarker proteins is also of interest for bacterial typing and the correlation to virulence factors, toxin production, and/or antibiotic resistance that could significantly improve the risk assessment in the food chain.

Shortening the culturing process has also been the focus of several studies based on vibrational spectroscopy. A real-time detection and identification approach to food pathogens has been described and was based on SERS measurement of labeled immunoassay reagents in cultural enrichment vessels while culturing is ongoing. The approach allowed sensitive detection of the pathogens *E. coli*, *Salmonella*, and *Listeria* in complex food matrices (Weidemaier et al., 2015).

Promising future prospects are those applications where the high specificity of molecular fingerprints is combined with the high selectivity of nanomaterials, such as aptamers, nanoparticles, or further recognition molecules. The use of magnetic nanoparticles allows the separation of target strains from the food matrix and nontarget strains and the posterior analysis by spectral fingerprinting.

Future trends will probably focus strongly on the analysis of a food product without any culturing step. An interesting study has been carried out aimed at the direct detection and quantification of the microbial load in meat and milk, using MALDI-TOF analysis (Nicolaou et al., 2012). Such rapid screening tests are essential for an effective risk assessment in the food control sector and are required for the fast and accurate detection of food contaminants, such as bacterial pathogens, as well as mycotoxins, drug residues, heavy metals, etc. To realize onsite analysis and field tests, portable and automated instruments are the objectives of ongoing research. A portable and automated optofluidic SERS system has been successfully applied to detect food and water contaminants (Yazdi and White, 2012).

Finally, the spectral fingerprinting techniques described can be combined with further analytical techniques, since this can give much more complete information as a single technique, especially in the field of bacterial strain characterization. As an example, vibrational spectroscopy is commonly implemented into microscope systems to recover cells after enrichment on agar and following spectral analysis. More interestingly, the combination of LIBS, vibrational spectroscopy, and mass spectrometry has great potential to unite all the information and create a "whole-organism spectral fingerprint."

REFERENCES

Alexander, T.A., Le, D.M., 2007. Characterization of a commercialized SERS-active substrate and its application to the identification of intact *Bacillus* endospores. Appl. Opt. 46 (18), 3878–3890.

Al-Holy, M.A., Lin, M., Al-Qadiri, H., Cavinato, A.G., Rasco, B.A., 2006. Classification of foodborne pathogens by fourier transform infrared spectroscopy and pattern recognition techniques. J. Rapid Methods Automat. Microbiol. 14 (2), 189–200.

Alm, R., Johansson, P., Hjerno, K., Emanuelsson, C., Ringnér, M., Häkkinen, J., 2006. Detection and identification of protein isoforms using cluster analysis of MALDI-MS mass spectra. J. Proteome Res. 5 (4), 785–792.

Al-Qadiri, H.M., Lin, M., Al-Holy, M.A., Cavinato, A.G., Rasco, B.A., 2008. Detection of sublethal thermal injury in *Salmonella enterica* serotype typhimurium and *Listeria monocytogenes* using Fourier transform infrared (FT-IR) spectroscopy (4000 to 600 cm(-1). J. Food Sci. 73 (2), M54–M61.

Al-Qadiri, H.M., Lin, M., Cavinato, A.G., Rasco, B.A., 2006. Fourier transform infrared spectroscopy, detection and identification of *Escherichia coli* O157:H7 and *Alicyclobacillus* strains in apple juice. Int. J. Food Microbiol. 111 (1), 73–80.

Alvarez-Buylla, A., Culebras, E., Picazo, J.J., 2012. Identification of *Acinetobacter* species: is Bruker biotyper MALDI-TOF mass spectrometry a good alternative to molecular techniques?. Infect. Genet. Evol. 12 (2), 345–349.

Amiali, N.M., Golding, G.R., Sedman, J., Simor, A.E., Ismail, A.A., 2011. Rapid identification of community-associated methicillin-resistant *Staphylococcus aureus* by Fourier transform infrared spectroscopy. Diagn. Microbiol. Infect. Dis. 70 (2), 157–166.

Ayyadurai, S., Flaudrops, C., Raoult, D., Drancourt, M., 2010. Rapid identification and typing of *Yersinia pestis* and other *Yersinia* species by matrix-assisted laser desorption/ionization time-of-flight (MALDI-TOF) mass spectrometry. BMC Microbiol. 10 (1), 285.

Baldauf, N.A., Rodriguez-Romo, L.A., Yousef, A.E., Rodriguez-Saona, L.E., 2006. Differentiation of selected *Salmonella enterica* serovars by Fourier transform mid-infrared spectroscopy. Appl. Spectrosc. 60 (6), 592–598.

Barbuddhe, S.B., Maier, T., Schwarz, G., Kostrzewa, M., Hof, H., Domann, E., et al., 2008. Rapid identification and typing of *Listeria* species by matrix-assisted laser desorption ionization-time of flight mass spectrometry. Appl. Environ. Microbiol. 74 (17), 5402–5407.

Barnett, C., Bell, C., Vig, K., Akpovo, A.C., Johnson, L., Pillai, S., et al., 2011. Development of a LIBS assay for the detection of *Salmonella enterica* serovar Typhimurium from food. Anal. Bioanal. Chem. 400 (10), 3323–3330.

Bessède, E., Solecki, O., Sifré, E., Labadi, L., Mégraud, F., 2011. Identification of *Campylobacter* species and related organisms by matrix assisted laser desorption ionization–time of flight (MALDI-TOF) mass spectrometry. Clin. Microbiol. Infect. 17 (11), 1735–1739.

Betz, J.F., Cheng, Y., Rubloff, G.W., 2012. Direct SERS detection of contaminants in a complex mixture: rapid, single step screening for melamine in liquid infant formula. Analyst 137 (4), 826–828.

Bille, E., Dauphin, B., Leto, J., Bougnoux, M.E., Beretti, J.L., Lotz, A., et al., 2012. MALDI-TOF MS Andromas strategy for the routine identification of bacteria, mycobacteria, yeasts, *Aspergillus* spp. and positive blood cultures. Clin. Microbiol. Infect. 18 (11), 1117–1125.

Bizzini, A., Durussel, C., Bille, J., Greub, G., Prod'hom, G., 2010. Performance of matrix-assisted laser desorption ionization-time of flight mass spectrometry for identification of bacterial strains routinely isolated in a clinical microbiology laboratory. J. Clin. Microbiol. 48 (5), 1549–1554.

Böhme, K., Fernández-No, I.C., Barros-Velázquez, J., Gallardo, J.M., Cañas, B., Calo-Mata, P., 2012a. SpectraBank: an open access tool for rapid microbial identification by MALDI-TOF MS fingerprinting. Electrophoresis 33 (14), 2138–2142.

Böhme, K., Fernández-No, I.C., Pazos, M., Gallardo, J.M., Barros-Velázquez, J., Cañas, B., et al., 2013. Identification and classification of seafood-borne pathogenic and spoilage bacteria: 16S rRNA sequencing versus MALDI-TOF MS fingerprinting. Electrophoresis 34 (6), 877–887.

Böhme, K., Morandi, S., Cremonesi, P., Fernández No, I.C., Barros-Velázquez, J., Castiglioni, B., et al., 2012b. Characterization of *Staphylococcus aureus* strains isolated from Italian dairy products by MALDI-TOF mass fingerprinting. Electrophoresis 33 (15), 2355–2364.

Branquinho, R., Sousa, C., Lopes, J., Pintado, M.E., Peixe, L.V., Osório, H., 2014. Differentiation of *Bacillus pumilus* and *Bacillus safensis* using MALDI-TOF-MS. PLoS One 9 (10), e110127.

Carbonnelle, E., Beretti, J.-L., Cottyn, S., Quesne, G., Berche, P., Nassif, X., et al., 2007. Rapid identification of Staphylococci isolated in clinical microbiology laboratories by matrix-assisted laser desorption ionization-time of flight mass spectrometry. J. Clin. Microbiol. 45 (7), 2156–2161.

Cattaneo, T., Holroyd, S., 2013. Review: the use of near infrared spectroscopy for determination of adulteration and contamination in milk and milk powder: updating knowledge. J. Near Infrared Spectrosc. 21 (5), 341–349.

Chae, E.J., Lee, J.H., Oh, B.K., Choi, J.W., 2013. Label-free nanobiosensor to detect infectious bacterica based on SERS. J. Biomed. Nanotechnol. 9 (4), 659–663.

Chen, X., Wu, D., Guan, X., Liu, B., Liu, G., Yan, M., et al., 2013. Feasibility of infrared and Raman spectroscopies for identification of juvenile black seabream (*Sparus macrocephalus*) intoxicated by heavy metals. J. Agric. Food Chem. 61 (50), 12429–12435.

Cheng, J.-H., Dai, Q., Sun, D.-W., Zeng, X.-A., Liu, D., Pu, H.-B., 2013. Applications of non-destructive spectroscopic techniques for fish quality and safety evaluation and inspection. Trends Food Sci. Technol. 34 (1), 18–31.

Cho, I.-H., Bhandari, P., Patel, P., Irudayaraj, J., 2015. Membrane filter-assisted surface enhanced Raman spectroscopy for the rapid detection of E. coli O157:H7 in ground beef. Biosens. Bioelectron. 64, 171–176.

Christner, M., Trusch, M., Rohde, H., Kwiatkowski, M., Schlüter, H., Wolters, M., et al., 2014. Rapid MALDI-TOF mass spectrometry strain typing during a large outbreak of Shiga-Toxigenic *Escherichia coli*. PLoS One 9 (7), e101924.

Clark, A.E., Kaleta, E.J., Arora, A., Wolk, D.M., 2013a. Matrix-assisted laser desorption ionization-time of flight mass spectrometry: a fundamental shift in the routine practice of clinical microbiology. Clin. Microbiol. Rev. 26 (3), 547–603.

Clark, C.G., Kruczkiewicz, P., Guan, C., McCorrister, S.J., Chong, P., Wylie, J., et al., 2013b. Evaluation of MALDI-TOF mass spectroscopy methods for determination of *Escherichia coli* pathotypes. J. Microbiol. Methods 94 (3), 180–191.

Craig, A.P., Franca, A.S., Irudayaraj, J., 2013. Surface-enhanced Raman spectroscopy applied to food safety. Annu. Rev. Food Sci. Technol. 4, 369–380.

Deak, E., Charlton, C.L., Bobenchik, A.M., Miller, S.A., Pollett, S., McHardy, I.H., et al., 2015. Comparison of the Vitek MS and Bruker Microflex LT MALDI-TOF MS platforms for routine identification of commonly isolated bacteria and yeast in the clinical microbiology laboratory. Diagn. Microbiol. Infect. Dis. 81 (1), 27–33.

Dieckmann, R., Malorny, B., 2011. Rapid screening of epidemiologically important *Salmonella enterica* subsp. enterica serovars by whole-cell matrix-assisted laser desorption ionization-time of flight mass spectrometry. Appl. Environ. Microbiol. 77 (12), 4136–4146.

Dieckmann, R., Helmuth, R., Erhard, M., Malorny, B., 2008. Rapid classification and identification of Salmonellae at the species and subspecies levels by whole-cell matrix-assisted laser desorption ionization-time of flight mass spectrometry. Appl. Environ. Microbiol. 74 (24), 7767–7778.

Dieckmann, R., Strauch, E., Alter, T., 2010. Rapid identification and characterization of *Vibrio* species using whole-cell MALDI-TOF mass spectrometry. J. Appl. Microbiol. 109 (1), 199–211.

Diedrich, J., Rehse, S.J., Palchaudhuri, S., 2007. Pathogenic *Escherichia coli* strain discrimination using laser-induced breakdown spectroscopy. J. Appl. Phys. 102 (1), 014702.

Du, Z., Yang, R., Guo, Z., Song, Y., Wang, J., 2002. Identification of *Staphylococcus aureus* and determination of its methicillin resistance by matrix-assisted laser desorption/ionization time-of-flight mass spectrometry. Anal. Chem. 74 (21), 5487–5491.

Duan, N., Chang, B., Zhang, H., Wang, Z., Wu, S., 2015. *Salmonella typhimurium* detection using a surface-enhanced Raman scattering-based aptasensor. Int. J. Food Microbiol. 218, 38–43.

Dubois, D., Leyssene, D., Chacornac, J.P., Kostrzewa, M., Schmit, P.O., Talon, R., et al., 2010. Identification of a variety of *Staphylococcus* species by matrix-assisted laser desorption ionization-time of flight mass spectrometry. J. Clin. Microbiol. 48 (3), 941–945.

Dupont, C., Sivadon-Tardy, V., Bille, E., Dauphin, B., Beretti, J.L., Alvarez, A.S., et al., 2010. Identification of clinical coagulase-negative staphylococci, isolated in microbiology laboratories, by matrix-assisted laser desorption/ionization-time of flight mass spectrometry and two automated systems. Clin. Microbiol. Infect. 16 (7), 998–1004.

Ellis, D.I., Broadhurst, D., Kell, D.B., Rowland, J.J., Goodacre, R., 2002. Rapid and quantitative detection of the microbial spoilage of meat by fourier transform infrared spectroscopy and machine learning. Appl. Environ. Microbiol. 68 (6), 2822–2828.

Erler, R., Wichels, A., Heinemeyer, E.A., Hauk, G., Hippelein, M., Reyes, N.T., et al., 2015. VibrioBase: a MALDI-TOF MS database for fast identification of *Vibrio* spp. that are potentially pathogenic in humans. Syst. Appl. Microbiol. 38 (1), 16–25.

Fan, C., Hu, Z., Mustapha, A., Lin, M., 2011. Rapid detection of food- and waterborne bacteria using surface-enhanced Raman spectroscopy coupled with silver nanosubstrates. Appl. Microbiol. Biotechnol. 92 (5), 1053–1061.

Fan, Y.X., Lai, K.Q., Huang, Y.Q., 2014. Application of surface-enhanced Raman spectroscopy to the determination of trace chemical hazards in food products. Guang Pu Xue Yu Guang Pu Fen Xi 34 (7), 1859–1864.

Farfour, E., Leto, J., Barritault, M., Barberis, C., Meyer, J., Dauphin, B., et al., 2012. Evaluation of the andromas matrix-assisted laser desorption ionization–time of flight mass spectrometry system for identification of aerobically growing gram-positive bacilli. J. Clin. Microbiol. 50 (8), 2702–2707.

Fernández-No, I.C., Böhme, K., Gallardo, J.M., Barros-Velázquez, J., Cañas, B., Calo-Mata, P., 2010. Differential characterization of biogenic amine-producing bacteria involved in food poisoning using MALDI-TOF mass fingerprinting. Electrophoresis 31 (6), 1116–1127.

Fernández-No, I.C., Böhme, K., Calo-Mata, P., Cañas, B., Gallardo, J.M., Barros-Velázquez, J., 2012. Isolation and characterization of *Streptococcus parauberis* from vacuum-packaging refrigerated seafood products. Food Microbiol. 30 (1), 91–97.

Fernández-No, I.C., Böhme, K., Díaz-Bao, M., Cepeda, A., Barros-Velázquez, J., Calo-Mata, P., 2013. Characterisation and profiling of *Bacillus subtilis*, *Bacillus cereus* and *Bacillus licheniformis* by MALDI-TOF mass fingerprinting. Food Microbiol. 33 (2), 235–242.

Gaia, V., Casati, S., Tonolla, M., 2011. Rapid identification of *Legionella* spp. by MALDI-TOF MS based protein mass fingerprinting. Syst. Appl. Microbiol. 34 (1), 40–44.

Green, G.C., Chan, A.D.C., Luo, B.S., Hanhong, D., Min, L., 2009. Identification of *Listeria* species using a low-cost surface-enhanced Raman scattering system with wavelet-based signal processing. IEEE Trans. Instrum. Meas. 58 (10), 3713–3722.

Grosse-Herrenthey, A., Maier, T., Gessler, F., Schaumann, R., Böhnel, H., Kostrzewa, M., et al., 2008. Challenging the problem of clostridial identification with matrix-assisted laser desorption and ionization-time-of-flight mass spectrometry (MALDI-TOF MS). Anaerobe 14 (4), 242–249.

Harz, M., Rösch, P., Peschke, K.D., Ronneberger, O., Burkhardt, H., Popp, J., 2005. Micro-Raman spectroscopic identification of bacterial cells of the genus *Staphylococcus* and dependence on their cultivation conditions. Analyst 130 (11), 1543–1550.

Hazen, T.H., Martinez, R.J., Chen, Y., Lafon, P.C., Garrett, N.M., Parsons, M.B., et al., 2009. Rapid identification of *Vibrio parahaemolyticus* by whole-cell matrix-assisted laser desorption ionization-time of flight mass spectrometry. Appl. Environ. Microbiol. 75 (21), 6745–6756.

He, L.D., Deen, B., Pagel, A.H., Diez-Gonzalez, F., Labuza, T.P., 2013. Concentration, detection and discrimination of *Bacillus anthracis* spores in orange juice using aptamer based surface enhanced Raman spectroscopy. Analyst 138 (6), 1657–1659.

He, S., Xie, W., Zhang, W., Zhang, L., Wang, Y., Liu, X., et al., 2015. Multivariate qualitative analysis of banned additives in food safety using surface enhanced Raman scattering spectroscopy. Spectrochim. Acta A Mol. Biomol. Spectrosc. 137, 1092–1099.

Hemati Farsani, M., Ghezelbash, M., Darbani, S.M.R., Eslami Majd, A., Soltanolkotabi, M., 2014. Determination of trace elements in some types of Iranian rice using laser induced breakdown spectroscopy. J. Mazandaran Univ. Med. Sci. 24 (118), 24–32.

Holland, R.D., Rafii, F., Heinze, T.M., Sutherland, J.B., Voorhees, K.J., Lay Jr., J.O., 2000. Matrix-assisted laser desorption/ionization time-of-flight mass spectrometric detection of bacterial biomarker proteins isolated from contaminated water, lettuce and cotton cloth. Rapid Commun. Mass Spectrom. 14 (10), 911–917.

Hsieh, S.-Y., Tseng, C.-L., Lee, Y.-S., Kuo, A.-J., Sun, C.-F., Lin, Y.-H., et al., 2008. Highly efficient classification and identification of human pathogenic bacteria by MALDI-TOF MS. Mol. Cell. Proteomics 7 (2), 448–456.

Hu, Y., Feng, S., Gao, F., Li-Chan, E.C.Y., Grant, E., Lu, X., 2015. Detection of melamine in milk using molecularly imprinted polymers–surface enhanced Raman spectroscopy. Food Chem. 176, 123–129.

Huang, H., Yu, H., Xu, H., Ying, Y., 2008. Near infrared spectroscopy for on/in-line monitoring of quality in foods and beverages: a review. J. Food Eng. 87 (3), 303–313.

Jackson, K.A., Edwards-Jones, V., Sutton, C.W., Fox, A.J., 2005. Optimisation of intact cell MALDI method for fingerprinting of methicillin-resistant *Staphylococcus aureus*. J. Microbiol. Meth. 62 (3), 273–284.

Jadhav, S., Sevior, D., Bhave, M., Palombo, E.A., 2014. Detection of *Listeria monocytogenes* from selective enrichment broth using MALDI-TOF Mass Spectrometry. J. Proteomics 97, 100–106.

Janbu, A.O., Møretrø, T., Bertrand, D., Kohler, A., 2008. FT-IR microspectroscopy: a promising method for the rapid identification of *Listeria* species. FEMS Microbiol. Lett. 278 (2), 164–170.

Jimaré Benito, M.T., Bosch Ojeda, C., Sanchez Rojas, F., 2008. Process analytical chemistry: applications of near infrared spectrometry in environmental and food analysis: an overview. Appl. Spectrosc. Rev. 43 (5), 452–484.

Josten, M., Reif, M., Szekat, C., Al-Sabti, N., Roemer, T., Sparbier, K., et al., 2013. Analysis of the matrix-assisted laser desorption ionization-time of flight mass spectrum of *Staphylococcus aureus* identifies mutations that allow differentiation of the main clonal lineages. J. Clin. Microbiol. 51 (6), 1809–1817.

Khot, P.D., Fisher, M.A., 2013. Novel approach for differentiating *Shigella* species and *Escherichia coli* by matrix-assisted laser desorption ionization-time of flight mass spectrometry. J. Clin. Microbiol. 51 (11), 3711–3716.

Kim, G., Kwak, J., Choi, J., Park, K., 2012. Detection of nutrient elements and contamination by pesticides in spinach and rice samples using laser-induced breakdown spectroscopy (LIBS). J. Agric. Food Chem. 60 (3), 718–724.

Kirkwood, J., Ghetler, A., Sedman, J., Leclair, D., Pagotto, F., Austin, J.W., et al., 2006. Differentiation of group I and group II strains of *Clostridium botulinum* by focal plane array Fourier transform infrared spectroscopy. J. Food. Prot. 69 (10), 2377–2383.

Klaenhammer, T.R., Pfeiler, E., Duong, T., 2007. Genomics and proteomics of foodborne microorganisms. In: Doyle, M.P., Beuchat, L.R. (Eds.), Food Microbiology: Fundamentals and Frontiers, third ed. ASM Press, Washington, D.C, pp. 935–951.

Kolínská, R., Drevínek, M., Jakubů, V., Zemlicková, H., 2008. Species identification of *Campylobacter jejuni* ssp. jejuni and *C. coli* by matrix-assisted laser desorption/ionization time-of-flight mass spectrometry and PCR. Folia Microbiol. (Praha) 53 (5), 403–409.

Kuhm, A.E., Suter, D., Felleisen, R., Rau, J., 2009. Identification of *Yersinia enterocolitica* at the species and subspecies levels by Fourier transform infrared spectroscopy. Appl. Environ. Microbiol. 75 (18), 5809–5813.

Kuhns, M., Zautner, A.E., Rabsch, W., Zimmermann, O., Weig, M., Bader, O., et al., 2012. Rapid discrimination of *Salmonella enterica* serovar Typhi from other serovars by MALDI-TOF mass spectrometry. PLoS One 7 (6), e40004.

Lasch, P., Drevinek, M., Nattermann, H., Grunow, R., Stämmler, M., Dieckmann, R., et al., 2010. Characterization of *Yersinia* using MALDI-TOF mass spectrometry and chemometrics. Anal. Chem. 82 (20), 8464–8475.

Lee, K.M., Herrman, T.J., Bisrat, Y., Murray, S.C., 2014. Feasibility of surface-enhanced Raman spectroscopy for rapid detection of aflatoxins in maize. J. Agric. Food Chem. 62 (19), 4466–4474.

Liu, Y., Delwiche, S.R., Dong, Y., 2009. Feasibility of FT–Raman spectroscopy for rapid screening for DON toxin in ground wheat and barley. Food Addit. Contam. Part A Chem. Anal. Control Expo. Risk Assess 26 (10), 1396–1401.

Lopez-Roldan, R., Tusell, P., Cortina, J.L., Courtois, S., 2013. On-line bacteriological detection in water. Trends Anal. Chem. 44, 46–57.

Lu, X., Al-Qadiri, H., Lin, M., Rasco, B., 2011. Application of Mid-infrared and Raman spectroscopy to the study of bacteria. Food Bioproc. Technol. 4 (6), 919–935.

Mandal, P.K., Biswas, A.K., Choi, K., Pal, U.K., 2011. Methods for rapid detection of foodborne pathogens: an overview. Am. J. Food Technol. 6, 87–102.

Mandrell, R.E., Harden, L.A., Bates, A., Miller, W.G., Haddon, W.F., Fagerquist, C.K., 2005. Speciation of *Campylobacter coli, C. jejuni, C. helveticus, C. lari, C. sputorum*, and *C. upsaliensis* by matrix-assisted laser desorption ionization-time of flight mass spectrometry. Appl. Environ. Microbiol. 71 (10), 6292–6307.

Männig, A., Baldauf, N.A., Rodriguez-Romo, L.A., Yousef, A.E., Rodríguez-Saona, L.E., 2008. Differentiation of *Salmonella enterica* serovars and strains in cultures and food using infrared spectroscopic and microspectroscopic techniques combined with soft independent modeling of class analogy pattern recognition analysis. J. Food Prot. 71 (11), 2249–2256.

Marcos-Martinez, D., Ayala, J.A., Izquierdo-Hornillos, R.C., de Villena, F.J., Caceres, J.O., 2011. Identification and discrimination of bacterial strains by laser induced breakdown spectroscopy and neural networks. Talanta 84 (3), 730–737.

Martiny, D., Busson, L., Wybo, I., El Haj, R.A., Dediste, A., Vandenberg, O., 2012. Comparison of the Microflex LT and Vitek MS systems for routine identification of bacteria by matrix-assisted laser desorption ionization-time of flight mass spectrometry. J. Clin. Microbiol. 50 (4), 1313–1325.

Matsumura, Y., Yamamoto, M., Nagao, M., Tanaka, M., Machida, K., Ito, Y., et al., 2014. Detection of extended-spectrum-β-lactamase-producing *Escherichia coli* ST131 and ST405 clonal groups by matrix-assisted laser desorption ionization-time of flight mass spectrometry. J. Clin. Microbiol. 52 (4), 1034–1040.

Mazzeo, M.F., Sorrentino, A., Gaita, M., Cacace, G., Di Stasio, M., Facchiano, A., et al., 2006. Matrix-assisted laser desorption ionization-time of flight mass spectrometry for the discrimination of food-borne microorganisms. Appl. Environ. Microbiol. 72 (2), 1180–1189.

Meisel, S., Stöckel, S., Rösch, P., Popp, J., 2014. Identification of meat-associated pathogens via Raman microspectroscopy. Food Microbiol. 38, 36–43.

Mohaidat, Q., Palchaudhuri, S., Rehse, S.J., 2011. The effect of bacterial environmental and metabolic stresses on a laser-induced breakdown spectroscopy (LIBS) based identification of Escherichia coli and *Streptococcus viridans*. Appl. Spectrosc. 65 (4), 386–392.

Moliner, C., Ginevra, C., Jarraud, S., Flaudrops, C., Bedotto, M., Couderc, C., et al., 2010. Rapid identification of *Legionella* species by mass spectrometry. J. Med. Microbiol. 59 (Pt 3), 273–284.

Multari, R.A., Cremers, D.A., Dupre, J.A., Gustafson, J.E., 2013a. Detection of biological contaminants on foods and food surfaces using laser-induced breakdown spectroscopy (LIBS). J. Agric. Food. Chem. 61 (36), 8687–8694.

Multari, R.A., Cremers, D.A., Dupre, J.M., Gustafson, J.E., 2010. The use of laser-induced breakdown spectroscopy for distinguishing between bacterial pathogen species and strains. Appl. Spectrosc. 64 (7), 750–759.

Multari, R.A., Cremers, D.A., Scott, T., Kendrick, P., 2013b. Detection of pesticides and dioxins in tissue fats and rendering oils using laser-induced breakdown spectroscopy (LIBS). J. Agric. Food. Chem. 61 (10), 2348–2357.

Müller, C., David, L., Chiş, V., Pînzaru, S.C., 2014. Detection of thiabendazole applied on citrus fruits and bananas using surface enhanced Raman scattering. Food Chem. 145, 814–820.

Nicolaou, N., Goodacre, R., 2008. Rapid and quantitative detection of the microbial spoilage in milk using Fourier transform infrared spectroscopy and chemometrics. Analyst 133 (10), 1424–1431.

Nicolaou, N., Xu, Y., Goodacre, R., 2012. Detection and quantification of bacterial spoilage in milk and pork meat using MALDI-TOF-MS and multivariate analysis. Anal. Chem. 84 (14), 5951–5958.

Novais, Â., Sousa, C., de Dios Caballero, J., Fernandez-Olmos, A., Lopes, J., Ramos, H., et al., 2014. MALDI-TOF mass spectrometry as a tool for the discrimination of high-risk *Escherichia coli* clones from phylogenetic groups B2 (ST131) and D (ST69, ST405, ST393). Eur. J. Clin. Microbiol. Infect. Dis. 33 (8), 1391–1399.

Pahlow, S., Meisel, S., Cialla-May, D., Weber, K., Rösch, P., Popp, J., 2015. Isolation and identification of bacteria by means of Raman spectroscopy. Adv. Drug Deliv. Rev. 89, 105–120.

Patel, I.S., Premasiri, W.R., Moir, D.T., Ziegler, L.D., 2008. Barcoding bacterial cells: a SERS based methodology for pathogen identification. J. Raman Spectrosc. 39 (11), 1660–1672.

Pathak, A.K., Kumar, R., Singh, V.K., Agrawal, R., Rai, S., Rai, A.K., 2012. Assessment of LIBS for spectrochemical analysis: a review. Appl.Spectrosc. Rev. 47 (1), 14–40.

Pennanec, X., Dufour, A., Haras, D., Réhel, K., 2010. A quick and easy method to identify bacteria by matrix-assisted laser desorption/ionisation time-of-flight mass spectrometry. Rapid Commun. Mass Spectrom. 24 (3), 384–392.

Putnam, R.A., Mohaidat, Q.I., Daabous, A., Rehse, S.J., 2013. A comparison of multivariate analysis techniques and variable selection strategies in a laser-induced breakdown spectroscopy bacterial classification. Spectrochimica Acta Part B: Atomic Spectrosc. 87, 161–167.

Quintela-Baluja, M., Böhme, K., Fernández-No, I.C., Morandi, S., Alnakip, M.E., Caamaño-Antelo, S., et al., 2013. Characterization of different food-isolated *Enterococcus* strains by MALDI-TOF mass fingerprinting. Electrophoresis 34 (15), 2240–2250.

Raemy, A., Meylan, M., Casati, S., Gaia, V., Berchtold, B., Boss, R., et al., 2013. Phenotypic and genotypic identification of streptococci and related bacteria isolated from bovine intramammary infections. Acta Vet. Scand. 55, 53.

Rajakaruna, L., Hallas, G., Molenaar, L., Dare, D., Sutton, H., Encheva, V., et al., 2009. High throughput identification of clinical isolates of *Staphylococcus aureus* using MALDI-TOF-MS of intact cells. Infect. Genet. Evol. 9 (4), 507–513.

Ravindranath, S.P., Mauer, L.J., Deb-Roy, C., Irudayaraj, J., 2009. Biofunctionalized magnetic nanoparticle integrated mid-infrared pathogen sensor for food matrixes. Anal. Chem. 81 (8), 2840–2846.

Rebuffo, C.A., Schmitt, J., Wenning, M., von Stetten, F., Scherer, S., 2006. Reliable and rapid identification of *Listeria monocytogenes* and *Listeria* species by artificial neural network-based Fourier transform infrared spectroscopy. Appl. Environ. Microbiol. 72 (2), 994–1000.

Rehse, S.J., Mohaidat, Q.I., Palchaudhuri, S., 2010. Towards the clinical application of laser-induced breakdown spectroscopy for rapid pathogen diagnosis: the effect of mixed cultures and sample dilution on bacterial identification. Appl. Opt. 49 (13), C27–C35.

Rehse, S.J., Salimnia, H., Miziolek, A.W., 2012. Laser-induced breakdown spectroscopy (LIBS): an overview of recent progress and future potential for biomedical applications. J. Med. Eng. Technol. 36 (2), 77–89.

Reil, M., Erhard, M., Kuijper, E.J., Kist, M., Zaiss, H., Witte, W., et al., 2011. Recognition of *Clostridium difficile* PCR-ribotypes 001, 027 and 126/078 using an extended MALDI-TOF MS system. Eur. J. Clin. Microbiol. Infect. Dis. 30 (11), 1431–1436.

Risch, M., Radjenovic, D., Han, J.N., Wydler, M., Nydegger, U., Risch, L., 2010. Comparison of MALDI TOF with conventional identification of clinically relevant bacteria. Swiss Med. Wkly.

Ryzhov, V., Fenselau, C., 2001. Characterization of the protein subset desorbed by MALDI from whole bacterial cells. Anal. Chem. 73 (4), 746–750.

Sandrin, T.R., Goldstein, J.E., Schumaker, S., 2013. MALDI TOF MS profiling of bacteria at the strain level: a review. Mass Spectrom. Rev. 32 (3), 188–217.

Sandt, C., Madoulet, C., Kohler, A., Allouch, P., De Champs, C., Manfait, M., et al., 2006. FT-IR microspectroscopy for early identification of some clinically relevant pathogens. J. Appl. Microbiol. 101 (4), 785–797.

Sauer, S., Freiwald, A., Maier, T., Kube, M., Reinhardt, R., Kostrzewa, M., et al., 2008. Classification and identification of bacteria by mass spectrometry and computational analysis. PLoS One 3 (7), e2843.

Schabauer, L., Wenning, M., Huber, I., Ehling-Schulz, M., 2014. Novel physico-chemical diagnostic tools for high throughput identification of bovine mastitis associated gram-positive, catalase-negative cocci. BMC Vet. Res. 10, 156.

Siegrist, T.J., Anderson, P.D., Huen, W.H., Kleinheinz, G.T., McDermott, C.M., Sandrin, T.R., 2007. Discrimination and characterization of environmental strains of *Escherichia coli* by matrix-assisted laser desorption/ionization time-of-flight mass spectrometry (MALDI-TOF-MS). J. Microbiol. Methods 68 (3), 554–562.

Singh, W., Aslam, F., Atif, M., Tawfik, W., Alsalhi, M.S., Singh, J.P., 2014. Study of bacterial samples using laser induced breakdown spectroscopy. Plasma Sci. Technol. 16 (12), 1141.

Smole, S.C., King, L.A., Leopold, P.E., Arbeit, R.D., 2002. Sample preparation of Gram-positive bacteria for identification by matrix assisted laser desorption/ionization time-of-flight. J. Microbiol. Methods 48 (2-3), 107–115.

Sousa, C., Botelho, J., Silva, L., Grosso, F., Nemec, A., Lopes, J., et al., 2014. MALDI-TOF MS and chemometric based identification of the *Acinetobacter calcoaceticus-Acinetobacter baumannii* complex species. Int. J. Med. Microbiol. 304 (5-6), 669–677.

Sparbier, K., Weller, U., Boogen, C., Kostrzewa, M., 2012. Rapid detection of *Salmonella* sp. by means of a combination of selective enrichment broth and MALDI-TOF MS. Eur. J. Clin. Microbiol. Infect. Dis. 31 (5), 767–773.

Stephan, R., Ziegler, D., Pflüger, V., Vogel, G., Lehner, A., 2010. Rapid genus- and species-specific identification of *Cronobacter* spp. by matrix-assisted laser desorption ionization-time of flight mass spectrometry. J. Clin. Microbiol. 48 (8), 2846–2851.

Stephan, R., Cernela, N., Ziegler, D., Pflüger, V., Tonolla, M., Ravasi, D., et al., 2011. Rapid species specific identification and subtyping of *Yersinia enterocolitica* by MALDI-TOF Mass spectrometry. J. Microbiol. Methods 87 (2), 150–153.

Suarez, S., Ferroni, A., Lotz, A., Jolley, K.A., Guérin, P., Leto, J., et al., 2013. Ribosomal proteins as biomarkers for bacterial identification by mass spectrometry in the clinical microbiology laboratory. J. Microbiol. Methods 94 (3), 390–396.

Sundaram, J., Park, B., Kwon, Y., Lawrence, K.C., 2013. Surface enhanced Raman scattering (SERS) with biopolymer encapsulated silver nanosubstrates for rapid detection of foodborne pathogens. Int. J. Food Microbiol. 167 (1), 67–73.

Szabados, F., Woloszyn, J., Richter, C., Kaase, M., Gatermann, S., 2010. Identification of molecularly defined *Staphylococcus aureus* strains using matrix-assisted laser desorption/ionization time of flight mass spectrometry and the Biotyper 2.0 database. J. Med. Microbiol. 59 (7), 787–790.

Tito, N.B., Rodemann, T., Powell, S.M., 2012. Use of near infrared spectroscopy to predict microbial numbers on Atlantic salmon. Food Microbiol. 32 (2), 431–436.

van Veen, S.Q., Claas, E.C.J., Kuijper, E.J., 2010. High-throughput identification of bacteria and yeast by matrix-assisted laser desorption ionization-time of flight mass spectrometry in conventional medical microbiology laboratories. J. Clin. Microbiol. 48 (3), 900–907.

Wan, X., Wang, J.Y., Ye, J.H., Wang, P., Zhang, Z.M., 2013. Analysis of distribution and contents of heavy metal pollution in fish body with laser-induced breaddown spectroscopy. Guang Pu Xue Yu Guang Pu Fen Xi 33 (1), 206–209.

Wang, Q.H., Liu, Y.L., Ma, M.H., Wang, H., 2015. Quantitative approach to melamine detection in egg white with Surface-Enhanced Raman spectroscopy. Guang Pu Xue Yu Guang Pu Fen Xi 35 (4), 919–923.

Weidemaier, K., Carruthers, E., Curry, A., Kuroda, M., Fallows, E., Thomas, J., et al., 2015. Real-time pathogen monitoring during enrichment: a novel nanotechnology-based approach to food safety testing. Int. J. Food Microbiol. 198, 19–27.

Welker, M., Moore, E.R.B., 2011. Applications of whole-cell matrix-assisted laser-desorption/ionization time-of-flight mass spectrometry in systematic microbiology. Syst. Appl. Microbiol. 34 (1), 2–11.

WHO, 2015. Food safety. Fact sheet N 399: World Health Organization (WHO). Available at: < http://www.who.int/mediacentre/factsheets/fs399/en/ >.

Wijaya, W., Pang, S., Labuza, T.P., He, L., 2014. Rapid detection of acetamiprid in foods using surface-enhanced Raman spectroscopy (SERS). J. Food Sci. 79 (4), T743–T747.

Wolters, M., Rohde, H., Maier, T., Belmar-Campos, C., Franke, G., Scherpe, S., et al., 2010. MALDI-TOF MS fingerprinting allows for discrimination of major methicillin-resistant *Staphylococcus aureus* lineages. Int. J. Med. Microbiol. 301 (1), 64–68.

Woodcock, T., O'Donnell, C., Downey, G., 2008. Review: better quality food and beverages: the role of near infrared spectroscopy. J. Near Infrared Spectrosc. 16 (1), 1–29.

Xie, Y., Xu, L., Wang, Y., Shao, J., Wang, L., Wang, H., et al., 2013. Label-free detection of the foodborne pathogens of Enterobacteriaceae by surface-enhanced Raman spectroscopy. Anal. Methods 5 (4), 946–952.

Yang, D., Ying, Y., 2011. Applications of Raman spectroscopy in agricultural products and food analysis: a review. Appl. Spectrosc. Rev. 46 (7), 539–560.

Yang, H., Irudayaraj, J., 2003. Rapid detection of foodborne microorganisms on food surface using Fourier transform Raman spectroscopy. J. Mol. Struct. 646 (1–3), 35–43.

Yazdi, S.H., White, I.M., 2012. Optofluidic surface enhanced Raman spectroscopy microsystem for sensitive and repeatable on-site detection of chemical contaminants. Anal. Chem. 84 (18), 7992–7998.

Zautner, A.E., Masanta, W.O., Tareen, A.M., Weig, M., Lugert, R., Groß, U., et al., 2013. Discrimination of multilocus sequence typing-based *Campylobacter jejuni* subgroups by MALDI-TOF mass spectrometry. BMC Microbiol. 13, 247.

Zhai, C., Li, Y., Peng, Y., Xu, T., Dhakal, S., Chao, K., et al. Research on identification and determination of mixed pesticides in apples using surface enhanced Raman spectroscopy. 94880R-94880R-7.

Zheng, J., He, L., 2014. Surface-enhanced Raman spectroscopy for the chemical analysis of food. Compr. Rev. Food Sci. Food Saf. 13 (3), 317–328.

Zhu, S., Ratering, S., Schnell, S., Wacker, R., 2011. Matrix-assisted laser desorption and ionization time-of-flight mass spectrometry, 16S rRNA gene sequencing, and API 32E for identification of *Cronobacter* spp.: a comparative study. J. Food Prot. 74 (12), 2182–2187.

Chapter 3

Creating a Food Defense/ Response Plan in Food Processing Facilities

C.L. Lorenzen[1] and C.N. Cutter[2]

[1]*University of Missouri, Columbia, MO, United States,* [2]*Pennsylvania State University, University Park, PA, United States*

> *For the life of me, I cannot understand why the terrorists have not attacked our food supply, because it is so easy to do.*
>
> Tommy Thompson, Secretary of Health and Human Services, 2004.

Food safety and food defense programs are designed to protect food products. However, it is important to understand that food safety programs are developed to prevent *unintentional* contamination from items such as metal, plastic, pathogens, pesticides, or sanitizers entering the food supply. Food defense programs are intended to prevent *intentional* contamination by individuals who deliberately contaminate food products and cause harm to the company or to the consumer. This chapter will discuss the creation and implementation of food defense/response plans in food processing facilities. Discussion of food defense plan requirements by the Food and Drug Administration is current with the Federal Register Notice for Mitigation Strategies to Protect Food Against Intentional Adulteration 5/27/16, which is the final rule (Federal Register, 2016).

3.1 INTRODUCTION TO DEVELOPING A FOOD DEFENSE PLAN

3.1.1 What Are the Benefits of Developing a Food Defense Plan?

Having a food defense plan can reduce the risk of intentional contamination to production or processing operations and may ultimately benefit their bottom lines.

Food Protection and Security. DOI: http://dx.doi.org/10.1016/B978-1-78242-251-8.00003-5

Writing a food defense plan helps employees identify steps that can be taken to reduce the risk that food in a facility can be harmed by intentional contamination. In addition, thinking through the processes used in an operation while developing the food defense plan can help pinpoint inefficiencies and redundancies that are costing operations time and money.

A well-developed response plan also helps family members, employees, and disaster response personnel respond appropriately to a suspected intentional contamination incident. A plan will map out a way to contain the damage and get operations back to normal production levels more quickly. By helping a company avoid a prolonged period of nonproduction, a food defense plan increases a business's chance of surviving a negative event.

All told, a food defense plan will help companies provide safe, high-quality products to their customers, keep employees safe and well informed, and protect the economic viability of the business.

3.1.2 What Operations Are Required to Have a Food Defense Plan?

Food defense plans are either required or highly recommended, but not required, by the following agencies for the specified food operations or products at this time. The *United States Department of Agriculture (USDA)* is responsible for the protection of preharvest agriculture, *USDA Food Safety and Inspection Service (FSIS)* is responsible for regulating and ensuring the protection of meat, poultry, eggs, and catfish, while the *Food and Drug Administration (FDA)* is responsible for regulating foods other than those covered by FSIS. As of July 2016 FDA covered facilities are required to implement food defense plans on the following schedule: large plants must be in compliance by July 2019; small plants (with less than 500 employees) must be in compliance by July 2020; and very small plants (those with less than $10 million in receipts averaged over 3 years) by July 2021. In addition, any operation supplying food for USDA feeding programs (e.g., school lunch program) are legally required to have a food defense plan.

Over the last several years, FSIS has been conducting periodic surveys of inspected plants to determine the percent voluntary compliance, before deciding whether to seek regulations requiring meat and poultry slaughter or processing facilities to have a food defense plan. The most recent results from 2015 demonstrate 85% compliance by meat and poultry plants, 92% compliance by processed egg product plants, and 85% compliance by import inspection establishments (http://www.fsis.usda.gov/wps/portal/fsis/topics/food-defense-defense-and-emergency-response/preparation-and-prevention/food-defense-plan-survey-results/food-defense-plan-survey-results accessed 6/2/2016).

3.1.3 Assembling a Food Defense Team

Writing and managing the food defense plan for an operation should not be a one-person job. Instead, consider assembling a team made up of the key personnel who are familiar with most aspects of the operation. These personnel also should be in positions to take responsibility for enacting and helping manage the plan and must be trained in food defense awareness. University extension personnel, insurance agents, county emergency managers, or members of the law enforcement community can also be included on the initial team to help with vulnerability assessments. However, the team that manages the plan should consist of employees from the operation.

3.1.4 What Other Documentation and Supplemental Information Should Operations Gather for Developing the Food Defense Plan?

When ready to begin the plan, gather the following documents: a detailed and labeled map of the facility; all written operational procedures, such as Hazard Analysis Critical Control Point (HACCP) plans, Sanitation Standard Operating Procedures (SSOPs), Good Agricultural Practices (GAP), Good Manufacturing Practices (GMPs), Quality Assurance Plans and Standard Operating Procedures (SOPs); and procedures related to the workforce, such as preemployment screening and security training. Supplemental publications are available to help ensure the food defense team considers all of the important areas of the operation and can be found through USDA FSIS, FDA, university extension, and food industry websites. These websites offer potential worksheets, food defense planning exercises, and model plans.

3.2 ASSESS VULNERABILITIES (FSIS: CONDUCT FOOD DEFENSE ASSESSMENT; FDA: VULNERABILITY ASSESSMENT)

The first step in developing a food defense plan is to conduct a vulnerability assessment. Look for areas of the operation that are accessible to someone wanting to intentionally contaminate products. Remember that these vulnerabilities may arise from either internal (people employed or contracted with the operation) or external threats (e.g., people not employed by the operation, such as truck drivers or organized terrorist or activist groups) (FSIS, 2008). To find the vulnerabilities in an operation, think like a disgruntled worker, a member of a political group, or anyone wanting to harm the business, cause illness or death, or make a statement for a cause or disrupt the food supply chain. Consider various people who have access to the operation, such as workers, delivery employees, contract cleaners, and/or visitors. Think about the processes and procedures used in the operation, such as food processing, receipt of shipments and marketing and which processes and procedures might be completed without supervision or in locations that are not visible. In the facility, identify locations where contamination would be easily distributed through normal operations, such as a feed or ingredient mixer. Consider points in the process or plant where a contaminant could be added and mixed easily. Also, identify critical areas that are not locked, as well as areas that are not visible to other employees, or where access is not limited. There are three specific elements in a vulnerability assessment that must be considered: (1) potential impact on public health; (2) physical access to the product; and (3) ability to successfully contaminate the product (Federal Register, 2016). When the vulnerability assessment is complete, keep the results and resulting plan confidential and shared only within the food defense team, to prevent it from being used as a tool for intentional contamination.

There are a number of vulnerability assessment tools, ranging from a series of questions to guide one through the planning process, to a more detailed approach, known as CARVER + Shock (Criticality, Accessibility, Recuperability, Vulnerability, Effect, and Recognizability). Establishments developing food defense plans will have the flexibility to choose the assessment tool that works best for them.

CARVER + Shock is an assessment tool that was commissioned by the Homeland Security Council, USDA-FSIS, and FDA for use in food processing facilities. The vulnerability assessment is a prioritization tool used in and by the food sector. It allows personnel to think like an attacker by identifying attractive targets at a food processing facility and then determining those vulnerable points in the infrastructure that personnel can focus on in order to prevent intentional contamination. The benefit of CARVER + Shock is that it standardizes the entire assessment process. So, any food processing facility can use this approach. CARVER + Shock also provides a scale for each characteristic, thereby facilitating a quantitative assessment. This assessment tool also provides for examination of public health, economic, and psychological consequences of an intentional attack, while breaking down the process into critical steps and evaluation of each node.

There are a number of key steps in CARVER + Shock:

Step 1. Establish Parameters

- Select the system to access.
- Develop an attacker profile—pick a worst-case scenario—inside versus outside threat.
- Identify an agent—consider biological, physical, chemical, or radionuclear agent, and pick the type of agent that takes the process and product into account.

Step 2. Assemble Subject Matter Experts

Choose individuals who understand the facility or system being assessed.

- Identify members of food defense team (see previous comments about suggested membership).

- Include external auditors for the team.
- Consider a number of individuals who can provide an overarching perspective on the process and can participate objectively in the activity.

Step 3. Detail Food Supply Chain

- Prepare a flow diagram with every step in the process.
- Validate the flow diagram by conducting a walk-through on the production floor.

Step 4. Assign Scores to Each Step in Flow/Node

- For each step in the process, evaluate the seven CARVER + Shock attributes and use this information to calculate an overall score for each step or node.
- Those steps with the highest score are those that are potentially the most vulnerable.
- Allocate financial and personnel resources to ensure the steps are protected from intentional contamination.

Again, the goal of using CARVER + Shock is to identify those critical nodes that are the most likely targets for a terrorist attack and allow personnel to design measures or mitigation strategies to reduce the risk. By using CARVER + Shock, personnel develop a systematic approach by going through each step in the flow diagram and assigning a number to each step, which can assist in prioritization of mitigation strategies; the higher the number, the more vulnerable the step. Based on the assessment, personnel can consider mitigation strategies that are inexpensive and can be done relatively quickly. Anything that requires more financial resources or more personnel involvement can be prioritized and implemented as time and finances allow.

With the facility map and operational and workforce procedures on hand, the food defense team should complete a comprehensive vulnerability assessment. The team should consider the security of the listed elements and additional elements specific to the operation. For broader food defense concerns, operations may find it easier to respond to a list of questions that cover vulnerable areas of concern (Table 3.1). Answer questions as "yes" if all the elements are secure; answer "no" if any elements are not secure, and indicate each insecure element and those not applicable (N/A) if the question does not apply to the operation. Only the questions that are answered "no" will need to be included in the food defense plan. The broad vulnerability assessment is designed to increase overall security and prevent intentional contamination in the most general sense.

The FDA has completed the CARVER + Shock process for the food industry in order to determine focused mitigation strategies. The more focused portion of the food defense plan should address the processes involved with producing high-risk foods, which would have large public health impacts. The characteristics of high-risk foods include having a likelihood of uniform mixing, being produced in high volumes, or having process

TABLE 3.1 Vulnerability Assessment Areas and Questions to Consider

Questions	Examples of Elements to Consider
Is the outside perimeter secure?	• Fencing restricts entry, within reason, and is inspected regularly. • Gates are locked when not in use, to limit access to the operation. • Locks are located on exterior doors (deadbolts with a minimum throw of 1.5 inches are recommended), windows, and other access points. • Vulnerable areas are well lit to make them more easily observable. • Cameras have been installed to make areas visible in a different way and to deter potential wrongdoers. • Exterior doors are metal or metal-clad and have tamper-resistant locking mechanisms. • Signage limits access to authorized persons or gives instructions for secure entry. • Area for vehicles is controlled and identification of vehicles is used.
Is access within the operation limited?	• Interior doors are locked to restrict access to sensitive areas. • Key inventory is kept up to date. Keys are returned by terminated employees. Keys are not left in machinery stored outside of buildings. • Exterior ladders used to access rooftops or storage bins are secured to prevent unauthorized access. • Interior windows are secured, as necessary, to limit access to sensitive areas. • Interior vents are locked, as necessary, to limit access to sensitive areas. • Interior signage limits access to sensitive areas. • A visitor log is maintained to record visitors' identification and the date and time of their visit. • Visitors park in a designated area that is monitored. • Computer system is password-protected, has limited access, and is protected from viruses. (Wrongdoers accessing an unprotected system can alter records to conceal tampering.)
Are processes or procedures secure?	• Procedures, in general, limit access to sensitive areas and ensure vulnerable production activities are observed by one or more employees, at all times. • Machines have locked lids or secure openings, or are observed by employees to prevent tampering.

(Continued)

TABLE 3.1 (Continued)

Questions	Examples of Elements to Consider
	• Production lines are enclosed, where possible, and observed at key points to limit opportunities for tampering. • Suppliers have a food defense plan. Contracts have been negotiated with suppliers, requiring seals or locks and a procedure for checking them upon delivery. • Uniforms do not leave the operation at any time, unless with a laundry service. • Laundry service can describe the security of their operation, as well as their pickup and delivery procedures. • In-house laundry facilities are secure and have procedures for daily uniform collection and distribution. Visitor and employee personal items are not taken into production or other sensitive areas. • Visitors are supervised by an appropriate employee at all times.
Is the shipping and receiving system secure?	• Loading area has limited access and procedures to deal with security issues, such as sealing loads and recording seal numbers. • Unloading area has procedures to deal with unscheduled deliveries, checking delivery invoices, and moving deliveries into storage. • A designated employee checks package integrity before supplies are placed in storage. • Trucks and trailer bodies within the facility are secured, even when empty. • Contracts have been negotiated with carriers so that liability is with the carrier while goods or products are in their possession.
Is there an inventory system for stored materials?	• Hazardous production inputs are secured when not in use, to prevent their being used to damage or intentionally contaminate the operation. • Inventory of raw materials is reconciled with shipping invoices to identify overages or shortages, which might be an indicator of contamination. • Inventory of packaging materials is reconciled with delivery invoices. • Chemical inventories are reconciled with records of delivery and usage. • Pharmaceutical usage is noted and reconciled with inventory.

(Continued)

TABLE 3.1 (Continued)

Questions	Examples of Elements to Consider
Is access to the water supply limited?	• Water source is tamper-resistant, wellhead is locked, and external water pipes do not have openings. • Inside water lines either have locks on access points or have access points that are easily observed by multiple employees. • Ice-making facilities have limited access. For facilities separate from processing areas and not easily observed, steps have been taken to increase observation or otherwise limit the opportunity for tampering.
Is mail opened away from sensitive areas?	• Mail is opened in a room separate from production areas with a separate ventilation system.
Are there screening and training procedures for the workforce?	• Before an employee is hired, background, reference, and credit checks are run. • Employees receive basic security training on how to recognize and deal with suspicious activities and to whom to report such activities.
Is access to sensitive areas limited?	• Storage area access is limited by locked doors, entry logs, or employee observation. • Processing and packaging area access is limited by locked doors, signage that restricts access, or employee badges or color-coded uniforms that designate work areas. • Chemical storage area access is limited by locked doors, signage, entry logs, or chemical usage logs. • Maintenance area access is limited by locked doors, signage or color-coded uniforms for maintenance employees.

Source: Adapted from Lorenzen, C.L., Hendrickson, M.K., Weaber, R.L., Clarke, A.D., Shannon, M.C., and Savage-Clarke, K.L. 2010. Food Defense: Protecting the Food Supply From Intentional Harm. University of Missouri Extension MP912.

steps that are easily accessible. The FDA has designated foods that include one or more of the following processes as high risk: bulk liquid receiving and loading, liquid storage and handling, secondary ingredient handling, or mixing and similar activities. To conduct this assessment, evaluate agents of intentional contamination and identify significant vulnerabilities, which are steps that are the most vulnerable; these are termed “actionable process steps” and will become the heart of the food defense plan.

3.3 WRITE THE FOOD DEFENSE PLAN

Once an operation's vulnerabilities and actionable process steps have been identified, it is time to write the food defense plan. The general areas addressed in a food defense plan include: processing, storage, shipping and receiving, and water and ice security (FSIS, 2008). Using the questions listed in Table 3.1 will reduce these broad areas into smaller, addressable vulnerabilities. In the food defense plan, address each vulnerable element and determine whether a simple, practical, and/or economical countermeasure could be implemented to make the element more secure. For each vulnerability considered practical to address, write down a countermeasure and indicate who is responsible for implementing it and by what date. Once the countermeasure has been implemented, have the person responsible date and initial the plan.

In addition to the vulnerability assessment, it may be helpful to develop a map and operational and workforce procedures available to work through the Food Defense Work Sheet. Once the work sheet is completed and assessed, a food defense plan that can stand alone or be added to any HACCP or other plans will be available.

3.3.1 Developing Countermeasures and Mitigation Strategies

Remember, the purpose of a food defense plan is to reduce the risk of intentional contamination in an operation. Countermeasures are actions taken to make vulnerable elements of the operation more secure. These countermeasures protect employees and customers, product, reputation and livelihood, and the business, property and assets. The goal of this exercise is to provide protection in the most economical ways possible.

As a general rule, procedural changes are the most economical. For example, checking references of potential employees is easy and inexpensive. Eligibility of new hires and validity of their Social Security numbers can be checked using the free E-Verify system (http://www.uscis.gov/e-verify). Another example is repositioning employees so that they face or are seen by other employees to reduce the chance of an intentional contamination.

The next most economical option may be inclusion of technology. Will technology, such as dusk-to-dawn lighting or a lock, reduce the risk of intentional contamination? If not, additional personnel may be needed, which may become the least economical option.

Be freethinking and creative in countermeasure development, and keep in mind the three Ls suggested by the National Food Processors Association: light it, lock it, and limit access (Hollingsworth, 2002).

Mitigation strategies are required under the Food Safety Modernization Act and there is guidance from both FDA and USDA. Mitigation strategies are designed to minimize chances of adulteration by minimizing accessibility

TABLE 3.2 Example of Focused Mitigation Strategies

Actionable Process Step	Mitigation Strategy
Outside security	• Secure the facility perimeter from unauthorized persons using a locked gate. • Locate visitor parking away from the main production facility. • Establish procedures for issuing, tracking, and retrieving keys to equipment and facilities.
Inside security	• Do not allow personal items into the production, storage, and loading areas. • Limit access to in-plant laboratories to authorized personnel only. • Ensure firewalls are built into the computer network.
Ingredient preparation area	• Examine packaging and containers for signs of tampering. • Maintain access log for ingredient preparation area. • Use peer monitoring for those handling ingredients.
Mixing process	• Ensure adequate lighting around mixer. • Use clean-in-place equipment when possible. • Conduct a visual inspection before mixing.
Blast freezer	• Install surveillance cameras in the freezer area. • Secure all cleaning supplies. • Ensure that the freezer equipment is cleaned and sanitized between uses.

Source: Mitigation strategies adapted from USDA and FDA websites (FDA. 2014. Vulnerability Assessment Software. Accessed 2/22/14 at: <www.fda.gov/food/fooddefense/toolseducationalmaterials/ucm295900.htm>. and FSIS. 2008. Developing a Food Defense Plan for Meat and Poultry Slaughter and Processing Plants. United States Department of Agriculture Food Safety and Inspection Service.

of an attacker to the product, reducing the opportunity for an attacker to successfully contaminate a product, or both, and mitigation strategies should be specific and customized to the product and process. These strategies need to be effective even when an attacker has legitimate access to the facility. Examples of mitigation strategies are found in Table 3.2 and more comprehensive suggestions can be found through both the FDA and USDA websites. The mitigation strategies need to be justified and documented as part of the food defense plan.

3.3.2 Marketing Challenges

Once the more obvious vulnerabilities and countermeasures have been addressed, one will need to address some of the challenges presented by marketing. The biggest challenge marketing presents is assignment of liability; that is, determining who is responsible for protecting produce or food products from contamination at each step along the way from operation to the consumer's table. For example, one will need to determine who is liable for the food product while it is in transit or awaiting auction. Liabilities must be considered for those operations that encompass niche marketing, contract marketing, or direct marketing.

A two-pronged approach to liability is recommended. Consider the double Cs approach: check and challenge. *Check* the contract, whether oral or written, and negotiate the liability. The goal here is to make sure that companies are liable only when the food is in their possession. *Challenge* those who might want to contaminate the product by making contamination more difficult with physical barriers, such as packaging or a lock on a trailer in transit, or procedural barriers, such a supervising visitors during tours.

3.4 PREPARE A RESPONSE PLAN

The countermeasures that are developed can reduce the risk of intentional contamination, but cannot prevent it. Companies still need to prepare to deal with an intentional contamination incident so that, should one occur, the organization can quickly and efficiently contain the damage and get the operation back to normal production levels. Getting back into production as quickly as possible is key to keeping the business afloat.

In preparing a response plan, have the facility map on hand, as well as contact information for all suppliers, customers and local emergency responders. Companies also may need to refer to operational plans, such as HACCP, GAPs, GMPs, and SOPs, which may contain information valuable to the response plan, such as regulatory agency phone numbers, emergency protocols or recall plans.

To contain and minimize an emergency situation, understanding what needs to happen and in which order is essential. In the case of possible intentional contamination, the steps that need to be addressed immediately include: containment, diagnosis, recall, and disposal. Each of these steps needs to be addressed in the response plan.

3.4.1 Containment

As soon as an intentional contamination incident is suspected, isolate all product that may have been contaminated. In the food defense plan, identify a location within the facility where potentially contaminated food can be quarantined, separate from uncontaminated products. Facilities need to determine what procedures will be used to contain contaminated food products and the exact location for the containment.

3.4.2 Diagnosis

To respond to the emergency appropriately, facilities need to know what contaminant was used and how. As soon as possible, contact the appropriate person to diagnose the contaminant. If there are issues, a food processing plant will need to call a food inspector: FSIS for meat, poultry, eggs, or catfish; FDA for other foods. In the plan, include a list of emergency telephone numbers. The numbers on the list will vary, depending on the operation and its location.

3.4.3 Recall

In the event of an intentional contamination incident, contaminated food that has already left the facility will need to be recalled and contained. To effectively recall these types of products, the organization must know where all of the food or livestock has gone. Keeping reliable contact information for suppliers, customers, and processing lots will make this process much easier. Include all contact information in the plan. HACCP plans or similar operational documents will contain information related to trace forward/trace back procedures, which is a requirement for food processors. Also, because recalls often result from contamination that has been unwittingly passed on to the company by suppliers, organizations will need to prepare for that possibility in the response plan. In addition, a recall may require e-notification of customers and communicating with specific media outlets.

3.4.4 Disposal

Contaminated food must not be allowed to enter the food chain, so the response plan must include a plan for disposal of contaminated livestock or food products and possible decontamination of the facility. Regulatory agencies such as FSIS or FDA are valuable sources to help determine what type of disposal will be needed and who will need to sign off on the plan before contaminated food can be disposed of.

3.4.5 Decontamination of Facilities

A specific plan for general decontamination of the facility, including chemicals and fumigants used to clean the facility, also is needed. Areas that may require decontamination include equipment, vehicles, facilities, personnel, and grounds. Decontamination procedures, beyond general procedures, will be directed by emergency responders and regulatory authorities.

3.4.5.1 *Facility Map*

A map of the operation or facility will be vital to emergency responders in any situation. The map should provide contact information for the owner or operator of the facility and show the following:

- The facility in relationship to other properties, structures, or environmental landmarks, such as streams;
- Road access, transportation routes, perimeter boundaries, and gates, including their dimensions;
- Locations of utilities, septic, and sewer systems; and
- Buildings, with doors and windows marked, and outbuildings, as well as building systems, such as ventilation, air conditioning, and heating.

3.4.5.2 *Emergency Phone List*

When compiling an emergency or other contact lists, be sure to include the area code, even with local telephone numbers. During an emergency, calls may be made from a nonlocal phone. The numbers on an emergency phone list will vary by location and type of operation, but in general should include the following categories:

- Emergency responders, including sheriff, highway patrol, police, fire, hospital, and poison control;
- Utilities, including electricity, water, phone, and gas;
- Regulatory groups, including FSIS for meat, poultry, eggs, and catfish; FDA for other food; Animal and Plant Health Inspection Service (APHIS) for animals (the responding vet will likely start the chain of phone calls); and
- Other state agencies, including the state's Department of Health and Senior Services, Emergency Management Agency, and Department of Homeland Security.

Supplier/customer phone list: To effectively respond to an emergency that is unfolding at a fast pace, maintain a list containing the names and contact information of all suppliers and customers.

Employee emergency contacts: Maintain an emergency contact list for all employees that includes their phone numbers and addresses. Keep the list where it can be accessed quickly in an emergency.

3.5 MANAGING THE FOOD DEFENSE PLAN

Store copies of the completed food defense and response plans in more than one secure location. Keep one copy on the facility's premises and a second in a secure but accessible location outside of the operation, such as a home. Also consider saving a copy of the food defense plan online using a virtual document storage service.

Once the food defense plan has been written and implemented, the team needs to consider how the plan will be managed for the long term. Managing the plan may include periodic tests and annual reviews to see if the plan is still effectively reducing the risk of intentional contamination or if it needs to be updated to reflect changes in the operation. In addition to changes in the operation, a critical contamination event at another operation may prompt a test or review of the plan to ensure that the organization has sufficient countermeasures in place to reduce the risk of a similar incident. The food defense plan team should determine practical guidelines for managing the plan. Once the guidelines are in place, the food defense coordinator will be responsible for notifying the team when action is required.

3.5.1 Employee Training

Management of the food defense plan must also include ongoing employee training. New employees must receive basic instruction about their responsibilities with regard to the food defense plan. Training is a requirement by the FDA, along with documentation of the training. However, as of June 2016, the specific training has not been defined. All employees need to know what type of suspicious individuals or activities should be reported, who they should report suspicious individuals or activities to, which employee will be responsible for calling the authorities in a case of suspected intentional contamination, and what each employee's responsibilities are regarding security procedures, such as locking up or filing inventory or accessing log sheets at the end of the day. The team should set up procedures to ensure all employees are updated on changes to the food defense plan and to record employee food defense training activities. A record should be used to track employee food defense training and ensure these records are kept with the food defense plan.

The objective of a food defense plan is to help establishments provide a safe, high-quality product to their customers, keep employees safe and well informed, and protect the economic health of the business. A well-thought-out management plan will help the food defense plan work for companies for the long term.

3.5.2 Corrective Actions

Corrective actions are taken when a focused mitigation strategy has not been implemented properly. Currently, they do not apply to broad mitigation strategies. Corrective actions need to be detailed for each focused mitigation strategy in the food defense plan and require documentation, when implemented. Corrective actions are those activities focused on the proper implementation or the proper action to take when a focused mitigation

TABLE 3.3 Examples of Corrective Actions for Mitigation Strategies

Actionable Process Step	Mitigation Strategy	Corrective Action
Ingredient preparation area	• Examine packaging and containers for signs of tampering. • Maintain access log for ingredient preparation area. • Use peer monitoring for those individuals handling ingredients.	• Reeducate employees on the importance of examining packaging, maintaining log, peer monitoring, and recheck to determine if packaging is now being examined, access log is now maintained, and that peer monitoring is now being done.
Mixing process	• Ensure adequate lighting around mixer. • Use clean-in-place equipment, when possible. • Conduct a visual inspection before mixing.	• Increase lighting until it is adequate. • When purchasing new equipment, purchase equipment that can be cleaned in place. • Inspect product after mixing for signs of contamination.
Blast freezer	• Install surveillance cameras in the freezer area. • Secure all cleaning supplies. • Ensure that the freezer equipment is cleaned and sanitized between uses.	• Install surveillance camera. • Resecure all cleaning supplies. • Empty freezer and inspect for signs of contamination on the product and then clean and sanitize freezer.

Source: Mitigation strategies adapted from USDA and FDA websites.(FDA. 2014. Vulnerability Assessment Software. Accessed 2/22/14 at: <http://www.fda.gov/food/fooddefense/toolseducationalmaterials/ucm295900.htm> and FSIS. 2008. Developing a Food Defense Plan for Meat and Poultry Slaughter and Processing Plants. United States Department of Agriculture Food Safety and Inspection Service).

strategy is not sufficient, such as tampered packaging, which would be addressed by GMPs. Corrective actions for food defense plans, unlike HACCP, do not require that the affected food be tested for food safety, due in part to the low frequency of intentional contamination events (Table 3.3).

3.5.3 Verification

3.5.3.1 Plan Reviews

Reviews of the food defense plan must be conducted every 3 years, at a minimum, but also can be triggered by changes in the operation, such as a new product line or category of livestock, change of supplier, expanded customer base, addition of new technology, newly developed or updated procedures, or change of food defense coordinator. The review should answer the following questions:

- Are the countermeasures continuing to reduce the risk of intentional contamination in vulnerable areas?
- Do new products or categories require additional countermeasures to reduce the risk of intentional contamination?
- Do new or updated procedures require additional countermeasures?
- Has supplier, customer, and employee contact information been updated?

3.5.3.2 Plan Tests

Tests of the food defense plan can be conducted randomly or scheduled two to four times a year, as determined by the food defense team, which should select an interval that is practical for the operation. The general purpose of these tests is to determine if the countermeasures are reducing the risk of intentional contamination. If the countermeasures are not adequately reducing risk, then new countermeasures should be developed and implemented. Tests that might be used include exercises in mock tampering, product quarantine, product recall, random food security checks, and computer system challenges. Specific areas to be checked include entry points to ensure that they are locked or secured, signage to ensure it is in place and legible, procedural compliance regarding uniforms and employee personal items, inventory log sheets to ensure they are being filed and properly maintained, and entry logs maintained for sensitive areas to be sure they are accurate and up to date.

3.5.3.3 Conducting a Food Defense Audit

Audits can be as simple as "...an official inspection of...an organization's accounts, typically by an independent body" or detailed such that they entail "...a systematic examination and verification of...accounts, transactions or other relevant documents, and physical inspection...by qualified auditors." Just like yearly reassessments of HACCP plans, periodic assessments or audits of food defense plans should be completed to ensure that the process and plan are sound. When it comes to auditing food defense plans, it is recommended that internal as well as external audits be performed with some degree of frequency. For internal audits, establishments may want to consider quarterly evaluation by management, while external audits may need to be conducted annually by an outside agency or third party. Either way, the goals of the audit(s) should be to ensure that the food defense plan remains relevant to the operation. It is important that audit information and frequency also be included in the food defense plan. There are several third-party auditors or audit programs, including self-assessments that can be applied to food manufacturing and/or processing facilities (see Table 3.4).

TABLE 3.4 Food Defense Self-Assessment or Checklists

US Food and Drug Administration	http://www.fda.gov/downloads/Food/GuidanceRegulation/ucm125192.pdf
USDA-FSIS: Industry Self-Assessment Checklist for Food Security	http://www.fsis.usda.gov/shared/PDF/Self_Assessment_Checklist_Food_Security.pdf
New York Department of Health	http://www.health.ny.gov/publications/7079.pdf
Ohio Department of Agriculture	http://medinahealth.org/images/company_assets/d98a6e31-3e37-43ff-bc1a-ecc84e8f1117/SelfAssessmentChecklist_4e66.PDF
USDA-FSIS Slaughter and Processing Establishments	http://www.fsis.usda.gov/wps/wcm/connect/63b6a057-ee99-41a0-813a-557cfb7f1c05/Slaughter_Plant_Checklist.pdf?MOD=AJPERES
Connecticut Department of Public Health Food Protection Program	http://ccthd.org/documents/foodoperatorsguide122310.pdf

3.5.4 Record Keeping

Record keeping starts with developing and writing the food defense plan. Establishments will want to have at least two copies of the food defense plan: one that will be kept in the food processing facility and one that will be kept offsite in the event that an intentional contamination or other emergency occurs and the team does not have access to the facility. The food defense plan will include the food defense team, vulnerability assessment with justification, mitigation strategies with documentation of implementation and justification, and monitoring procedures, including frequency, corrective actions, and verification activities. The second part of record keeping, which should be kept in a separate notebook or file in the facility, would include records relating to monitoring activities, corrective actions, and verification. Records have to be maintained for 2 years to comply with US government regulations; however, records must be accessible within 24 h when requested by regulatory agencies. Examples of methods for retaining records include keeping the original document, scanning the original document into an electronic database, or collecting data in an electronic form.

REFERENCES

Federal Register. 2016. Mitigation Strategies to Protect Food Against Intentional Adulteration: Final Rule. 81(103) part IV, May 27, 2016.

FDA. 2014. Vulnerability Assessment Software. Accessed 2/22/14 at: <http://www.fda.gov/food/fooddefense/toolseducationalmaterials/ucm295900.htm>.

FSIS. 2008. Developing a Food Defense Plan for Meat and Poultry Slaughter and Processing Plants. United States Department of Agriculture Food Safety and Inspection Service.

Hollingsworth, P., 2002. Hot topics address terrorism, fickle consumers, and obesity. Food Technol. 58 (8), 48, 50-52.

Lorenzen, C.L., Hendrickson, M.K., Weaber, R.L., Clarke, A.D., Shannon, M.C., and Savage-Clarke, K.L. 2010. Food Defense: Protecting the Food Supply From Intentional Harm. University of Missouri Extension MP912.

Chapter 4

Creating a Food Defense and Response Plan in Complex Food Production Systems

M. Seeger[1], T. Sellnow[2] and E.L. Petrun[3]
[1]*Wayne State University, Detroit, MI, United States,* [2]*University of Central Florida, Orlando, FL, United States,* [3]*University of Maryland, College Park, MD, United States*

4.1 INTRODUCTION

Food safety is a recurring technical and management challenge, which is further complicated by constantly evolving public perceptions. Currently, the food supply in the United States remains one of the safest in the world. The Centers for Disease Control and Prevention (CDC), however, estimates that 76 million people get sick, more than 300,000 are hospitalized, and 5,000 Americans die each year from foodborne illness (Mead et al., 2000). More than 200 known diseases are transmitted through food. The food supply is susceptible to both unintentional and intentional contamination by a wide range of other agents. Fortunately, there are only a handful of documented instances where the food supply has been intentionally adulterated, although the threat of such contamination is very real. The continued industrialization of food systems, including the development of long, complex supply chains and distribution channels, has complicated efforts to guarantee a safe food supply.

Efforts to understand crises in industrial systems draw heavily on the principles of complex systems theory or chaos theory (Perrow, 1984; Seeger et al., 2003). These approaches emphasize the dynamic and nonlinear nature of highly complex systems. As systems become centralized, increasingly complex and tightly coupled, the probability of unforeseen interactions increases. These interactions carry the potential to create a crisis, or what system theorists call bifurcation. In bifurcation, a system is fundamentally altered in some dramatic way. Thus, a system designed to distribute safe food may function to spread contamination.

Food Protection and Security. DOI: http://dx.doi.org/10.1016/B978-1-78242-251-8.00004-7

Although the modern food production system is very safe, it is increasingly efficient, dynamic, integrated, tightly coupled, and complex. Interestingly, the US food supply is exceedingly centralized, as food companies become larger, more vertically integrated, and fueled by added producers who provide escalating inputs to existing organizations. Today, food travels through agricultural production supply chains on both domestic and international farms, to orchards and ranches, through transportation, to processing in industrial settings, to distribution, into wholesale and retail outlets, and on to the consumer and storage. Modern food production is very susceptible to systemic breakdowns during any number of these steps. This extended chain of production, often expressed with the phrases "from farm to fork," or "from seed to shelf," inherently creates vulnerabilities. Industrial, mass production of food products and width of distribution has added to the complexity and increased the chances that an adverse event will be quite widespread.

Greater emphasis on efficiency and smaller profit margins may also serve to reduce slack resources and buffers that may have served to contain crises. The use of technology, such as automated production, while reducing some threats, has introduced others and further enhanced overall complexity. Finally, globalization of food production and distribution has added additional levels of intricacy and reduced levels of predictability. In the global food market, food is produced under a very wide range of regulatory, cultural and economic contexts. These features of the food production and distribution system are all illustrated in the cases presented here.

Insuring the safety and reliability of food production systems is a multistage process involving appropriate risk awareness, communication, mitigation resources, and appropriate response strategies. This chapter describes the difficulty of providing food defense within complex, globalized, and highly dynamic food production systems. The high reliability organization and mindfulness framework is proposed as a useful approach for managing and mitigating risks (Weick and Sutcliffe, 2007). The CDC's Crisis and Emergency Risk Communication (CERC) crisis planning template is presented as a resource to help prepare for and respond to an event (Reynolds and Seeger, 2012).

4.2 FRAMING RISKS AND RESPONSES

Many efforts to address issues of risks and their manifestation in the form of a crisis focus on the stages of development. These approaches seek to identify the structure of a crisis both to enhance understanding and to facilitate effective management of risk. Phase or stage models describe how a crisis will develop and evolve over time in a relatively predictable pattern. The result is the description of a series of relatively general and discrete stages or phases that describe the unfolding of crises generally regardless of the industry, organization, or cause. These approaches also allow for anticipating communication and informational needs over the life cycle of a crisis. As such, they are particularly useful frameworks for risk and crisis management.

Next, we introduce food defense and response plans by discussing a simple three-phase model of precrisis, crisis, and postcrisis. This model is used widely by organizational crisis theorists and is probably the most widely used framework in part due to its simplicity. During precrisis, an emerging threat or risk develops and interacts with some other aspects of a system. This process is typically described as an incubation or gestation period where the magnitude of a threat grows and interacts in unanticipated ways. Often, this incubation involves a risk judged by managers as relatively minor interacting in a nonlinear and disproportional way with other factors. Sometimes, organizations have failed to conduct adequate risk assessment. In other cases, threats converge or connect and interact with new deficiencies or fallacious assumptions about risk. Another common interaction concerns the level of threat preparation interacting with other system needs. In the case of Peanut Corporation of America, for example, what was perceived as relatively minor issues of sanitation interacted with poor inspection procedures resulting in a massive salmonella outbreak in 2008 and 2009. The recall of Peanut Corporation of America was complex and extensive because the company's product, peanut paste, was an ingredient in dozens of other food products. The outbreak was eventually associated with almost 700 reported illnesses and 9 deaths.

The second stage, or the crisis stage, begins with the trigger event and a general recognition that a crisis has occurred. Most often the trigger event is some dramatic, sudden occurrence that signals a severe disruption of the system and onset of harm or the potential for harm. In other cases, the realization that a crisis is occurring is a slow realization. The crisis stage continues until the harm is contained and the organization has returned to some relatively normal operation. One of the challenges in the case of contaminated food, however, is that an outbreak of foodborne illness is usually not immediately identified. Tracking the outbreak back to the source usually takes even longer. Disease surveillance systems have become much more sophisticated in the last decades and advances in epidemiology allow for rapid strain typing of bacteria (CDC, 2012). In the case of the 1993 Jack in the Box outbreak of *Escherichia coli* O157:H7 bacteria, even with good local surveillance, it took 39 days to determine that a serious outbreak was happening. By the time the outbreak was identified, traced to the source, and contained, the deaths of 9 children and 600 illnesses were associated with eating undercooked hamburgers from Jack in the Box (Bottemiller, 2013).

The final stage, postcrisis, begins when the harm, drama, confusion, and uncertainty of the crisis dissipate and some sense of order is reestablished. Postcrisis is generally accompanied both by a sense of relief and recognition of the loss that has occurred. This stage typically involves an intense investigation and analysis that includes efforts to create plausible explanations of what went wrong to answer: why, how, who is to blame, and what should be done to prevent future crises. Apologies may be warranted to those who were harmed and further reflection could provide lessons learned from the incident.

In August 2008, the Canadian food company Maple Leaf Foods announced a recall for several products linked to a listeria outbreak. The outbreak would eventually be associated with five deaths. Maple Leaf Foods acted very quickly to close the plant where the products were produced and recalled products. By taking this aggressive, proactive stance, the company demonstrated sincerity and commitment to consumers. On August 23, 2008, Maple Leaf CEO, Michael McCain, issued a public apology accepting responsibility for the outcome. In the video posted on YouTube, he acknowledged that the company was at fault and described what he was doing. He expressed his deep personal expressed sympathy for those affected: "... our best efforts (to keep customers safe) failed and we are deeply sorry." McCain noted that his company exists in a "culture of food safety" and that "We have a unwavering commitment to keep our food safe." Finally, McCain acknowledged that, "We know this has shaken your confidence in us. I commit to you that our actions are guided by putting your interest first" (http://www.youtube.com/watch?v = zIsN5AkJ1AI).

The postcrisis stage can provide direction for both food defense systems and crisis response plans. Defense of the food system is a precrisis process and activity that involves risk analysis, assessment, developing appropriate process structures, norms, capacities, and values to insure the ongoing production of a safe product (Novak and Sellnow, 2009). Crisis planning is also a preevent activity that is manifest in the crisis and postcrisis stages. Crisis planning establishes procedures for operating in response to the unique conditions created by a crisis and is necessary but not always sufficient for an effective response.

In the following sections we describe processes and procedures for food defense as an ongoing process of risk assessment and analysis. We describe general methods and procedures for crisis planning that will aid a food organization to respond more effectively to a crisis.

4.3 PROCESSES AND PROCEDURES FOR FOOD DEFENSE: HIGH RELIABILITY ORGANIZATIONS

In every aspect of the food industry, organizations rely on well-planned routines to collect, transport, clean, process, package, and distribute their products. These routines provide both efficiency and safety for consumers. Yet, at every step of the process the routines of food production are susceptible to pathogens that can enter the production process and contaminate consumer products. To further complicate the issue, food production procedures are vulnerable to intentional attacks ranging from sabotage to full-blown terrorism (Doeg, 2005; Mohtadi and Murshid, 2009). The routines of production warrant constant review and updating.

The routines that operate in the food industry should be reviewed mindfully. A mindless approach occurs when routines are applied with such calloused regularity that workers and managers forsake the conscious observation and reporting of failures that is necessary to preserve high standards of food safety. By contrast, mindfulness is characterized by workers and managers whose "attention naturally goes to what is different and out of balance" (Langer, 2009, p. 13). In other words, food industry workers are mindful when they notice minor failures throughout the production process that could result in the contamination of products. Ideally, every member of the organization exhibits a "high level of sensitivity to errors, unexpected events, and—more generally—to subtle cues suggested by the organization's environment or its own processes" (Levinthal and Rerup, 2006, p. 503).

Mindfulness is an essential characteristic in maintaining high reliability in organizations and requires effective communication about risks. Weick (1993) suggested that organizations begin the high reliability process by making sense of the minor failures they observe. This sensemaking process includes creatively solving problems, clearly defining roles among employees, continually questioning assumptions—even those with a long history. A willingness to encourage employees to voice their concerns and to follow up on issues is also important. More recently, Weick and Sutcliffe (2007) advocated a series of specific steps or commitments based on effective communication.

High reliability organizations (HROs) typically encounter risk on a regular basis. They respond to those risks through the mindful use of techniques for observing failures, gathering the best information available, and engaging in corrective action. Novak and Sellnow (2009) found that the mindful participation of employees at every stage of food production can reduce the risk of both unintentional and intentional contamination. They observed that food production workers do notice production problems and, if a supportive environment exists, are willing to report problems to supervisors. The HRO model offers a series of specific suggestions for anticipating and containing problems (Weick and Sutcliffe, 2007).

4.4 ANTICIPATING PROBLEMS

HROs anticipate problems by displaying a preoccupation with failure, reluctance to simplify, and sensitivity to their operations. When organizations use these strategies and open communication to accompany them, they increase their potential to identify and address a risk before it becomes a crisis. Anticipation is based on scrutinizing all potential vulnerabilities. This scrutiny helps workers and supervisors to avoid oversimplifying tasks to a point of encouraging a mindless application of routines. Food processing organizations should anticipate problems at every stage ranging from harvest to delivery of a packaged product to retail outlets.

4.4.1 Preoccupation With Failure

The observation, reporting, and analysis of even minor failures is critical to the HRO approach. Without a willingness to reconsider routines based on minor failures, organizations often develop a false sense of confidence. In fact, many organizations interpret near misses as a sign of resilience when in fact a crisis was averted by luck or chance rather than by procedural effectiveness (Tinsley et al., 2012). To counter this preoccupation with failure, the HRO model suggests embracing failure more than success (Weick and Sutcliffe, 2007). In short, a preoccupation with failure assumes that risk mistaken for safety is far more threatening than safety mistaken for risk.

4.4.2 Reluctance to Simplify

Efficiency and simplicity are not always synonymous. A simple system may be efficient. If, however, that simplistic system allows harmful bacteria to contaminate a food product, the ensuing harm may be quite complex. The HRO management philosophy resists the temptation to simplify. Organizations should constantly create new categories for interpreting the risks and procedural challenges (Weick and Sutcliffe, 2007). Operations or procedures will function best when differentiating categories or labels are accepted by employees. Rather than categorizing all risks into one category, HRO uses subcategories and scrutinizes "examples that fit the category imperfectly to see what *new* category they suggest" (p. 58). This reluctance to simplify enables HROs to create a language for risk communication that is more sensitive to the dynamic nature of risk. It also allows the HRO to see differences in kinds of risks that might otherwise be overlooked.

4.4.3 Sensitivity to Operations

Simply put, organizations display sensitivity to operations when they attend to "the messy reality inside most organizations" (Weick and Sutcliffe, 2007, p. 59). Routine procedures can lead employees to mindlessly assume that their intentions and expectations reflect reality. In truth, these assumptions can create a false sense of confidence that inhibits the recognition of failures. Sensitivity is the antithesis of these mindless assumptions. Sensitivity to operations is accomplished when organizations "focus on actual work rather than intentions, define actual work by its relationships rather than its parts, and treat routine work as anything but automatic" (p. 62).

4.5 CONTAINMENT

Anticipation strategies are designed to help organizations see new and emerging risks and to respond to these emerging problems before they result in crises. The HRO model accepts that not all risks can or will be observed before they manifest into threatening conditions. Thus, the model introduces containment as a means for responding to events "mindfully and swiftly" after they have occurred (Weick and Sutcliffe, 2007, p. 65). Strategies for doing so include a commitment to resilience and deference to expertise.

4.5.1 Commitment to Resilience

From an HRO perspective, organizations are resilient when they are "mindful of events *that have already occurred* and ... correct them before they worsen and cause more serious harm" (p. 68). Doing so requires organizations to engage in precrisis planning (Seeger, 2006). The National Center for Food Protection and Defense explains that such precrisis planning involves establishing communication networks, assigning communication roles, and having the resources needed before a crisis event occurs (Sellnow and Vidoloff, 2009). Organizations committed to resilience prepare themselves for crises so that, despite the uncertainty, they have the strategies and resources in place to continue to function, recover quickly, and adapt their operations after such events.

4.5.2 Deference to Expertise

The HRO model does not prioritize organizational hierarchies in the management of risk. In fact, such hierarchies can preclude an organization's decision makers from receiving the most accurate and informative input regarding failure. Deference to expertise occurs when organizations allow information to flow in all directions throughout the organization so that expertise is shared both upward and downward. For example, regular feedback from line workers can help supervisors regularly revise routine procedures to adjust to evolving risks. Organizations that diminish the role of worker input at any level increase the potential for minor failures to grow into serious crises.

4.6 PLANNING A RESPONSE TO A FOOD-RELATED CRISIS

While the HRO approach can significantly reduce the occurrence of crises in the food industry, some will occur. Eventually, food products contaminated with serious pathogens will be consumed by the public resulting in illness. Some of these illnesses will be serious and several will result in deaths. In these cases crisis preparation, including a crisis plan, is essential. What is crisis preparation and how can a food company be prepared? Preparing for a crisis is a multifaceted and ongoing process which involves a variety of steps. Preparation includes understanding how a crisis is likely to evolve, training personnel, cultivating crisis response capacity, and developing an operations plan alongside a communication plan.

While the HRO approach can significantly reduce the occurrence of adverse events in the food industry, crises do occur. Contaminated products may be distributed and consumed by the public resulting in illness. Some reported cases will be very serious and invariably consumer death is always a possibility. In these cases, crisis preparation is essential. Food companies can contribute to their preparation by assembling a crisis communication plan. While a crisis plan does not insure an effective response, it significantly increases the chances that the response will be carefully thought out and timely, and that critical skills and resources are available.

Crisis preparation is a complex and ongoing process. To be successful, communicators will need to understand how crises evolve, the role of training, how to establish response capacities, and how to develop an operations plan and a communication plan.

4.6.1 Crisis Phases

The CDC developed a five-stage model to help identify specific communication activities associated with a public health emergency. These include: (1) precrisis, (2) initial event, (3) maintenance, (4) resolution, and (5) evaluation (Reynolds and Seeger, 2012). Crisis preparation and planning can take advantage of these stages to indicate what will need to happen during each stage.

4.6.1.1 Precrisis

Planning and preparation can help an organization be ready once a crisis emerges. There are certain challenges that are unique to the food industry that should be anticipated. For example, there should be a process in place to determine the source of a contamination. Organizations should anticipate predictable crises and form appropriate responses in advance. The precrisis phase affords organizations time to plan for common crisis events, develop chains of command, assign tasks, select spokespersons, train staff, form alliances with partner organizations, draft messages, and collect resources.

4.6.1.2 Initial Event

When a crisis emerges, the time to plan has passed. Responders will need to activate existing plans as quickly as possible while still verifying facts about the event. Once a basic understanding of the event is available, the communication plan can be activated and amended to fit the current situation. During the initial stage the public needs to know how current risks will impact them personally, what they should do to protect themselves, where to get more information, which agencies are tasked with responding to the event, and ultimately who will be responsible for fixing the problem. Food events often result in recalls which need to be communicated in a timely manner and designed to reach the target audience. The public will also want to know the basics of who is at fault, what happened, where and when the crisis occurred, and why current procedures or policy failed to protect them.

4.6.1.3 Maintenance

Crises generally enter the maintenance stage once the immediate danger from the event is contained. Once the shock of the initial event subsides, responders will need to answer questions about fault, the likelihood that the crisis could have been prevented, what will be done to ensure the crisis never happens again, and finally what the organization will do differently in the future. Instead of simply reporting facts, the organization may need to respond to questions and criticism from the media. In some cases, organizations have apologized for the harm caused by a contaminated product. Moreover, although the crisis may seem to be contained, communication surrounding the event could escalate as new details emerge.

4.6.1.4 Resolution

There is no clear-cut moment that defines a shift to the resolution phase. The exact amount of time depends on how quickly responsibility is defined (i.e., if an investigation is necessary) and how quickly those affected recover. Be prepared to truly examine what went wrong, improve organizational capabilities to control future risks, engage in communication to bolster public support, and potentially draw attention to systemic failures outside of organizational control that need to be addressed (e.g., a food safety policy that needs to be changed).

4.6.1.5 Evaluation

It is important to allocate time to revisit the communication plan and make note of what worked well and what failed. Ignoring lessons learned will increase the chances that an organization will repeat a mistake again in the future. Consider archiving communication documents along with a final report which reflects collective understanding of the event for future use.

4.6.2 The Unique Nature of Food Crises

Food safety professionals can anticipate some of the challenges unique to their industry. The first is determining the source of contamination. For example, in 2007 ConAgra responded to a *Salmonella* outbreak in Banquet brand frozen poultry pot pies. At first, ConAgra concluded that consumers were in fact undercooking the pot pies (Sellnow and Petrun, 2009).

Later, the company learned that in fact prescribed cooking instructions for the pot pies were incorrect. Unfortunately, this misunderstanding implied that ConAgra was simply trying to shift the blame for the contamination to consumers instead of assuming responsibility. If possible, it is helpful to anticipate who will be in charge of an initial investigation, who would need to be notified about a contamination (e.g., other industry partners, distributors, consumers), and how each of those groups could be reached.

Food illness can be difficult to identify at first and different types of information may need to be communicated at different times. Organizations will also need to work with the US Food and Drug Administration and/or the US Department of Agriculture. Recalls are usually voluntary and conducted by the manufacturing organizations, although in some instances if a company has failed to identify a contamination, recalls may be mandated. During the precrisis phase organizations should anticipate the process of both communicating with the public, reporting to regulatory agencies, and releasing internal updates and calls to action for employees.

4.6.3 Training and Developing Crisis Response Capacity

Crisis preparation should begin by assessing current crisis response capabilities. Weick and Sutcliffe (2007) suggest performing "conscious audits" to help anticipate current capabilities (p. 83). Knowing what response capacity is available, for example, means that organizations should know what needs to go right and how things can go wrong. Companies can also assess mindfulness to see where employees, departments, and leadership fall short. Identifying organizational strengths and weaknesses will help in developing effective crisis plans.

Several training options are available to bolster capabilities during the precrisis phase. For example, managers can help employees learn crisis procedures by holding drills. Drills typically test a part of the crisis plan. Another option is to facilitate an exercise that simulates a larger-scale organizational response within a realistic scenario. Exercises allow responders to test policies and procedures under pressure and become familiar with operations. Exercises and drills should be conducted when crisis response operations are created or changed, and typically at least once a year.

Additionally, planning should anticipate surge capacity. For example, during a national or international contamination, additional communication staff, public health professionals, spokespersons, administrative support, among other skills, may be needed. Anticipating surge capacity should assess what internal staff can manage, when contracted or additional staff would be called, and how other organizational partners could contribute to the response. While it is usually impossible to know initially how widespread a recall will be, organizations need to be ready to assume the greatest level of damage has happened so they can be ready to respond should a worst-case scenario occur.

Finally, organizations will need to cultivate relationships with external audiences prior to including them in a crisis plan. Strategic partners could include industry groups, employees, investors, media, or nongovernment organizations, or others. Alerting these key groups to crisis plans in advance, and sharing finalized crisis-planning documents with them, will allow them to be prepared to work with a responding organization. Ideally, organizations will work together to disseminate consistent messages and avoid communication breakdowns (Seeger, 2006).

4.6.4 The Emergency Operations Plan and Communication Plan

An emergency operations plan outlines procedures for mitigating the harm surrounding a crisis. In the case of food systems, this may include halting production and shipment, inspecting, cleaning, and repairing equipment and operations, securing information and records, and notifying appropriate agencies, among other activities. An emergency operations plan will specify who is responsible for these activities and specify procedures and steps. Typically the plan specifies the team members who will manage the crisis response. Alongside outlining technical procedures, the communication plan will detail policy and procedures to communicate with implicated stakeholder groups.

Ideally, a communication plan will be developed alongside the emergency operations plan. While information may overlap in some areas, the communication plan will delineate roles, responsibilities, and resources to reach the public, media, government, and nongovernment organizations during a crisis. While communication plans will vary organization to organization, several foundational elements are essential including (Reynolds and Seeger, 2012):

A note from leadership. An introductory note from leadership endorsing a crisis plan signifies support for a plan and inspires confidence in those who will follow the plan.

Public information team responsibilities. Each team should be listed in the document, along with an outline of which team will be held accountable for assigned tasks.

Information verification and clearance policy. Clearance procedures for information should be established in the precrisis stage. Names of who approves what and when should provide a detailed overview of the clearance process. This information is imperative to ensure information is not released without verification and approval.

Media contact lists. Compile any contact that could be needed from local, state, and national media.

Coordination information. Discuss how multiple organizations would work together during a response. Include a point of contact for emergency response partners.

Spokespersons. Identify approved and media-trained spokespersons.

Emergency response team member's contact information. Include each team member's full contact information so they can be reached day or night.

Procedures to acquire resources. Overview how additional resources, including space, equipment, and personnel, should be acquired.

How information will be disseminated. Crisis response can use many methods to release information, spanning from traditional newspapers and television to new media and other digital tools. Plan what channels will work best to reach anticipated stakeholder groups.

Identified list of key stakeholder groups. Finally, define stakeholder groups and list known points of contact. Stakeholder could include anyone that might contact the organization (Coombs, 2012).

Communication planning will yield a document unique to each organization. There is no prescribed length or design format; rather, plans can be composed in a style and format consistent with other organizational documents. It is likely that no two organizations will have the exact same plan. Plans should be revisited, updated, and revised as organizations adapt to current events and new threats.

4.7 CONCLUSION

Creating a food defense and response plan in complex food production systems is an ongoing process that should be matched to the evolving risks an organization faces. Risks in the food industry are continuously changing as products, production systems, consumers, and regulations change. The high reliability approach is useful in maintaining strategic risk awareness. Regardless of vigilance, food companies still face some likelihood that crises will occur. In these cases, the crisis planning process and a crisis plan can be critical to mounting an effective response.

REFERENCES

Bottemiller, H., 2013. Outbreak detection since Jack in the box: a public health evolution. Food Safety News. Available at: <http://www.foodsafetynews.com/2013/02/outbreak-detection-since-jack-in-the-box-a-public-health-evolution/#.UssLDKkeXdl>.

Coombs, W.T., 2012. Ongoing Crisis Communication: Planning, Managing, and Responding, Third ed. Thousand Oaks, CA: SAGE.

Centers for Disease Control and Prevention. (September 24, 2012). Foodborne Illness, Foodborne Disease, <http://www.cdc.gov/foodsafety/facts.html#tracking>.

Doeg, C., 2005. Crisis Management in the Food and Drinks Industry: A Practical Approach, Second ed. Springer-Science + Media, Inc, New York.

Langer, E.J., 2009. Counterclockwise. Ballantine Books, New York.

Levinthal, D., Rerup, C., 2006. Crossing an apparent chasm: bridging mindful and less-mindful perspectives on organizational learning. Organi. Sci. 17, 502–513.

Mead, P.O.S., Slutsker, L., Dietz, V., McCaig, L.F., Shapiro, C., Griffin, P.M., et al., 2000. Food related illness and death in the United States. Emerg. Infect. Dis. 5, 5. Online: Downloaded June 15, 2005 Available: <http://www.cdc.gov/ncidod/eid/vol5no5/mead.htm>.

Mohtadi, H., Murshid, A.P., 2009. Risk analysis of chemical, biological, or radionuclear threats: implications for food security. Risk Anal. 29 (9), 1317–1335. Available from: http://dx.doi.org/10.1111/j.1539-6924.2009.01260.x.

Novak, J.M., Sellnow, T.L., 2009. Reducing organizational risk through participatory communication. J. Appl. Commun. Res. 37 (4), 349–373.

Perrow, C., 1984. Normal Accidents. Basic Books, New York.

Reynolds, B., Seeger, M.W., 2012. Crisis and Emergency Risk Communication, 2012 edition. Centers for Disease Control and Prevention, Atlanta, GA. Available: <http://emergency.cdc.gov/cerc/pdf/CERC_2012edition.pdf>.

Seeger, M., Sellnow, T., Ulmer, R.R., 2003. Communication and Organizational Crisis. Book. Quorum Press.

Seeger, M.W., 2006. Best practices in crisis communication: an expert panel process. J. Appl. Commun. Res. 34, 232–244. Available from: http://dx.doi.org/10.1080/00909880600769944.

Sellnow, T.L., Petrun, E.L., 2009. ConAgra: audience complexity in risk communication. In: Sellnow, T.L., Ulmer, R.R., Seeger, M.W., Littlefield, R.S. (Eds.), Effective Risk Communication: A Message-Centered Approach. Springer Science + Business Media, LLC, New York, NY, pp. 119–129.

Sellnow, T.L., Vidoloff, K.G., 2009. Getting crisis communication right: eleven best practices for effective risk communication can help an organization navigate the slippery path through a crisis situation. Food Technol. 63 (9), 40–45.

Tinsley, C., Dillon, R., Cronin, M., 2012. How near-miss events amplify or attenuate risky decision making. Manag. Sci. 58 (9), 1596–1613.

Weick, K.E., 1993. The collapse of sensemaking in organization: the Mann Gulch disaster. Admin. Sci. Q. 38 (4), 628–652.

Weick, K.E., Sutcliffe, K.M., 2007. Managing the Unexpected: Resilient Performance in an Age of Uncertainty, Second ed. Jossey-Bass, San Francisco, CA.

Chapter 5

A Data-Driven Approach to Food Safety Surveillance and Response

N.P. Greis and M.L. Nogueira

University of North Carolina at Chapel Hill, Chapel Hill, NC, United States

5.1 INTRODUCTION

Maintaining a safe and secure food supply is critical to the well-being of millions around the world. An increasingly global food chain—in which products are sourced from locales far from the end consumer—has increased the potential for contamination. These pressures have only increased the sense of urgency in addressing gaps in the food safety system. In particular, early detection and rapid response are challenges that must be met to minimize the impact of a contamination event—whether due to unintentional failure of the food chain or due to an intentional terrorist act. This chapter explores the potential of data-driven informatics tools to provide situational awareness and decision-making intelligence for an intrinsically complex and dynamic process—the detection of and response to a foodborne illness outbreak. A data-driven approach is introduced that builds situational awareness by coalescing real-time data fusion of both traditional and nontraditional sources, analytics based on tools of data science, visualization using a Common Operating Picture (COP), and real-time collaboration across stakeholders of the system to reduce the latency in detecting an emerging contamination event. By reducing the latency of detection, responses such as medical alerts and product recalls can be accelerated, thereby saving lives and cost. These principles of situational awareness were used to develop a prototype software tool for the State of North Carolina, the North Carolina Foodborne Events Data Analysis Tool or NCFEDA. Latencies reductions in surveillance and response are illustrated using a typical example—a cluster of unspecified illness cases reported with symptoms of gastrointestinal distress that may (or may not) indicate a possible foodborne disease outbreak.

Food Protection and Security. DOI: http://dx.doi.org/10.1016/B978-1-78242-251-8.00005-9

5.2 CHALLENGES OF FOOD SAFETY

Recent high-profile contamination events have elevated the need to adopt a data-driven approach to assure the safety of the country's food system. One of the most widely reported contamination events was the recent closure of more than 40 restaurants belonging to Chipotle, a US-based fast-food restaurant chain, in Washington and Oregon in October 2015 due to an *Escherichia coli* contamination. The Centers for Disease Control and Prevention (CDC) reported that 45 people were sickened by the *E. coli O26* outbreak strain and, of those, 43 reported eating at Chipotle. Sixteen people were hospitalized although no deaths were reported. In February 2016, the CDC concluded their investigation, unable to find the source of the *E. coli* contaminations.

The CDC has also linked the Chipotle outbreak in the US Pacific Northwest with other reported *E. coli* cases in California, Ohio, New York, and Minnesota. And only a few months before, Chipotle had been linked to two other cases of foodborne contamination and resulting illness—a norovirus outbreak in California in August and cases of *Salmonella* in Minnesota that have been traced to tomatoes from out-of-state farms. Then, in December 2015, 80 individuals were sickened after eating at a Chipotle restaurant in Massachusetts. And before the chain of outbreaks, Chipotle had taken the step of removing pork from its restaurant menus when one of the company's suppliers failed to follow animal welfare standards.

Despite recent efforts and the passage of the Food Safety Modernization Act (FSMA) in 2012, foodborne infections continue to be an important public health problem in the United States. Federal data released by the Foodborne Diseases Active Surveillance Network (FoodNet) in 2015 showed little improvement in terms of foodborne illnesses when compared with data collected between 2006 and 2008, and between 2011 and 2013. The data indicated that illness due to *Campylobacter*—usually caused by consuming undercooked poultry—has risen by 13%. In addition, illnesses from two strains of *Salmonella*, javiana and infantis, typically found in undercooked eggs, milk, and meat, have more than doubled. And *Listeria*, the likely culprit in this year's massive Blue Bell Creameries outbreak in the United States, was responsible for the most deaths of any strain last year. Of the 118 people who were diagnosed with listeriosis, 18 of them died.

The problems experienced by Chipotle and other food purveyors are emblematic of the challenges faced by today's food industry. Our food supply chains are dynamic and complex—with an array of governmental agencies at different jurisdictional levels charged with regulating and supervising the safety of millions of food products produced by thousands of companies across the globe. Assuring safe food depends critically on our ability to collect, interpret, and disseminate electronic and other information across organizational and jurisdictional boundaries. The lack of visibility due to interoperability across the stakeholders of the food chain makes it difficult to quickly determine when a contamination has taken place. And once a contamination has been confirmed, the lack of visibility makes it difficult to trace contaminated food products back to the farm or country where they were produced—and also forward to locations where similar products may be waiting to be sold.

In response to these challenges, the food industry has looked to data science and "big data" for insight and a way forward. Significant efforts are being made to marshal big data tools to the cause of food safety. The definition of big data remains in flux depending on industry and application, but it typically involves the digital generation of data, often passively produced and automatically collected and stored, but also actively generated through events that serve as a trigger for marshaling response to an emerging contamination event. The premise of data science and big data for improved food safety is that, when fusing multiple types and formats of data including new and nontraditional sources, new analytics will make it possible to enhance our visibility of the food system to better monitor and respond in (near) real time to contamination threats as they occur.

5.3 MOVING TO DATA-DRIVEN FOOD SAFETY

Major advances in many industries can be attributed to the convergence of multiple technological advances whose synergistic effects enable major transformation within that industry. Defined as the coming together of two or more disparate disciplines or technologies, convergence has been associated with advances from early in the industrial age—from firearms and sewing machines at the beginning of the 20th century to jet engines today. The fax revolution was produced by a convergence of telecommunications technology, optical scanning technology, and printing technology. Today fundamental shifts in our basic industries are emerging from a confluence of the internet and related communication technologies with technical advances in specific domains—including the food industry.

A fortuitous and simultaneous convergence of internet and communications technologies along with a new generation of sensors and analytical tools is reshaping the food industry—and its ability to reduce the risk of food contamination and resulting foodborne illness. The availability of low-cost sensors, scanners, and various mobile devices, along with new communications technologies linked to the internet, offers visibility across the food chain. When combined with data analytical tools capable of fusing extremely large quantities of data of different formats and extracting relevant information, these technologies are opening the door to real-time, end-to-end monitoring, and control of the movement of food products across the chain. And, as we will see later in this chapter, this convergence can be marshaled to make it possible to reduce the latencies in both detecting a food contamination event and responding to it.

The food supply chain starts at the farm and encompasses food transportation companies, processing facilities, distributors, retailers, brokers, importers, and governmental agencies responsible for overseeing and regulating the system—and ends at the consumer's table. Given the large-scale and distributed nature of the food system, it can be viewed as a "system of systems" whose components are complex, heterogeneous, self-organizing networks of systems that operate independently but are ultimately integrated into a dynamic, evolving "organism" that expertly manages the continuous production, distribution, and sale of food. Bringing these stakeholder systems together into an efficient and effective food safety network has been the signature challenge of regulatory agencies such as the US Food and Drug Administration (FDA).

Across many of these food chains today, sensors and other hardware are able to record a wide range of parameters—from location of a pallet or even item of food to its temperature while in transit from farm to fork. These sensors provide a level of granularity that was not available previously. A sensor attached to a carton of New Zealand milk will record the swings in temperature that accompany that carton as it moves from the New Zealand dairy farm by truck to airplane hold and by truck to retailer in China or elsewhere in Asia. This information, alone, can assist in identifying milk that might have spoiled before it is placed on the grocer's shelf. Temperature traces in route when combined with weather data, as well as shelf-life curves for that product, can also let retailers know what the remaining shelf-life is for that product.

In addition to preventing food spoilage and contamination, these new technologies are enabling better surveillance to determine the onset of foodborne illness. Although the specific authority varies from country to country, surveillance has typically been the purview of public health departments. Public health officials engage in surveillance activities to determine whether reported cases of foodborne illness are part of a large outbreak. Local public health departments are usually the first to pick up the signals of foodborne illness. These signals may correspond to isolated reports of illness or they may be causally linked and part of a larger outbreak. Or they may be uncorrelated and isolated cases that are not precursors of an emerging event.

When public health officials suspect a set of causally related cases, samples are sent to official laboratories such as the CDC for DNA "fingerprinting" to confirm that the illness is due to the same pathogen. Confirmation of the pathogenic source becomes the starting point for investigations by response teams to determine the specific food types that are responsible for the illness. Numerous delays occur in the surveillance and response processes. The promise of data science and big data is that these latencies can be reduced by timely fusion and interpretation of information on potential cases of foodborne illness.

Large amounts of data are already collected during the surveillance and response processes. What separates "big data" from "small data" in the food chain? Big data is distinguished by five characteristics referred to as the "5 Vs"—the volume, velocity, variety, veracity, and value of the data generation process. More data is being collected faster and in many different formats. The highly structured data that is typical of processing histories, shipment records, and lab reports is being augmented by data generated and/or transmitted from many nontraditional sources including wireless sensors such as RFID, temperature, and chemical sensors that monitor ambient conditions during transport, and mobile technologies—as well as satellite images, real-time data collected by drones, text data from telephone hotline calls, electronic medical data, and even social media.

Except for highly sensored food chains, the volume of data currently collected across a food chain is not extremely large when compared with other industrial processes such as aerospace where voluminous data is reported by aircraft in flight to ground stations for analysis. Similarly, the velocity with which the data is gathered is not extremely high compared with other domains such as financial systems. However, in both the food and agriculture industries, there is a proliferation of data variety with different levels of value. To build the capabilities necessary for improved surveillance and response, data must be collected and combined from the multiple and heterogeneous sources listed previously. And with increasing numbers of sources, there is inevitably a data quality and confidence problem—so veracity is an issue as well.

The proliferation of multiple data systems and tools that lack interoperability hinders effective information gathering and timely response to emerging but yet unconfirmed foodborne illness. As already noted, most of the public health and food safety informatics work in the United States—from early detection of food-related outbreaks by local and state health departments to confirmation by the CDC through "fingerprinting" of pathogenic contaminants—takes place at different local, state, and federal jurisdictional levels causing significant delays that have significant cost in terms of lives and dollars. A data-driven approach to food safety would reduce these latencies by bringing together: (1) traditional and new nontraditional data sources across all stakeholders in the food safety network; (2) new information and communication technologies for fusing and interpreting this data; and (3) new informatics and visualization tools capable of extracting knowledge that establishes "evidence" that can be used effectively by *all* the stakeholders across the food chain.

5.4 NEW FOOD SAFETY STAKEHOLDER MODEL

In the United States, the passage of the FMSA of 2012 was an attempt to bridge gaps in food safety surveillance and response activities by mandating the implementation of new information processes and informatics tools that reduce both the scale and scope of a food contamination event—whether unintentional or intentional. Not only did the passage of FSMA serve as a milestone in food safety law in the United States that sets the stage for a "big data" approach to assuring food safety, it also fundamentally changed the landscape of stakeholders that play a role in assuring safe food.

The FSMA signaled the emergence of a new food safety stakeholder model in which the private and public sectors, as well as the consumer, assume new roles in meeting the challenge of safe food. While the public sector has traditionally been the guardian of food safety, increasingly private sector enterprises and the consumer are playing an important role. The private sector is being given more responsibility for recording and providing information about its processes, suppliers, and customers (when requested by the FDA). And consumers have more opportunity to provide information to regulatory agencies and private sector enterprises about the quality and safety of their food.

Under FSMA new responsibilities fall on private sector companies. Food manufacturers are required to register and to examine their processing systems to identify possible ways that food products can become contaminated and to develop detailed plans to keep that from occurring. Companies must share those plans with the FDA, and provide the agency with records, including product test results, showing how effectively they can carry them out. The FDA was mandated to work with private sector companies on pilot projects to develop traceability systems that strike a balance between protecting public health and preventing any undue burden to businesses.

Increasingly the consumer is also a key stakeholder in the system. Previously, the consumer has had limited direct input into the food safety system. Official laboratory reports of cases of foodborne illness typically take many days, or even weeks, to find their way into the food safety system. Increasingly, however, consumer input into the surveillance and response processes is occurring through new channels. "Complaint" hotlines to food retailers and to public agencies provide real-time signals of possible foodborne illness. Consumers also "blog" information related to food using social media and other emerging technologies. Harnessing these sources of real-time consumer information can be critical in reducing delays in detecting foodborne illness.

Fig. 5.1 presents the new food safety stakeholder model comprised of the food safety system's four major stakeholders. They are: (1) a private sector that controls the production and commercialization of food products, the sale and distribution of potentially contaminated products, and participates in the recall of tainted products; (2) a public health system in charge of surveillance and management of outbreaks of disease caused by food contamination; (3) the governmental agencies pertaining to agricultural activities and the protection of the environment and natural resources, which regulate the production of food for human consumption by the agricultural and food

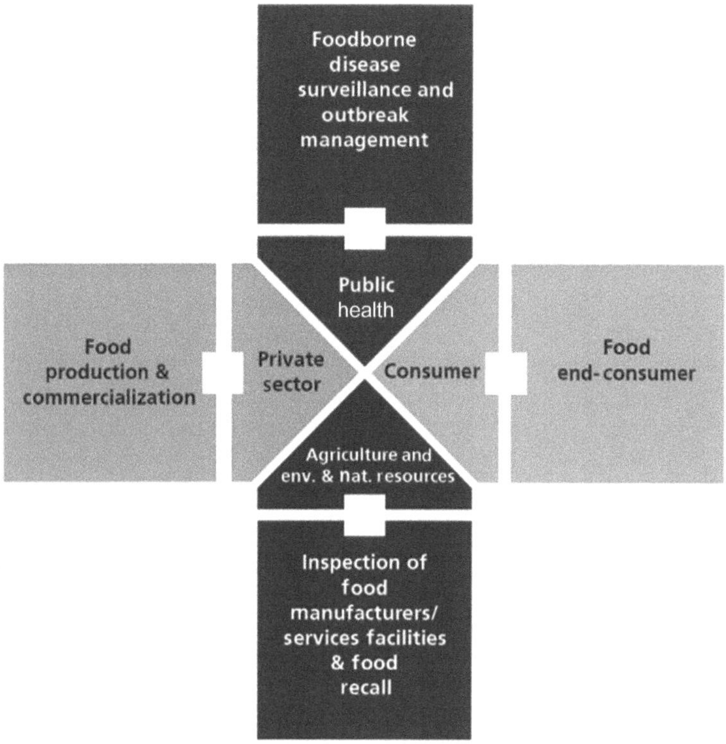

FIGURE 5.1 Food safety stakeholder model.

manufacturing sectors, oversee the safe use of natural resources and the environmental and sanitary conditions of establishments offering food services, as well as monitor and assist in food recall efforts; and (4) consumers of food products.

5.5 REDUCING LATENCY IN SURVEILLANCE AND RESPONSE

Continuous surveillance for early detection of foodborne outbreaks and rapid response to reduce the scale and scope of outbreaks are essential components of timely response for safeguarding our food supply. As noted earlier, our current ability to detect and respond to foodborne illness outbreaks is hampered by a number of gaps in the food safety system that create latencies in these processes. FSMA provides increased authority and resources for FDA to address many of the existing gaps in our food safety system. The law seeks to bridge some of the biggest gaps by mandating the implementation of new information processes and informatics tools that reduce both the scale and scope of a food contamination event.

An illustrative example of the degree of latency in responding to food contamination outbreaks was the 2008–09 *Salmonella Typhimurium* contamination of peanut butter produced by the now-defunct Peanut Corporation of America (PCA). The outbreak sickened 714 people in 46 states and may have contributed to nine deaths, according to the CDC. The illnesses began in January 2009 and ultimately prompted one of the largest food recalls in US history. This contamination triggered the most extensive food recall in US history up to that time, involving 46 states, more than 360 companies, and more than 3900 different products manufactured using PCA ingredients. The cost to food companies and the government was estimated to be more than $1 billion.

The timeline for the PCA outbreak is shown in Fig. 5.2. As shown in the figure, the first suspected contamination occurred in August 2008. It took almost 6 months to confirm that a foodborne outbreak had occurred, and another 6 months to locate all the contaminated products and remove them from retail shelves across the country. Nearly 6 years later, on September 21, 2015, the owner of the now-defunct PCA was sentenced to 28 years in prison for knowingly shipping out salmonella-contaminated peanut butter and hiding the evidence. This was the toughest punishment in US history to date for a producer in a foodborne illness case.

The key tasks associated with the surveillance and response processes are represented by four phases of the food safety wheel shown in Fig. 5.3. The right-hand side of the food safety wheel represents the surveillance phase. The left-hand side represents the response phase. During the first phase public health officials engage in detection activities to determine whether

Date	Event
2008	
August	First contamination?
September	First contamination?
October	First laboratory-confirmed cases reported
November	CDC confirms cluster of *Salmonella Typhimurium*
December	CDC holds national conference to confirm cases
2009	
January	CDC and FDA suggest peanut butter as source
February	FDA confirms source as PCA King Nut peanut butter; recall begins
March	PCA files for bankruptcy
April	Recall continues
May	Recall continues
June	Last contaminated PCA product recalled

FIGURE 5.2 Timeline of Peanut Corporation of America outbreak.

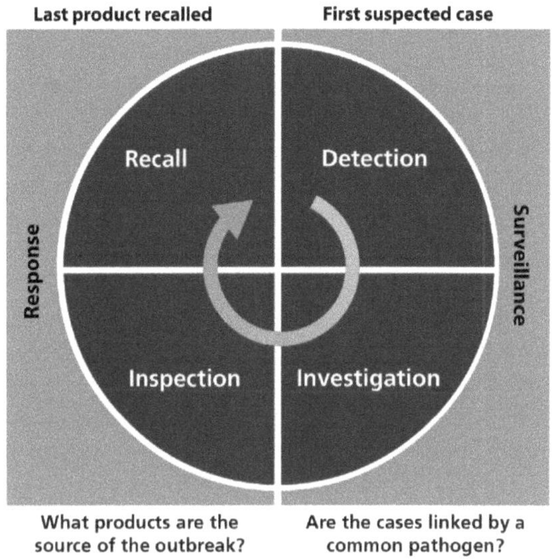

FIGURE 5.3 Key tasks in surveillance and response.

individual cases of foodborne illness are part of a larger outbreak. Laboratory testing to look for common pathogens (e.g., *Salmonella*) is used to confirm that an outbreak has occurred and that a cluster of cases has a common pathogen. Once a common pathogen has been identified and an outbreak has been confirmed, epidemiologists conduct interviews to discover the offending food types (e.g., tomatoes).

During the recall phase, the specific food products (e.g., Red Ripe Tomatoes) and facilities (e.g., Best Produce Company) are tested and inspected to identify specific product brands and/or production facilities. Once a source has been located, the difficult task of recalling all contaminated products in the food chain begins. The scale and scope of a food contamination event is directly related to the speed with which these tasks can be performed. Reducing the latencies associated with these events is crucial to saving lives and reducing costs.

Thus, we can reduce the costs of contamination significantly by reducing the surveillance period during which evidence is gathered and used to confirm the outbreak. In this section, we apply big data principles and techniques to the problem of reducing these food safety gaps.

5.6 BUILDING SITUATIONAL AWARENESS ACROSS THE FOOD CHAIN

The example of the PCA demonstrates the overarching need for capabilities that enhance situational awareness across the stakeholders in the food safety system. Theoretical frameworks for situational awareness are based on an understanding of cognitive processes of the human mind for decision making. The most common theoretical framework is provided by Endsley who defined situational awareness as the "perception of the elements in the environment within a volume of time and space, comprehension of their meaning and the projection of their status in the near future." We target four essential capabilities that contribute to enhanced situational awareness in food safety: (1) data integration; (2) visualization; (3) analytical tools; and (4) real-time collaboration.

Collectively, these four capabilities support an operational environment necessary for understanding, evaluating, and responding to foodborne outbreak events in an effective and timely manner. In continuously and rapidly changing environments such as public health, capabilities that support situational awareness maximize results of operational procedures, improve team collaboration, and enable better-informed decision making. The relevance of each capability to our end goal of reducing latencies in surveillance and response to foodborne illness outbreaks is described briefly in the following paragraphs.

Data Integration. Fusing data from all major food safety stakeholders can offer a more complete and clear picture of an emerging or ongoing (i.e., near real-time) event. In order to create situational awareness an informatics tool must provide a coherent representation of those data elements that are relevant to respective food safety stakeholders and that are essential to perceiving the status, attributes, and dynamics of any emerging or ongoing event. Currently, each major food safety stakeholder (c.f., public health official or private company) has only partial knowledge of what is happening based on that stakeholder's limits of responsibility and authority. Combining relevant information across all relevant food safety stakeholders into a single shared view, i.e., common operational picture, will create a more complete representation of present conditions that may allow faster recognition of existing problems and generate new knowledge that will contribute to latency reductions.

Visualization. A visualization tool not only provides a graphical representation of data that is more easily interpreted, but can also be used as a problem-solving tool. Trying to answer questions by examining large numerical tables or spreadsheets is typically more difficult and time-consuming than allowing a user to process the same data presented in graphs or maps or charts. Exploring different visual views of the same data facilitates analytical reasoning by taking advantage of human capabilities to process images. Benefits obtained from fusing diverse data sources can be augmented by adding visual analysis capabilities to the food safety system.

Analytical Tools. Analytics are broadly defined as a set of tools based on logic, statistics, or data science that are used to support decision making. In food safety, analytical tools can discover disease or exposure patterns that require further epidemiological investigation and will, as a result, speed up the process of identifying possible sources of contamination. For example, analytical tools can generate clusters based on similar foods consumed, places visited, or other common elements among data records that may help point out the source of contamination or uncover a totally new, still unreported, existing problem. Such tools can also assist in reducing latencies in the recall process by making the recall and effectiveness checks more efficient. Analytical tools can also be used to assess the likelihood of the emergence of a food safety event from fused data that can then be used to guide response.

Real-Time Collaboration. The need for better mechanisms for informal and formal communication among stakeholders is multifold and: (1) calls for a communication vehicle that enables exchange of information between participants; (2) offers 24/7 access; and (3) entails keeping a comprehensive roster of responders and public health officials at the state level including direct contact information and location, and an analogous roster of local healthcare providers' representatives and physicians at the local level. Such capability enables anytime, anywhere collaboration and exchange of ideas and information.

5.7 BUILDING A DATA ANALYTICS ENGINE FOR SURVEILLANCE

Savings lives and reducing the costs of a foodborne illness outbreak depend directly on the ability to reduce the latency with which a contamination event can be confirmed and the speed with which the offending products can be removed from the shelves of retail stores, as illustrated in Fig. 5.3. There are many cases of illness due to food consumption every day. Not all of them signal an impending food safety crisis. Individuals may react poorly to certain types of food. And, in other cases, food safety problems may be attributed to an individual's malfunctioning refrigerator. Distinguishing between these two cases is essential in responding effectively and efficiently to potential food contamination problems.

In making this assessment, the human decision-making process takes into account what is known by the decision maker. This includes *facts* describing the situation at hand and preestablished procedures/regulations, or *processes*, that dictate how that particular situation must be handled. The human processes this information through an activity known as *logical reasoning*, which allows the human to identify relationships among seemingly independent elements of a problem in the search for a solution. When it is not possible to apply any known processes to the known facts, humans can resort to using logical reasoning to link apparently unrelated facts to get a better understanding of the problem and to find an answer, or to delay any decision until more information is available or a new method is devised.

Connecting information, or finding the relationships among isolated facts, and selecting what is relevant to the task at hand is key to enabling humans to make better decisions in a timely and efficient manner. Today, representing facts and processes in a format so that they fit traditional execution models for computers is an ordinary task which makes it possible to automatically control many operations with these machines. Facts are well-suited to be represented in databases and processes as sequences of instructions for computer programs. In the case of food safety surveillance and response, these instructions are analogous to the thought processes that assist the decision-making process of the human.

In translating the cognitive processes by which we assess an emerging food safety event, we think of the food safety surveillance process as one in which many different bits of (big) data are being received in sequence. These data points contain information such as an admission to the emergency room with presenting gastroenteritis, a physician's report of a suspected foodborne illness case to public health authorities, personal blogs on social media that report illness after eating at a particular restaurant, FDA food product recalls, or even calls to government poison hotlines. These bits of data can be thought of as "events" that contain information that can help determine when a food contamination event has occurred and to distinguish that contamination event from an isolated case of food poisoning.

An illustrated example of such an events sequence is shown in Fig. 5.4. In the figure, a couple enjoys a meal at *MyFoodChain* restaurant and blogs

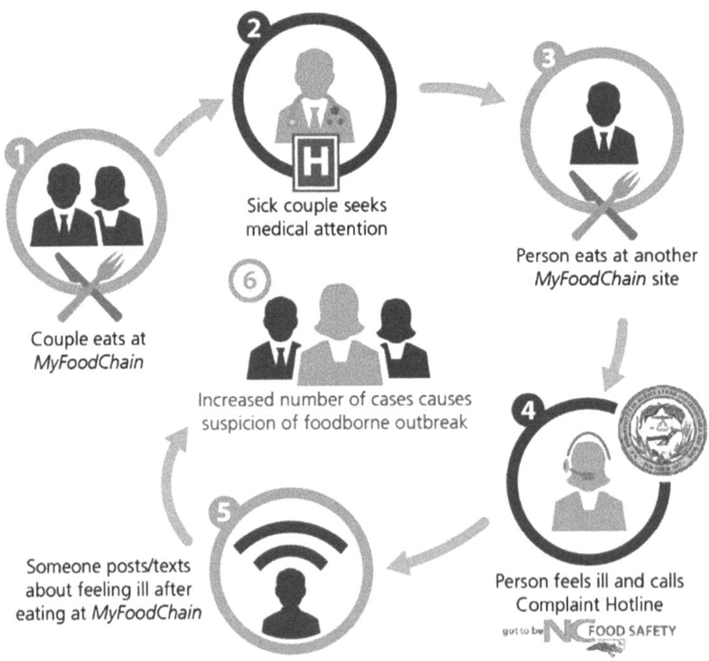

FIGURE 5.4 Events sequence for foodborne illness analysis.

about it to friends. They fall ill shortly thereafter and visit the local emergency room for treatment. A single report of gastroenteritis by a single couple does not, by itself, confirm a food safety event. However, soon thereafter another person eats at *MyFoodChain* and falls ill. Not sick enough to report to the hospital, the individual calls a state complaint hotline to report the problem. The likelihood that the illness is due to a contaminated food product at *MyFoodChain* has increased with the second report. The likelihood increases even more when a third person posts a blog about feeling ill after also eating at *MyFoodChain.*

The food safety challenge, then, is to develop analytics that interpret this sequence of events in real time and assess the likelihood that a foodborne illness outbreak is emerging. A number of data science methods can be adapted to look for clues in the various information events that are being received to determine the strength of the supporting evidence. Conceptually, the task is to "connect-the-dots" between possibly related pieces of information. As a new piece of evidence is observed (c.f., another hotline report of illness), it is compared to the available set of events to determine whether or not the newly received event increases the likelihood that the current situation signals an emerging foodborne illness outbreak.

Representing the necessary logical reasoning in such a way that it can be performed by computers, on the other hand, is not an easy endeavor because the relationships to be represented may require a complex set of rules that cannot be easily encoded in a database or a program with a well-defined flow of control. Usually logical reasoning is encoded as *inference rules* using some computer programming language and these rules are processed, together with facts, by another software application called a *reasoning engine* to produce answers. The analytics engine described herein performs rule-based predictive analytics and "reasons" about an existing situation as described by the known facts and encoded rules. The analytics engine deduces relationships among events to generate an *evidence set* of relevant events and information, which is shared with food safety stakeholders to improve their situational awareness and help in determining the likelihood that a food contamination event is emerging.

In addition to identifying relevant information concerning possible emerging events that can be "pushed" to users, an analytics engine can also rate the strength of the relationships among the events for users and compute a measure of the likelihood that the event under consideration is indeed an emerging event. In NCFEDA, the strength of the relationship among events in the evidence set is captured by the computation of the Event Likelihood Index (ELI) metric. This metric is based on the number and "connectedness" of the events that comprise the evidence set. The ELI metric is captured in an ordinal scale as shown in Fig. 5.5. In the example, seven possible levels of the ELI ratings scale range from "no relationship" at ELI = 1 and "highest likelihood" at ELI = 7.

ELI ratings	
Level	**Description**
1	No likelihood
2	Low likelihood
3	Some likelihood
4	Moderate likelihood
5	Significant likelihood
6	High likelihood
7	Highest likelihood

FIGURE 5.5 Event Likelihood Index (ELI).

5.8 NCFEDA—NORTH CAROLINA FOODBORNE EVENTS DATA INTEGRATION AND ANALYSIS TOOL

The North Carolina Foodborne Events Data Integration and Analysis (NCFEDA) prototype tool demonstrates the potential of improved situational awareness—created through real-time data fusion, analytics, visualization, and real-time communication—to reduce latency of response to foodborne illness outbreaks by North Carolina public health personnel. Data integration occurs across responding agencies—the North Carolina Department of Public Health (NCDPH), the North Carolina Department of Agriculture and Consumer Services (NCDA&CS), and the North Carolina Department of Environmental and Natural Resources (NCDENR)—as necessary for situational awareness. NCFEDA also includes new data sources from the private sector and the consumer. For example, on the private sector side, FDA recall alerts and enforcement reports provide information about contaminated food products as reported by manufacturing companies to the FDA and USDA. On the consumer side, consumer complaints collected by agencies' complaint hotlines are used as triggers for the NCFEDA system.

At its present state, the NCFEDA Analytics Engine processes triggering event data against other food safety data already stored in its databases and generates one or more possible "models" of the situation being evaluated. By definition, a model is a consistent set of knowledge assertions that the engine infers from the given inputs and the concepts it knows. The Analytics Engine data usage flow is illustrated by the diagram in Fig. 5.6 and indicates the types of results expected to be produced by the engine for two different use case scenarios.

Fig. 5.6 presents a high-level view of the major components that comprise NCFEDA's modular architecture including input data from stakeholders, analytical tools such as the Analytics Engine, and stakeholder dashboards. Input

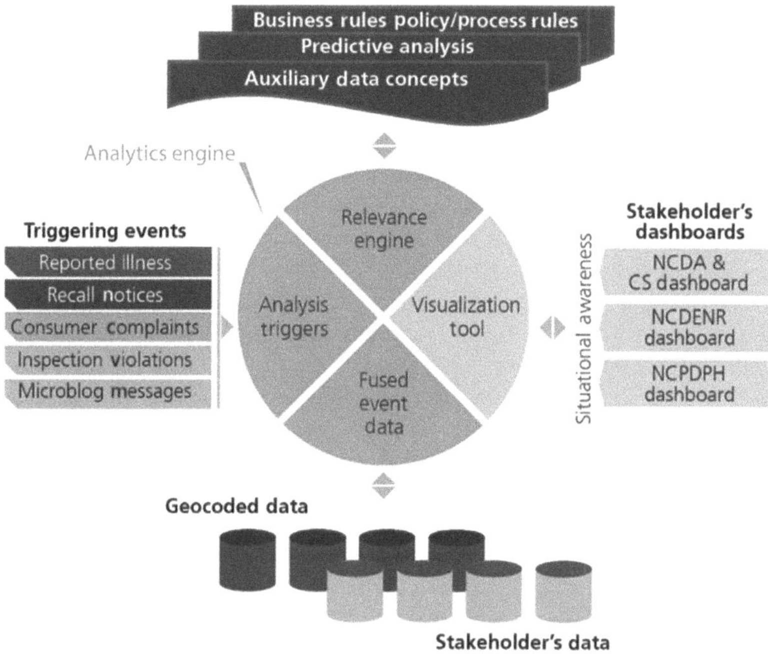

FIGURE 5.6 High-level view of NCFEDA's modular architecture.

data are observations associated with a food-related event that triggers NCFEDA analysis. These triggering events could be illness cases reported to public health officials, recall notices issued by FDA and USDA, or consumer complaints reported to NCDA&CS. These events are then provided to NCFEDA's Analytics Engine to determine whether they are relevant to the stakeholders' decision-making process and support the likelihood of an emerging foodborne illness. These new events can be thought of as signals that may indicate an emerging event or confirm an existing event.

When new events—or signals—arrive, they are interpreted by NCFEDA to determine whether they are relevant to other previously received data. Every new arrival may or may not activate one or more NCFEDA logical rules which are the basis for NCFEDA's Analytics Engine. If arriving event information is determined by the NCFEDA Analytics Engine to be relevant to a suspected emerging event, NCFEDA "pushes" that information to the appropriate stakeholder dashboard. When NCFEDA determines that the event may be relevant, the Analytics Engine computes the ELI—a measure of the likelihood that the suspected outbreak is real, which can assist public health officials in planning a response.

The major components of NCFEDA's modular architecture are described in the following paragraphs.

Analytics Engine. The Analytics Engine is the intelligent component of the system responsible for drawing conclusions about a given food safety situation. The engine is the core, domain-independent inference module and includes a set of inference rules that were created to define the food safety domain problem for North Carolina. NCFEDA reasoning capabilities are powered by formal logic. This means that the Analytics Engine "reasons" about food safety events by applying deductive reasoning, i.e., inference rules, to facts informed to the engine in order to infer (new) knowledge. The Analytics Engine's modules execute the following main functions: (1) analysis of the incoming triggering information events; (2) fusion of known event data previously acquired directly or indirectly from various stakeholder's surveillance and reporting systems; and (3) processing of new trigger information against the known data by using the relevance engine's deductive mechanisms together with various sets of rules, i.e., predictive analytics. A sample of the logical rule set that reasons to build the evidence set in NCFEDA is shown in Fig. 5.7.

Auxiliary Data Concepts. The auxiliary data concepts are a set of seven factual databases which store concepts of interest necessary for the task of reasoning about food safety events, and which are represented as logical knowledge for easy processing by the Analytics Engine. These concepts include four (simplified) ontologies for food, foodborne illness, and geographical information, as well as three databases which contain FDA's Food Code and the medical and consumer complaints codes utilized by the NCDA&CS to process consumer complaints about food products.

Stakeholder Databases. NCFEDA's databases store all event data obtained from both private and public sources and are the source of all information analyzed by the Analytics Engine and displayed on stakeholders' dashboards. All received event information is recorded in NCFEDA databases so that the databases are kept up to date. These event data constitute the history of food safety in North Carolina and are used by the Analytics Engine to support or refute possible conclusions regarding emerging and other food events.

The following data are provided to NCFEDA and stored in the NCFEDA databases:

- **Public Health Illness Data**. Records of patient illness reported to the North Carolina Division of Public Health containing, among other fields, the office visit date, probable diagnosis, and patient's county of residence.
- **Food Recall Notifications**. Recall notices of food products issued by FDA containing the recall issuing date, the product recalled, the company recalling the product, the cause for the recall (i.e., pathogen causing the contamination when available), and areas (states) where the product has been distributed.

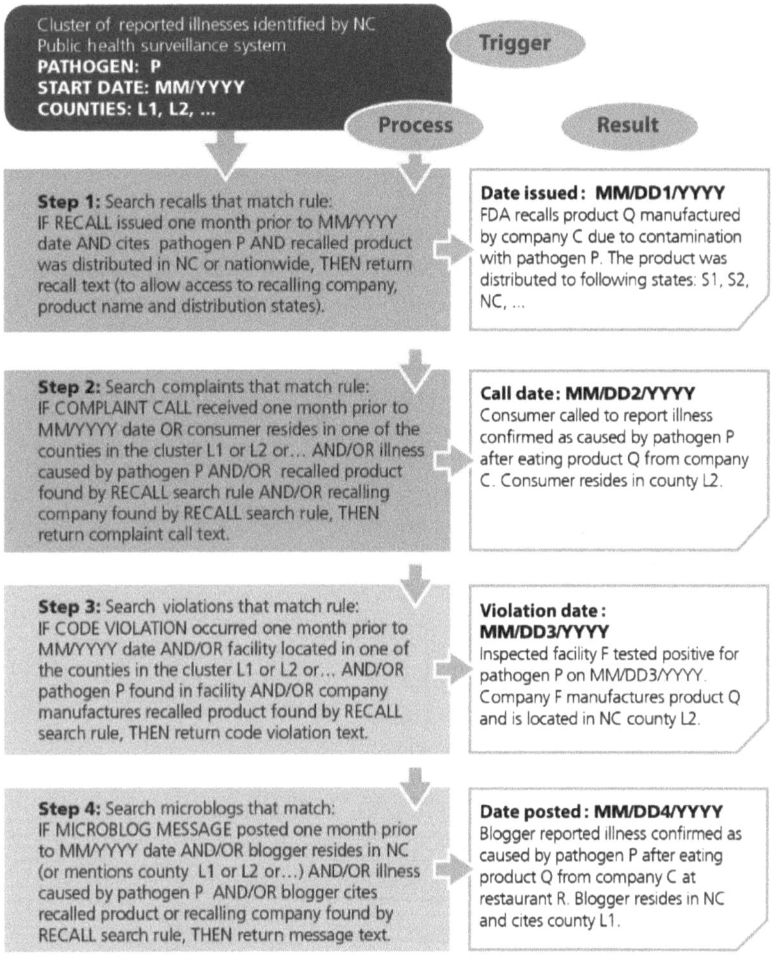

FIGURE 5.7 Building the evidence set by reasoning.

- **Consumer Complaints**. Consumer complaint calls to the NCDA&CS implicating a possible contaminated food product including date of the call, complainant county of residence, product implicated, retailer/manufacturer/food service provider implicated, complainant medical status (i.e., illness, hospitalization), diagnosis, and description of the complaint.

Visualization Dashboards. The visualization tolls in NCFEDA create visual representations of the results produced by the Analytics Engine for display on users' dashboards, increasing users' situational awareness by presenting information in a user-friendly interface. These dashboards are a set of dynamic graphical user interfaces that provide each agency-user and stakeholder with a COP and additional customized screens that, together, convey situational awareness to the various stakeholders.

5.9 PULLING IT ALL TOGETHER

In the following section we illustrate how NCFEDA works using a typical example of the progression of a foodborne illness event. The emerging event occurs over a 3-day period during which time a cluster of unspecified illness with symptoms of gastrointestinal problems is recorded by the system. This cluster may be an indication of an ongoing foodborne illness outbreak. Over the 3-day period, new information from various sources is provided daily to NCFEDA's Analytics Engine. As each new "event" is received, NCFEDA continuously evaluates this newly acquired information against knowledge previously acquired by the system to determine what information is "connected" and whether it belongs to the evidence set. NCFEDA also provides a measure of the likelihood that a foodborne illness or threat is occurring based on the strength of evidence contained in the evidence set, referred to as the ELI, or Evidence Likelihood Index. The 3-day simulation is summarized in Fig. 5.8.

Users connect to NCFEDA by accessing a login page, shown in Fig. 5.9, and then entering the name of the agency for whom they work, their user identification number, and a personal password to be verified by the system before any further access can be granted. The login page can also provide users with links to sites hosting relevant news related to food safety. For example, the login page provides a direct link to the latest recall issued by FDA, to the latest recall issued by USDA, and to an additional link to a site hosting recent food safety news.

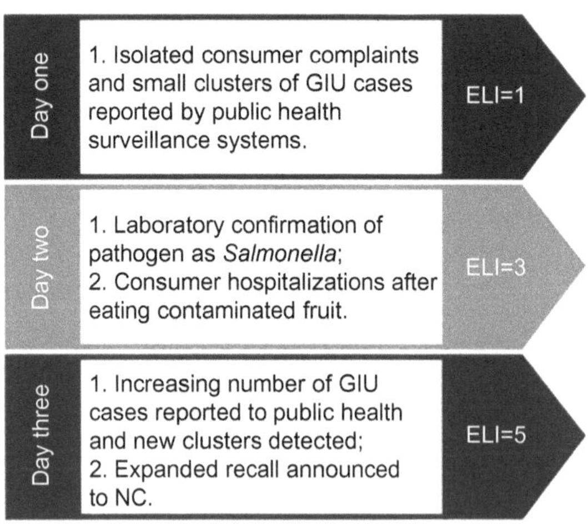

FIGURE 5.8 NCFEDA simulation timeline.

FIGURE 5.9 NCFEDA user login page.

The NCFEDA screen shown as Fig. 5.10 corresponds to a COP of all food-related events occurring in North Carolina and is intended to be used by all agencies as their primary NCFEDA work screen. Its goal is to increase user situational awareness across all users of ongoing events. The key areas of the COP are as follows:

1. The *North Carolina Map* provides the primary view of the COP. The map offers visual cues as to where "events" are occurring to help users assimilate the spatial distribution of possible food contamination threats.
2. The *Emerging Events Table* keeps a continuous record of any possible emerging event identified by the NCFEDA engine. The *Emerging Events Table* is "pushed" to users via separate pop-up windows. The pop-up window contains a short description of the Analytics Engine result and the corresponding ELI rating at any point in time.
3. The *New Incoming Reports/Information Relevant to Food Safety in NC* area in the middle of the screen displays three tables that contain key data fields from the three primary sources—consumer complaints received by NCDA&CS, illness cases reported to NCDPH, and food recalls issued by USFDA.
4. Finally, the *NCFEDA Searchable Database of Food Safety Reports* table at the bottom of the screen provides an easy mechanism for users to query NCFEDA databases by typing words of interest on dedicated search areas attached to each field.

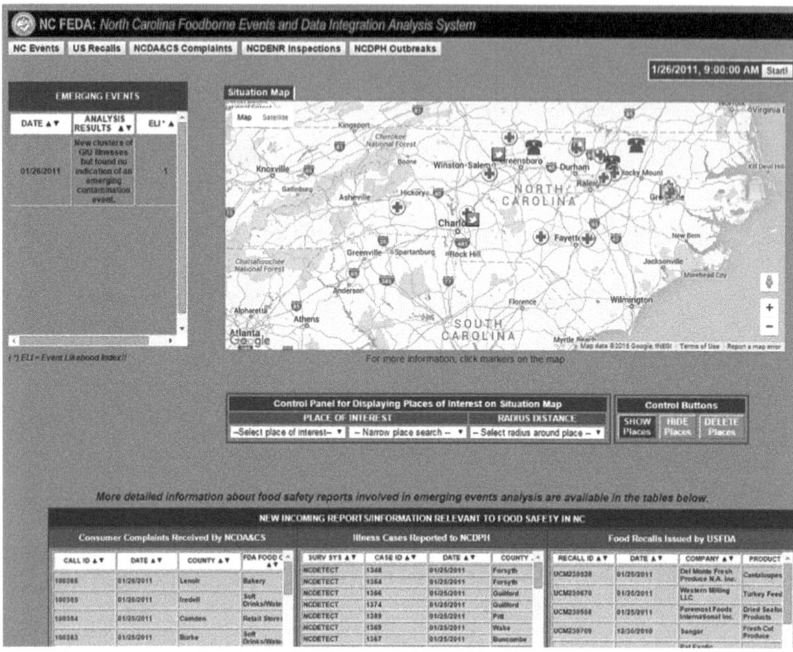

FIGURE 5.10 NCFEDA Common Operating Picture with ELI = 1.

5.9.1 Day One: Reports of Gastrointestinal Illness

On Day One a small cluster of illnesses with general symptoms of gastrointestinal ulceration (GIU) is reported by the public health department to NCFEDA. These cases are represented as icons on the North Carolina map. Without confirmatory test results, or a more precise diagnosis, no pathogen can be identified. These GIU records are also displayed in both the NCDPH records table and the Searchable Database table appearing on the COP.

The receipt of information about this cluster can be viewed as a trigger in NCFEDA. When this cluster is reported to NCFEDA, it is compared against other data "events" by the system, such as recent food product recalls and consumer complaint calls, to determine whether these events can be linked to these cluster cases. The locations of hospital visits (marked by an iconic red cross) and complainant counties of residence (marked by the icon of a green telephone) are plotted in NCFEDA's North Carolina map.

Using all information that is determined to be relevant and part of the evidence set, NCFEDA's engine computes the ELI rating for this situation and displays a short message in the Emerging Events Table to inform the user of its findings. As shown previously in Fig. 5.6, ELI ranges from 1 to 7 where a score of 7 indicates the highest likelihood. Without any confirmatory information to indicate an emerging foodborne illness outbreak, the ELI rating is computed to be 1 (ELI = 1). The emerging events for Day One are shown in Fig. 5.10.

5.9.2 Day Two: Lab Results and Consumer Complaint Calls

On Day Two a new cluster of illness cases is detected by public health officials. This new information is analyzed by NCFEDA to determine whether it is part of the evidence set, and thus increases the likelihood of an emerging event. Laboratory tests confirm that these cases are associated with the pathogen *Salmonella*. Given that a pathogen has now been positively identified, NCFEDA searches among both incoming and previously active recall notices to verify if any of those are also a result of contamination by *Salmonella*. NCFEDA also looks to see whether any products associated with these recalls are known to have been shipped to North Carolina. But no results are found.

NCFEDA also searches among incoming consumer complaint calls, and any complaints currently under investigation by NCDA&CS, for any illnesses confirmed to have been caused by *Salmonella*, or for implicated good products susceptible to this pathogen. NCFEDA searches its databases and locates a complaint call in which the caller reported being hospitalized because of possible consumption of contaminated fruit. Because fruit is susceptible to *Salmonella*—as documented by existing recall data—the NCFEDA relevance engine deduces that there is a possible emerging *Salmonella* contamination event and issues a warning to responsible agencies.

An emerging events map pops up in a separate window, as shown in Fig. 5.11, displaying the location of all events in the evidence set linked to this threat. When the user hovers the computer mouse over the map icons, detailed information about each reported case/complaint is displayed. In light

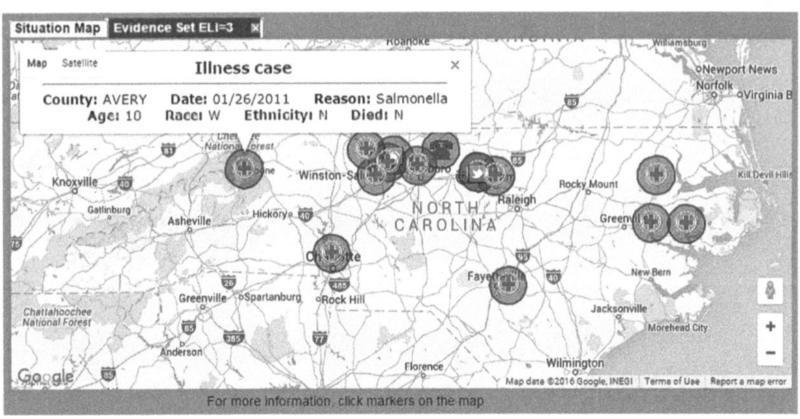

FIGURE 5.11 Evidence set for ELI = 3 on day two.

of the confirmatory evidence, the ELI for the event is computed to be 3 (ELI = 3) by the Analytics Engine and appears in the corner of this pop-up window. A screen shot of NCFEDA with ELI = 3 is shown in Fig. 5.11.

5.9.3 Day Three: Food Recall Issued

On Day Three, an increasing number of new illness cases are reported to NCFEDA from the public health system and new clusters are detected. Because we do not have personalized information about patient identity due to government HIPAA regulations and other privacy concerns, NCFEDA cannot deduce an exact relationship among patient cases beyond same county of residence.

However, the arrival of a new recall notice from FDA expanding the area of distribution of recalled cantaloupe to the state of North Carolina is thought to be linked to the cluster of *Salmonella* cases. The cantaloupe recall had previously been restricted to three states on the west coast of the United States. NCFEDA recognizes that cantaloupe is a fruit and that the *Salmonella* pathogen causing this recall is also the same pathogen causing a reported illness and hospitalization as reported by a consumer complaint call.

The emerging events map appears in the COP displaying all events linked to this threat, which now includes a recall notice (shown in the gray box on the bottom left corner of the pop-up window viewed in Fig. 5.12). The new ELI rating has been elevated to a score of 5 (ELI = 5) because more connections among the data have been discovered and confirmed by the relevance engine in the Analytics Engine.

Now that the threat level has been elevated to ELI = 5, indicating a high likelihood of an emerging threat, a warning message is pushed to users that

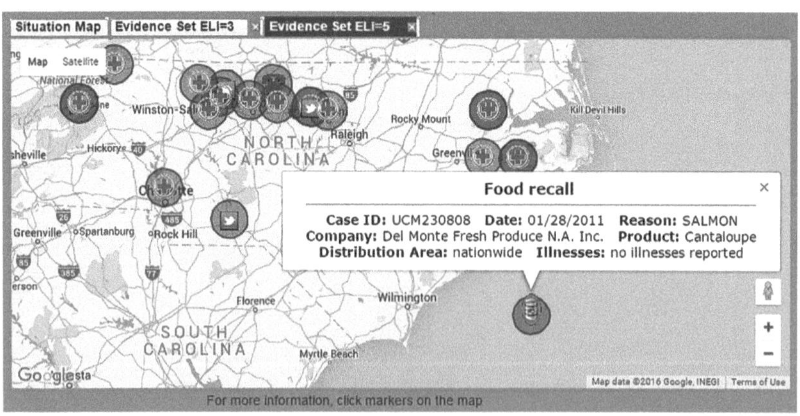

FIGURE 5.12 Evidence set for ELI = 5 on day three.

includes a complete set of information about the relevant events and possible threat including the suspect food product (cantaloupe) and the pathogen (*Salmonella*). The evidence set with ELI = 5 is shown in Fig. 5.12.

5.10 FUTURE TRENDS

Public agencies and private companies alike are working hard to adopt methods of data science in the interest of food safety. In 2013 the US FDA awarded a $50 million federal contract to Dynamics Research Corporation (now part of Engility, Inc.) to help move the agency into the big data era. And, given the prevalence of mobile apps and smart phones, it is not surprising that a number of efforts are mining social media data, much as Google did for influenza. The City of Chicago Department of Public Health is working with the Smart Chicago Collaborative to develop mobile applications that monitor Twitter for possible food poisoning references. The New York City Department of Health and Mental Hygiene is working with Columbia University to review restaurant-goer comments on Yelp for possible clues to a food contamination event or outbreak.

Many private sector companies are also contributing data-driven technologies and analytics to support the push to an integrated, data-driven approach to food safety. IBM recently announced a new predictive analytics technology that the company claims is capable of identifying contaminated products "within as few as 10 outbreak case reports." Like NCFEDA, the goal of IBM's technology is to reduce the time required to identify the likely contamination sources by days or even weeks. Predictive analytics and other algorithms look through petabytes of grocery store food sales data from retailers and distributors in search of patterns and relationships that may indicate contamination. Visualization techniques link the data to geographical information to connect suspected contaminations with clinical and lab reports, as well as other data. A pilot is being conducted with the Department of Biological Safety at the German Federal Institute. The project will process information from 1.7 billion supermarket items sold in each country.

These developments are the first steps toward the integration of the food chain within an Internet-of-Things (IoT) environment. Situational awareness of complex and lengthy food chains is currently constrained by difficulties associated with the timeliness of data collection and fusion—in fact, much data is still manually entered into systems. In an IoT environment, end-to-end data needed for both surveillance and response can be autonomously and automatically collected using the sensor-enabled network environment of the IoT. In this, hopefully, near-term future, all stakeholders in the supply chain from the farm to the consumer will have sensors and systems in place to monitor both the food as it moves through the food chain and the health data and laboratory data needed to identify and confirm foodborne illness—and most importantly the connections between them that enable the incidence of illness to be linked immediately with the offending products in the food chain and with its source.

5.11 FURTHER INFORMATION

Further information about the application of data science in food safety can be found in several disciplines. *Food Safety Magazine* (http://www.foodsafetymagazine.com/) offers many articles about the critical challenges of food safety and the application of new data-driven and digital tools to address those challenges. The CDC website offers up-to-date information about the current state of food safety and capabilities of surveillance and response (http://www.cdc.gov/foodsafety/fsma/index.html). Their website provides basic information about the current responsibilities and procedures for managing a foodborne disease outbreak. Two studies published by the CDC address the state of food safety in the United States. The CDC's annual food safety progress report measures foodborne illnesses from nine key germs and is produced from data compiled by the FoodNet. The National Outbreak Reporting System (NORS) publishes an annual summary of foodborne outbreaks reported to CDC by state and local health departments.

ACKNOWLEDGMENTS

This research was funded by the US Department of Homeland Security through the Research Triangle Institute—Institute for Homeland Security Solutions (IHSS) under Contract HSHQDC-08-C-00100.

REFERENCES

FoodNet Report Shows Mixed Bag of Foodborne Illness Trends, May 15, 2015, <http://www.foodsafetymagazine.com/news/foodnet-report-shows-mixed-bag-of-foodborne-illness-trends/>.

Beach, C., CDC declares Chipotle *E. coli* outbreaks over; cause unknown, Food Safety News, February 1, 2016, <http://www.foodsafetynews.com/2016/02/cdc-declares-chipotle-e-coli-outbreaks-over-cause-unknown/#.VrjH4v7bKig>.

Dube, L., LAbban, A., Moubarac, J.-C., Heslop, G., Ma, Y., 2014. A nutrition/health mindset on commercial big data and drivers of food demand in modern and traditional systems. Ann. N. Y. Acad. Sci. USA 1331, 278–295.

Doinea, M., Boja, C., Batagan, L., Toma, C., Popa, M., 2015. Internet of things based system for food safety management. Inf. Econ. 19 (1), 87–97.

Greis, NP, Nogueira, M, MacDonald, P, Wilfert, R., NCFEDA North Carolina Foodborne Events Data Integration and Analysis Tool: A New Informatics Tool for Food Safety in North Carolina, 2011. Prepared by RTI International–Institute for Homeland Security Solutions under contract HSHQDC-08-C-00100.

IBM. The four V's of big data, 2014. <www.ibmbigdatahub.com/infographic/four-vs-big-data>.

Keller, M., Blench, M., Tolentino, H., et al., 2009. Use of event-based unstructured reports or global infectious disease surveillance. Emerg. Infect. Dis. 15 (5), 689–695.

Manyika, J., Chui, M., Brown, B., et al., Big data: the next frontier for innovation competition, and productivity, 2011, <http://www.mckinsey.com/insights/business_technology/big_data_the_next_frontier_in_innovation>.

Piramuthu, S., Zhou, W., 2016. RFID and Sensor Network in the Food Industry. Wiley Blackwell.

Strawn, L. K., E. W. Brown, J.R.D. David, H.C. Den Barker, P. Vangay, F. Yiannas, and M. Wiedmann, Big Data in Food, Food Technol., 2015, 69(2), pp. 42–49, Retrieved at: <https://www.researchgate.net/publication/272238175>.

Wang, Y., Yang, B., Luo, Y., He, J., Tan, H., 2015. The Application of Big Data Mining in Risk Warning for Food Safety. Asian Agric. Res. 7, 83–86.

Chapter 6

Hygienic Design of Open Food Processing Equipment

F. Moerman[1] and K. Lorenzen[2]
[1]*Catholic University of Leuven – KU Leuven, Leuven, Belgium,* [2]*European Hygienic Engineering & Design Group, Frankfurt, Germany*

6.1 INTRODUCTION

Food legislation developed in many countries around the globe requires that microbiologically safe food shall be produced by means of process equipment that minimizes the risk of contamination and that is easily cleanable. Good hygienic engineering and design practices have become the tools to reduce or exclude microbial (e.g., pathogens), chemical (e.g., lubricating fluids, cleaning and disinfectant chemicals), or physical (e.g., glass, wood) contamination of food. Good hygienic equipment design also allows to eliminate food product "held up" within the process equipment that could deteriorate and affect product quality on rejoining the main product flow. So, also cross-contamination of one batch by a previous batch can be avoided by means of good hygienic design practices. Although initially more expensive than poorly designed equipment, hygienically designed equipment is more cost effective in the long term because it may reduce the downtime required for an item of process equipment to be cleaned, disinfected, or maintained.

Open processes include many different types of equipment according to the product (e.g., dairy, alcoholic and nonalcoholic drinks, ice cream, sweet oil, nutrient fat, coffee, sugar, cereals, vegetables, fruits, bakery products, ready meals, meat, and fish). During open processing, contamination may additionally occur from microorganisms present in the factory and so the operating environment also becomes an important factor. The type and level of product contamination and the stage of the manufacturing process must also be taken into account.

This chapter intends to inform food safety professionals and inspectors/auditors about the risks associated with poor hygienic design of open process equipment. Along with typical examples of poor hygienic design, the necessary technical and practical guidance is given to identify and control open

Food Protection and Security. DOI: http://dx.doi.org/10.1016/B978-1-78242-251-8.00006-0

food processing equipment-related food safety hazards. This chapter may help the food manufacturer to select the most suitable food processing equipment, to construct a food production line that meets all current and future hygienic requirements, and to set up an appropriate food safety management plan (e.g., HACCP) intended to eliminate or control all food safety hazards along the food chain.

In Section 6.2, an overview is given of the current legislation and standards dealing with the hygienic design of food processing equipment. Section 6.3 lists the basic hygienic design requirements that food processing equipment must meet to produce microbiologically safe food products. Section 6.4 describes the hygienic and food-grade materials that can be used in the manufacturing of food processing equipment; Section 6.5 outlines the requirements for the food contact surface finish. In the next sections, we discuss the hygienic design of several open food processing equipment (components) such as open vessels, containers and bins (Section 6.6), framework (Section 6.7), feet (Section 6.8) and casters (Section 6.9), belt conveyors (Section 6.10), motors (Section 6.11), covers and guards (Section 6.12), electrical equipment enclosures and cabling (Section 6.13), human interfaces (Section 6.14), and stairs, raised walkways, and platforms (Section 6.15).

6.2 LEGISLATION, STANDARDS AND GUIDELINES COVERING HYGIENIC DESIGN

6.2.1 Legislation

Annex I of the Machine Directive 2006/42/EC & 98/37/EC (formerly 89/392/EEC and its amendments 91/368/EEC & 93/44/EEC) and Annex V of Council Directive 93/43/EEC on the Hygiene of Foodstuffs require that all equipment used to handle food should be hygienically designed: (a) be so constructed, be of such materials, and be kept in such good order, repair and condition as to minimize any risk of contamination of the food; (b) with the exception of nonreturnable containers and packaging, be so constructed, be of such materials, and be kept in such good order, repair, and condition as to enable them to be kept thoroughly cleaned and, where necessary, disinfected, sufficient for the purposes intended; (c) be installed in such a manner as to allow adequate cleaning of the surrounding area.

6.2.2 US Standards and Guidelines

In the United States, the American Meat Institute (AMI), United States Department of Agriculture (USDA) and 3-A are considered to be the experts in sanitary design. “Sanitary design” in the United States has the same meaning as “hygienic design” in Europe.

6.2.3 European Standards and Guidelines

In Europe, the European Hygienic Engineering & Design Group (EHEDG) is the most experienced organization in the field of hygienic design. Besides these guidelines, many food and equipment manufacturers have developed their own hygiene standards for internal use.

6.3 BASIC HYGIENIC DESIGN REQUIREMENTS

In all stages of design, construction, installation and maintenance of food processing equipment, hygienic design aims to reduce the buildup of food material or microorganisms in individual items of equipment and the complete line, and to ensure that all detectable soil is removed after cleaning (and disinfection). According to European Standard EN1672-2, soil is "any matter, including product residues, microorganisms, residual detergents or disinfecting agents." Food processing equipment should at least meet the following basic hygienic requirements (Moerman & Kastelein, 2014):

- To avoid bimetallic corrosion, the right combination of steels, alloys or metals in the assembly of food processing equipment and food processing support systems must be used. So, piping and components should be constructed out of the same materials to prevent contact corrosion between dissimilar metals (Fig. 6.1). However, this is not always possible.
- Product contact surfaces (including the welds) must be smooth, enabling them to be easily cleaned.
- The design of food processing equipment must prevent bacterial ingress, survival, growth and reproduction on both product and nonproduct contact surfaces. The food processing equipment also must be constructed so as to ensure effective and efficient cleaning over the lifetime of the equipment (Fig. 6.2).

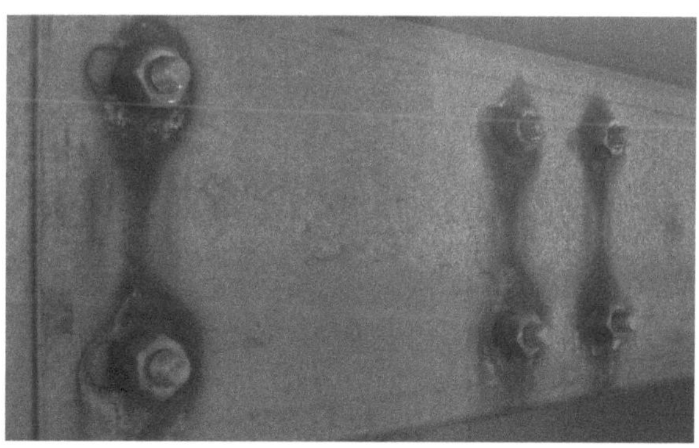

FIGURE 6.1 In this part of the food processing equipment, galvanized steel plate is combined with stainless steel bolts, giving rise to galvanic corrosion between the stainless and galvanized steel components.

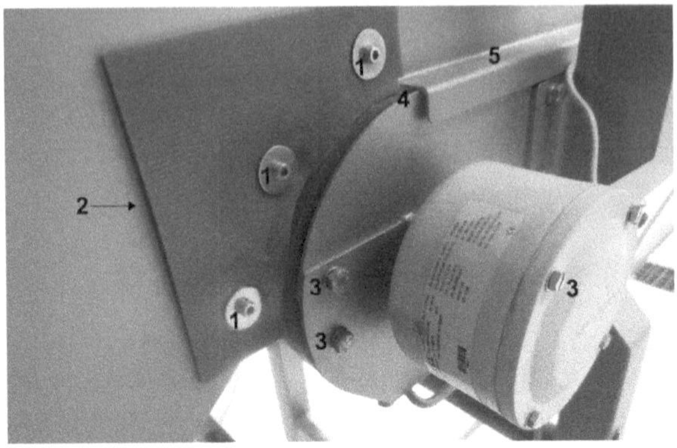

FIGURE 6.2 The blue sheet of plastic used to reduce product spillage is fixed on the equipment surface by means of rivets (1), leaving behind holes in which dirt may accumulate. Dirt built up behind the blue plastic sheet (2) cannot be removed during cleaning procedures. Bolt heads create a lot of crevices (3) where dirt may collect, while reducing the cleanability. Part of the conveyor belt is running beneath the product guide (4), prohibiting effective cleaning of the whole conveyor belt. The product guide also provides a horizontal ledge (5) on which product may lodge. *Frank Moerman, © 2016.*

FIGURE 6.3 Welds must be continuous and smooth, free of pits and cracks. Gaps, lap seams, bolts, and threads will accumulate dirt and will make this equipment not cleanable. *Photo left, courtesy of John Butts, Land O'Frost, © 2016; photo right, courtesy of Joe Stout, Commercial Food Sanitation LLC - Intralox, ©2016.*

- Welding or continuous bonding is preferred over fastenings. Avoid exposed screw threads, nuts, bolts, and rivets whenever possible, certainly in product contact areas. Alternative methods of fastening can be used where the washer has a rubber compressible insert to form a bacteria-tight seal.
- Welds must be continuous and smooth, free of pits and cracks (Fig. 6.3).

FIGURE 6.4 Penetration of hollow sections of equipment such as frames with bolts is not allowed. Hollow sections shall be hermetically sealed (Kold et al., 2016).

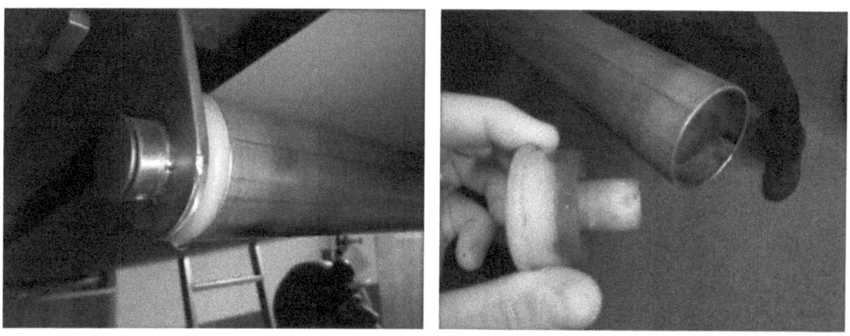

FIGURE 6.5 Hollow roller. *Organization Sanitary Design Workshop, © 2016.*

- In design, construction, installation, and maintenance, hollow areas of equipment such as frames (Fig. 6.4) and rollers (Fig. 6.5) must be eliminated or they shall be hermetically sealed. As such, studs, mounting plates, brackets, junction boxes, end caps, sleeves, and other such items must be continuously welded to the surface, and shall not be attached via drilled and tapped holes.
- Fastening of nameplates on the equipment should be avoided in favor of direct continuous welding (Fig. 6.6). As a further improvement, direct application of graphics on equipment components by laser engraving eliminates the need for identification plates.
- Niches (Figs. 6.2–6.5) such as pits, cracks, crevices, open seams, gaps, lap seams, inside threads, holes that may accumulate dirt and hamper the cleanability of the food processing equipment are not allowed.
- Avoid dead areas, pockets or other conditions that may trap food and harbor contamination. They prohibit effective cleaning and disinfection, and allow for cross-contamination.

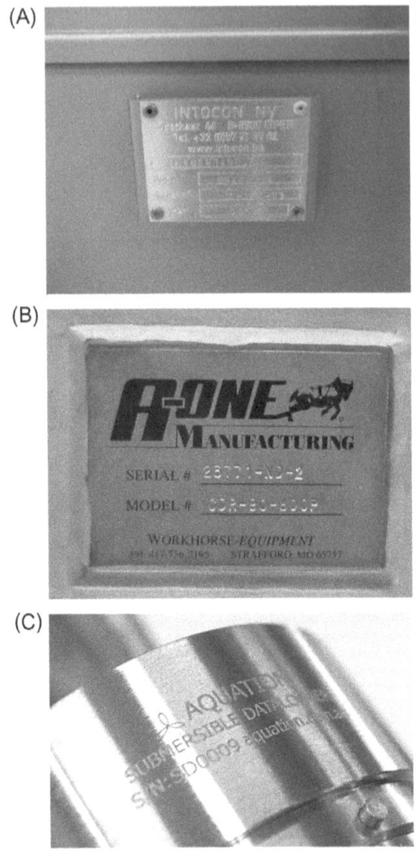

FIGURE 6.6 (A) Nameplates are often fastened to the surface of process equipment by means of rivets. (B) Continuous welding of nameplates onto the equipment surface is possible, but (C) direct application of graphics on equipment components by laser engraving is preferable.

- All inaccessible horizontal flat areas, ledges (Fig. 6.2), projections, protrusions, recesses, edges, etc. where product residues can accumulate should be eliminated.
- The exterior of indirect product contact surfaces should be so arranged that harboring of contamination in and on the equipment itself, as well as in its contact with other equipment, floors, walls, or hanging supports, is prevented.
- For the same reason and to facilitate cleaning, internal angles, and corners should be well radiused.
- All equipment surfaces in the product zone must be so arranged that they are self-draining (Fig. 6.7) to minimize contamination and corrosion risks when liquid food, cleaning and disinfection solutions, and rinsing water are retained during idle periods. Microbes can flourish in stagnant pools

FIGURE 6.7 Equipment surfaces in the product zone must be self-draining, and internal angles and corners well radiused. *Courtesy of Krones AG.*

of water, especially when nutrients are trapped in the internal pockets. Moreover, accumulated and pooling cleaning and disinfection solutions may contaminate food products.

- Equipment design, therefore, should not permit the formation of condensate that may enter the food zone and contaminate product or product-contact surfaces.
- All parts of the equipment shall be readily accessible for inspection (Fig. 6.8), so as to facilitate the detection of all potential contaminants on representative surfaces throughout the product contact zone. So, all surfaces in the product zone must be immediately visible for inspection, or the design of the equipment shall allow easy dismantling without use of any tools.
- Equipment surfaces must be readily accessible for manual cleaning and disinfection (Fig. 6.9), unless it can be demonstrated that the result of in-place cleaning and disinfection procedures without dismantling is equivalent to the result of dismantled and manual cleaning procedures. All potential obstructions to cleaning, disinfection, and maintenance should be avoided or minimized.
- Instruments not only must be hygienically designed, but also hygienically installed.
- Equipment design must ensure hygienic compatibility with other equipment and systems, such as hydraulics and electrical, steam, air, and water systems.
- Maintenance equipment enclosures and human machine interfaces, such as push buttons, valve handles, switches, and touchscreens, must be designed to ensure food product and water or liquid product do not penetrate in or accumulate on the enclosure or interface. The enclosures should be sloped to an outside edge to avoid use as storage areas. Doors,

FIGURE 6.8 Despite their general acceptance over the last two decades and their mechanical versatility, modular belts, relying as they do on hinges and pins, versatility, modular belts, relying as they do on hinges and pins, are difficult to clean. Especially, the pinholes in the hinges and pins are major dirt and bacteria traps. Hence, all parts of modular conveyor belts at the running side shall be readily accessible for inspection so as to facilitate the detection of all potential contaminants, as well as cleaning procedures.

covers, and panels must prevent entry and/or accumulation of soil. To facilitate cleaning, they should be easy to remove.

- Bearings should be mounted outside the product area to avoid contamination of food products by lubricants and to exclude the ingress of bacteria. When the bearing is within the product area, its design should allow the passage of cleaning fluid.
- Food grade lubricant (Fig. 6.10) should be used, and leaking of lubricant onto food product must be excluded (Fig. 6.11). To protect the product zone, a drip pan should be used, or motors driving equipment components such as belt drives should be placed outside the product area. If they are within the splash area, they should be protected by a removable cover.

FIGURE 6.9 Belt lifting device used to provide improved access for cleaning and disinfection of the conveyor bed (frame and wear strips), as well as the bottom side of the conveyor belt. *Courtesy of Intralox.*

FIGURE 6.10 The process equipment is overlubricated with a nonfood-grade red grease. *Courtesy of John Butts, Land O'Frost, © 2016.*

FIGURE 6.11 This motor without drip pan is positioned above the product stream, increasing the risk of lubricant dripping onto the food product. Dirt also may build up in the chain guard.

6.4 MATERIALS OF CONSTRUCTION

6.4.1 General Recommendations

Construction materials for open processing equipment should be as hygienic (smooth, nonabsorbent, nontoxic, easily cleanable, impervious and nonmold supporting), as chemical resistant (nondegrading and maintaining its original surface finish after sustained contact with product, process chemicals, cleaning agents), as physically durable and mechanically stable (resistant to steam, moisture, cold, the actions of cleaning and disinfecting agents, abrasion and corrosion resistant, resistant to chipping, unbreakable), and as easy to maintain (Hauser et al., 2004a; Partington et al., 2005) as possible. Table 6.1 gives an overview of the corrosion durability of the most frequently used materials in the construction of open equipment.

6.4.2 Use of Metals and Alloys

Nonferrous and ferrous metals and alloys are used in the construction of equipment and services for the food industry. Alloys for food contact may only contain aluminum, chromium, copper, gold, iron, magnesium, manganese, molybdenum, nickel, platinum, silicon, silver, tin, titanium, cobalt, vanadium, and carbon (EDQM, 2013). The austenitic chrome-nickel or chrome-nickel-molybdenum stainless steels are used for the construction of open equipment, as well as ancillary support systems in the food industry. Because AISI SS 304/304(L) suffers from some corrosion over a long time period, especially in the presence of chloride (e.g., salt, sodium hypochlorite), stainless steel AISI SS 316(L) is commonly used as construction material for food processing equipment. Galvanized steel should be avoided in the product contact area (the splash area included), as the zinc coating may peel off and is easily dissolved in diluted acids and bases, releasing zinc and traces of cadmium and lead. Painted steel (Fig. 6.12) should never be used in contact with food, because paints can crack or peel off. Brass, bronze, and copper quickly react with strong alkaline detergents, sodium hypochlorite, acid, and salty food, making them not suitable in the food contact zone. Aluminum is attacked by alkaline detergents, sodium hypochlorite, and acid food, but anodized aluminum is acceptable in the food contact area (Moerman & Partington, 2014).

6.4.3 Use of Plastics

Plastic materials may be used to preclude metal-to-metal contact (e.g., for bearing surfaces), because of their plasticity and corrosion resistance, and because of their smaller thermal mass as compared to metals. The latter may be advantageous to reduce the pull-down load in freezers. Plastics must have high mechanical strength (resistant to aging, creep, brittleness, fatigue,

TABLE 6.1 Corrosion Durability Classes

Class	Materials
1	• Stainless steel AISI 304(L), AISI 316(L) • Hastelloy B & C • Titanium • Polyvinyl chloride (PVC) • Teflon (PTFE) • Polypropylene (PP) • Polyethylene (LDPE, HDPE) • Polyvinylidene fluoride (PVDF) • Polysulfone (PES) • Polyetheretherketone (PEEK) • Polystyrene (PS) • Poly(methyl methacrylate) (PMMA) • Epoxy resin • Neoprene rubber • Ethylene propylene diene monomer (EPDM)-rubber
2	• Hard chromium plated steel • Nickel-plated steel • Nickel-plated brass • Anodized aluminum • Nickel • Acrylonitrile butadiene styrene (ABS) • Polyamide (PA) • Polyacetal plastics (POM) • Phenolic resins (PF) • Ureum and melamine resins (UF, MF) • Polyurethane rubber (PU) • Nitrile rubber (NBR)
3	• Galvanized, carbon and painted steel • Cast iron • Bronze and brass • Copper • Zinc • Aluminum • Polycarbonate

1 = highly durable, 2 = moderate, acceptable durability, 3 = sensitive to chemical attack.

etc.). Plastic components which shatter under adverse tensile or bending loads, or under impact, can cause food product contamination with sharp particles that are not detected by common in-line metal detectors. Such materials therefore represent a similar hazard to glass.

Most plastics (especially PA and ABS) can absorb moisture. In extreme cases, and in addition to simple porosity effects, swelling becomes visibly noticeable and the mechanical performance will degrade. When food

FIGURE 6.12 Paints can crack or peel off.

components diffuse into the plastic, they subsequently may leak back out into "later" food, causing a loss of the perceived quality of the food, such as changes in visual appearance or organoleptic qualities (sometimes called *tainting* of the flavor). Application of formaldehyde resins in the food contact area of food processing equipment is not at all recommended due to formaldehyde that may be released or migrate into the food products produced. Free phenolic compounds (phenol, cresol and/or tertiary-butyl phenol) present in phenolic resins may migrate into food products. Melamine-formaldehyde resins may release melamine and formaldehyde. Also plastics containing plasticizers (e.g., plasticized PVC) are not recommended and should only be used in the nonproduct contact zone. The use of glass-reinforced plastic products should be avoided as it is known that components of glass-reinforced plastic can react with certain wetting agents in detergents. Plastics applied outside the food contact area require no special approval, but they should be easy to clean and resistant to chemicals and temperatures occurring within their immediate installed environment. Porous plastics, which may harbor microorganisms, require special cleaning/disinfection procedures and periodic inspections.

When using a plastic material (belt, gaskets, electric cables, etc.), it is of utmost importance to make sure that the material is able to withstand all temperatures from −50°C to temperatures as high as 121°C (steam sterilization) without cracking or breaking. Moreover, the plastic material must be chemically resistant to alcohols, acid, alkaline, reducing and oxidizing agents, cleaning and disinfection agents, and corrosive food gases at these temperatures.

6.4.4 Use of Rubbers

Rubber products are widely used in food processing equipment, such as for seals, gaskets, hoses, conveyor belting, skirting, milk liners, and feather pluckers. Elastomers must be chemically resistant to fat, cleaning agents and disinfectants. Rubbers which are not in direct contact with food product and located outside the contact area do in principle not require special approval, but they should be easy to clean and resistant to the chemicals and temperatures occurring within their immediate installed environment. Preferably, gaskets and seals should be of a removable type, because they may be degraded by product or cleaning agents or damaged by excessive mechanical or thermal compression leading to severe deformation. In both cases, their cleanability will be adversely affected and ingress of liquids containing chlorides under the gaskets and seals may cause severe corrosion problems, even with stainless steel. Elastomers also must be abrasion resistant (e.g., rotary shaft seals, or seals in static applications subjected to abrasion) and retain their surface and conformational characteristics (no loss of elasticity, no embrittlement, no rubbed-off parts and crevices, etc.). PTFE (Teflon), EPDM, natural (white) silicone, neoprene, nitrile, and nitrile/butyl rubber are usually applied. Note, however, that PTFE elastomers do not have enough elasticity to assure resealing, and EPDM is sensitive to grease and oil.

6.4.5 Other Materials

As it can retain microorganisms which can subsequently grow in the presence of product nutrients, wood is not allowed within the product contact area. Exceptions are butcher's blocks and wooden barrels for ripening of cheese or for producing wines and vinegar. Take care of wood splinters which can result in foreign body contamination. Certain types of insulation are not allowed within the product contact area. To avoid their exposure to the outside, they must be permanently and tightly sealed off from the product zone. Glass may be used as a food contact surface, but its application is not recommended due to the potential for breakage. Ceramics can be applied in the coating of other stable materials and processing equipment for very sensitive food products. They are very hard, but may suffer from porosity and brittleness. All ceramic surfaces in direct contact with food must have smooth, unbroken and lead-free glassy surfaces, entirely free of crazing (small hairline cracks) and blemishes.

6.5 SURFACE FINISH

Product contact surfaces must be finished to a degree of surface roughness that is smooth enough to enable them to be easily cleaned and disinfected. The surface finish must have a roughness area R_a as low as practicable and without cracks, pits or cavities where water or soil might be retained. Although a surface finish $R_a < 0.3$ μm is a minimum requirement in the pharmaceutical industry, a surface roughness $R_a \leq 0.8$ μm is considered as acceptable for the food industry. Surface roughness, R_a, of enclosures in hygienic production areas should not exceed 2.5 μm. As surfaces deteriorate with time, their cleaning becomes more difficult (Hauser et al., 2004a).

The technique used for achieving the appropriate surface finish is of great importance. Although a surface roughness of Ra 0.8 μm can be achieved with different surface finish techniques (glass blasting, ceramic beats blasting, electropolishing, pickling), the topography/structure of the surface can differ a lot, giving different cleaning results.

6.6 HYGIENIC DESIGN OF OPEN VESSELS, CONTAINERS, AND BINS

6.6.1 Interior and Exterior Design of Open Vessels, Containers, and Bins

Appropriately designed and installed open vessels, containers, and bins shall meet the following recommendations (Moerman & Kastelein, 2014):

- Equipment without bottom outlets must be pivoted (Fig. 6.13) over an angle of at least 93 degrees for fully discharging product and cleaning solution. While fully drainable, contaminants from the exterior of the open vessel, container, or bin (e.g., dirt from casters) may not gain access to the food product being discharged.
- The vessel, container, and bin tipped for discharge must be designed for improved cleanability. Vessel corners should be well rounded and hinges must allow for maximum cleanability.

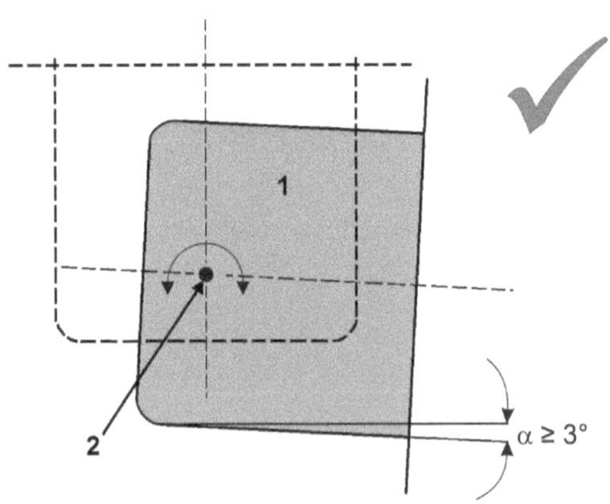

FIGURE 6.13 To fully empty containers without bottom outlet, they must tip over an angle of at least 93 degrees. The interior and exterior of the container must be designed to exclude any contamination of the food product when it is drained. Vessel should have well-rounded bottom corners, with hinges designed for maximum cleanability (Lelieveld et al., 2003; Hauser et al., 2004b).

- Open vessels, containers, and bins with bottom outlets must have their discharge outlet at the lowest level and their bottom shall be sloped (more than 3 degrees toward the outlet). Their corners shall be well-rounded, with a radius equal to or larger than 3 mm (Fig. 6.14).
- The design of the top rims of product-containing equipment (e.g., open tanks, chutes, boxes) must avoid ledges where product can lodge and which are difficult to clean (Figs. 6.15 and Fig. 6.16A). Open top rim designs must be rounded and sloped for drainage (Fig. 6.16B). If the top rim is welded to the wall, the weld must be flush and polished to provide a smooth surface and the rim must be totally closed. Any holes, therefore, must be sealed by welding or by fitting sealed caps (Fig. 6.16B).

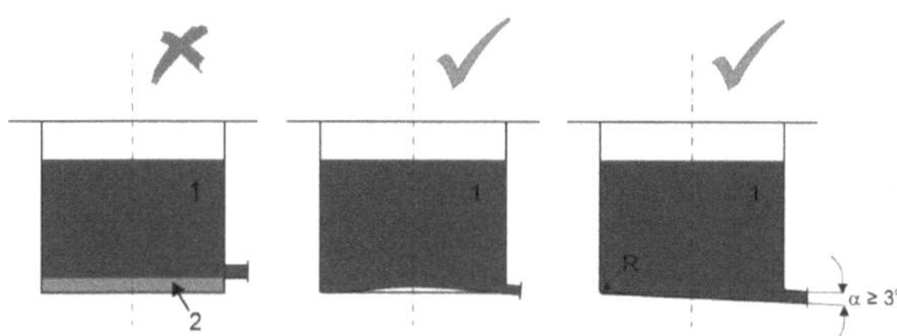

FIGURE 6.14 For good drainability and cleanability, open vessels, containers, and bins used in the processing of food (1) shall have their discharge outlet at the lowest level. Their bottom shall be sloped (more than 3 degrees toward the outlet), and their corners shall be well-rounded. Where food product and cleaning solutions are not allowed to drain, residual soil (2) will be left. Sharp corners (≤90 degrees) must be avoided (Lelieveld et al., 2003; Hauser et al., 2004b).

FIGURE 6.15 A badly designed rolled-over part of top rim provides a ledge where product debris can lodge. *Don Graham, Graham Sanitary Design Consulting LCC, ©2010.*

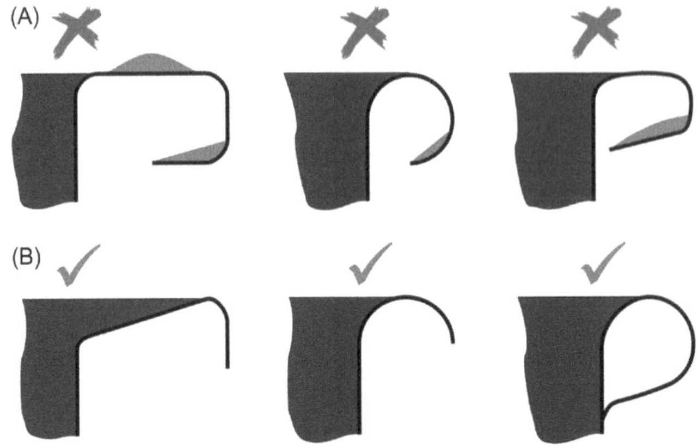

FIGURE 6.16 Top rims may impart rigidity to the construction. (A) However, a rim with an upper horizontal part provides a surface where debris may collect. When the rolled-over part of the rim is badly designed, it may provide a ledge where product debris can lodge. This soil can indirectly affect the product. (B) Open top rims must be rounded in a way that at one side the product drains back in the bulk of the product, while the more exterior part of the rim must allow drainage to the outside. Where preference is given to closed top rims, the top rim should be welded correctly to the wall over its full length. The weld must be flush and polished to provide a smooth surface. The rim must be totally closed and any holes sealed by welding or by fitting sealed caps (Hauser et al., 2004b).

6.6.2 Installation of Agitators in Open Vessels (e.g., Kettles)

Installation of agitators in open vessels should occur as follow (Moerman & Kastelein, 2014):

- Stirrers, homogenizers, or mixers installed via the bottom side or a side wall require sealing of the shaft at the product side. The problem is that seals may wear with time resulting in leakage of product to the outside and product contamination. Parts of the seal may get lost in the product as a foreign body contaminant (Fig. 6.17). Moreover, when the seal gets damaged, product buildup around the shaft will occur, providing nutrients for microorganisms to grow. These microorganisms may subsequently flow back into the product.
- Where mounting of the equipment outside the product zone is possible, the mixer used to mix open product should be fixed beside the equipment, not only to prevent the contamination of the product with dripping oil, but also to avoid the introduction of soil, and concomitantly spoiling microorganisms and pathogens into the product along with overhanging electrical cabling (Fig. 6.18).

6.7 FRAMEWORK

The number of support legs and cross-bracings should be reduced but shall be of sufficient number and strength and so spaced that the process

FIGURE 6.17 Stirrers, homogenizers, or mixers installed via a side wall require sealing of the shaft at the product side. When the seal is worn, however, leakage of product to the outside and product contamination may occur. Particles of the broken seal act as foreign body contaminants, and the food may be spoiled with microorganisms migrating out of the gap in which they could grow due to the presence of nutrients. *www.ourfood.com, courtesy of Karl Heinz Wilm, © 2016.*

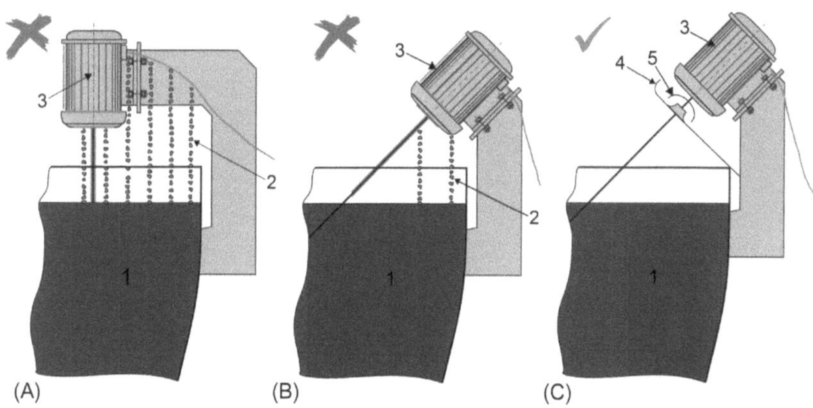

FIGURE 6.18 (A) A motor and cabling mounted over any exposed product (1) can contaminate it by soil, condensate, or lubricants (2). (B) The motor drive (3) and power line should be placed beside the recipient. But without drip protection, soil, condensate, and lubricants still can contaminate the product. (C) A self-draining protection sheet with "upstand" (4) in combination with a cowl (5) on the shaft must exclude any food safety risk. The bottom side of the thrower ring (cowl) should be made inspectable (Lelieveld et al., 2003; Hauser et al., 2004b; Moerman, 2011).

equipment will be adequately supported. Where applied, cross-bracers should be fitted in a diamond configuration. Solid cross-members as structural components are preferred over hollow section members, while sealed hollow section members are usually more preferable over open profile angle or channel sections (Fig. 6.19A). Round section members or square section members turned through 45 degrees provide sloping surfaces (Fig. 6.19B). For vertical

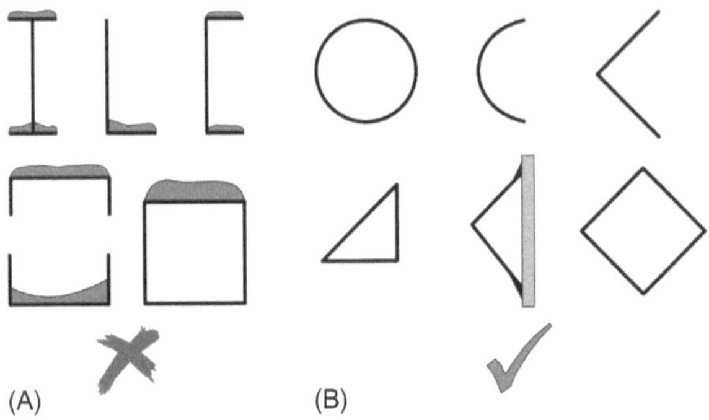

FIGURE 6.19 (A) In the horizontal plane, flat open and closed support members allow debris and liquids to lodge. (B) Structural components with round section and square section members turned through 45 degrees may prohibit buildup of dirt when installed in the horizontal plane. Open profile members with sloped surfaces should have the folding turned outward in both the horizontal and vertical plane, so as to facilitate cleaning (Hauser et al., 2004b).

FIGURE 6.20 A metal sheet frame and subframe ending in metal sheet legs may provide higher strength than hollow framework members. The open profile framework members of this spiral freezer are still perfectly cleanable as they are turned outward. *Courtesy of JBT FoodTech, a Frigoscandia freezer.*

parts of frames all the cross sections shown in Fig. 6.19 can be used. When legs and supports are designed with open profiles, the folding should be turned outward for easy cleaning.

Because small fatigue cracks can arise from vibration in hollow sections, allowing penetration of moisture, soil, and microorganisms, an outward turned open profile construction should be considered for framework exposed to continuous vibrations (Figs. 6.20 and 6.21), even when installed

FIGURE 6.21 Channelized frame construction eliminating hollow tubes and threaded leg adjusters. *Courtesy of Rudi Groppe, Heinzen Manufacturing International, © 2016.*

in the horizontal plane. Furthermore, in temperature conditions <0°C, water inside hollow sections may freeze to form ice. As ice expands, it may cause hollow sections to crack and split. From that point of view, it is important to install hollow sections in such a manner that liquids accumulated on the inside are allowed to drain.

Rolled hollow sections (e.g., legs) must be sealed with great care by welding that should be filled by foaming or made drainable away from the product zone. Allowing drainage to the floor via the bottom of the hollow sections is better than trying to seal them (e.g., plastic plugs). Hollow sections shall not be penetrated by fasteners, and hence drilled and tapped holes are not allowed. Preference should be given to welded plugs when fastening to hollow sections. Welded studs and tapping plates are not recommended. Welding framework members together is more preferable than bolted overlapping constructions.

6.8 FEET

Feet begin at the point where they attach to the leg or the body of the equipment and end at the support point on the floor. These feet are indirect-product contact surfaces but have a hygienic significance because they may become a harborage of soil and create a source of secondary contamination to the products (e.g., during high-pressure cleaning of equipment and especially floors, and dirt present on the feet may splash on the food contact surfaces).

Use a minimum number of support legs/floor mountings, because they are important obstacles for cleaning and service personnel. However, feet must be sufficient in number and strength and so spaced that the equipment

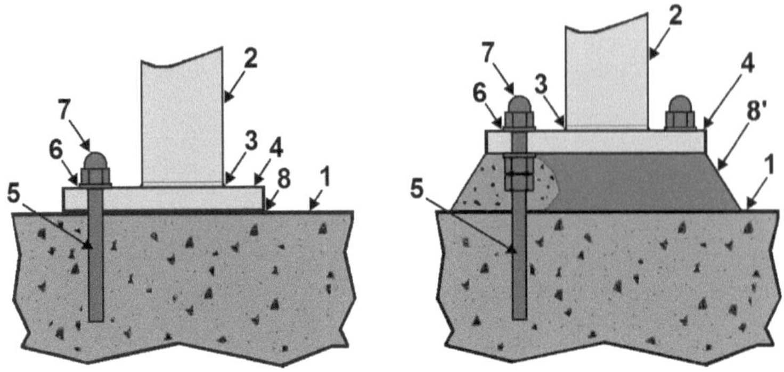

FIGURE 6.22 If the equipment is heavy, the contact face of the foot (2) with the floor (1) must be sufficient to absorb the pressure. To distribute the load, feet should be provided with a footplate (4) welded to the foot leg (3). The foot may be fastened to the floor by means of (a) stainless steel anchor bolt(s) (5) which must have (a) seal washer(s) (6) and (a) dome nut(s) (7) fitted. When the equipment must be bolted to the floor, floorplates shall be sealed with polymer material (e.g., epoxy) to the floor (figure left, 8). The bolting hole(s) must be sealed so that water and dirt are not allowed to leak into the hole(s). Grouting with concrete (figure right, 8′) is not recommended (Moerman & Kastelein, 2014).

will be adequately supported. The general rule is to minimize the floor contact area, but the contact face of the foot must be sufficient to absorb the pressure of the equipment. Skid-proof antislip feet or feet with a footplate fastened to the floor can be used. The latter are applied when the load of heavy equipment must be distributed over a larger surface of contact with the floor. When anchored to the floor, the equipment can't move from its designated position during operation. However, it is better to avoid fastening to the floor because of hygiene issues. The manner in which feet are fastened to the floor depends on the type of floor and the presence of equipment (e.g., machinery producing heat) or services (e.g., electricity) immediately below the surface. When the process equipment must be bolted to the floor requiring floor slab penetration, footplates shall be additionally sealed to the floor with polymer material (e.g., epoxy) (Fig. 6.22 left and Fig. 6.23). Grouting with concrete (Fig. 6.22, right) is not recommended, as practice in the food industry has proven that the concrete grouting can break, allowing food residue to accumulate and bacteria to find a niche in the cracks and crevices of the concrete. Chemical anchors without bolting (fixing to floors by means of a polymer seal) are more recommended. Care must be taken during installation to assure that the footplate does not span over cracks, grout lines, or other floor imperfections.

The footbase of foot ends may either have flat (not recommended) or sloped (recommended) surfaces. Provide only fixing holes where bolting to the floor is necessary, and avoid the use of extra brackets. Avoid all unnecessary (sharp) corners and edges, as well as crevices at the fixing point. When installed on the

FIGURE 6.23 By anchoring the process equipment to the floor, it cannot move from its designated position during operation. Foot anchors should be polymer sealed into the floor. Grouting foot anchors into a concrete floor may give rise to unhygienic conditions with time. *Courtesy of Surface Solutions, Inc.*

FIGURE 6.24 The foot has plenty of pits, folds and other imperfections where dirt and liquids may build up. In particular the exposed treads, bolt nuts and washers create a lot of crevices. The footbase forms a difficult-to-drain flat surface, and dirt and water may penetrate under the footplate. In this manner, they form a niche where microorganisms may grow. *Photo left, courtesy of Mondelez International, 2016; photo right, American Meat Institute, © 2016.*

machinery and within the specified load conditions, all exposed surfaces shall be free of pits, folds, cracks, crevices, and other imperfections in the final fabricated form. A smooth finish is required such that soil may be cleaned from the surface using manual cleaning techniques. Several examples of nonhygienically designed feet are shown in Figs. 6.24 and 6.25.

Equipment should be adequately located in position, with all its feet having a contact face that is even, so as to ensure complete contact with or to allow fixation to the floor. Where the floor is flat, hygienic nonadjustable feet can be used (Figs. 6.26 and 6.27).

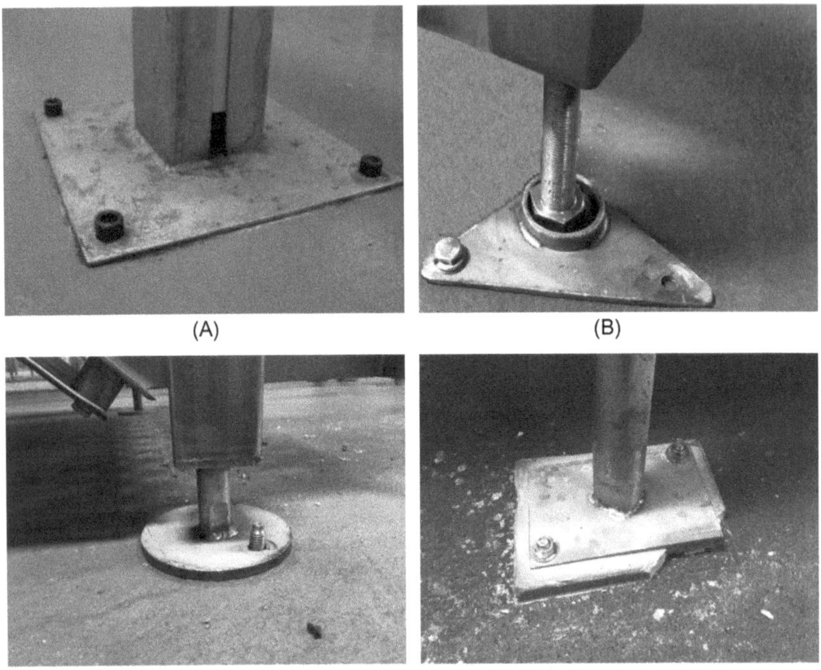

(A) (B) (C) (D)

FIGURE 6.25 Foot ends with a flat footbase (footplate) are not recommended. The use of a bush (welded to a footplate, and either open or closed) to insert the legs of the equipment or the foot spindle is a proof of bad hygienic design practice because debris and water may collect into the bush (A & B). Welds to fix the equipment leg or foot spindle onto the foot base must be smooth, without pits and folds where dirt and liquids may build up (C & D). Holes in the footplate should be omitted where bolting to the floor is not necessary. Exposed threads such as threaded foot spindles (B) and bolting screws without dome nut (C, D) are not allowed. Countersunk screws with slots or other drive configurations (A) and bolts with flat hexagon bolt heads (B) are not recommended. Bolts with hexagon dome nuts and seal washers must be used instead of flat open hexagon nuts (D). Because the footplate is not provided with an antislip rubber pad or polymer sealed onto the floor, dirt and water may penetrate under the footplate, forming a niche where microorganisms may grow. Concrete pads should not be used, as the concrete may crack (D).

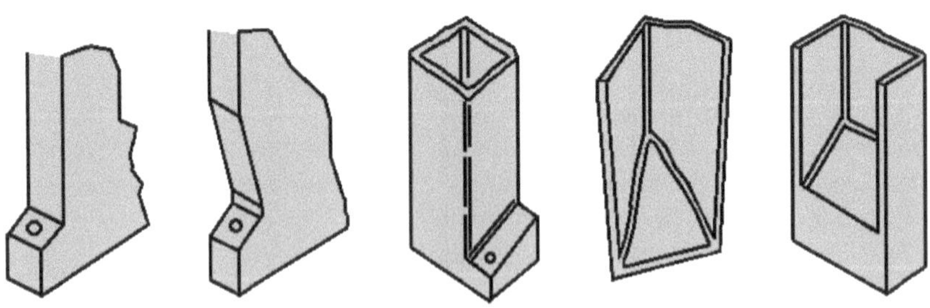

FIGURE 6.26 Nonadjustable feet with sloped surfaces, rounded corners and smooth welds for maximum drainability should be used (APV Baker, 2001).

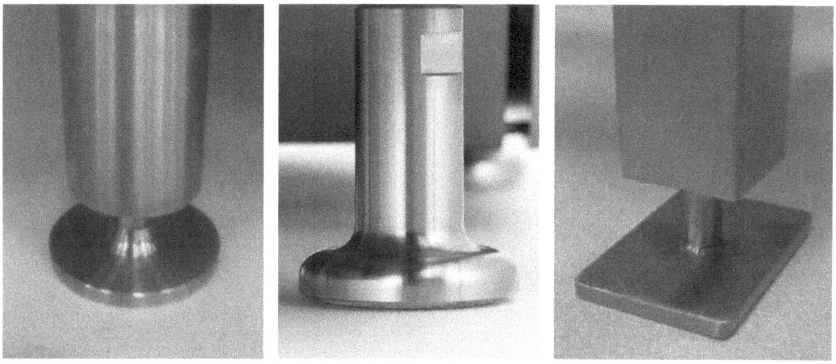

(A) (B) (C)

FIGURE 6.27 (A) Foot with a smooth radiused leg-footbase transition (Frank Moerman, © 2016). (B) Foot radiused down to the foot base. (C) Foot spindle smoothly welded to the foot-plate and the hermetically closed leg. *Photo right, courtesy of Joe Stout, Commercial Food Sanitation LLC - Intralox, © 2016.*

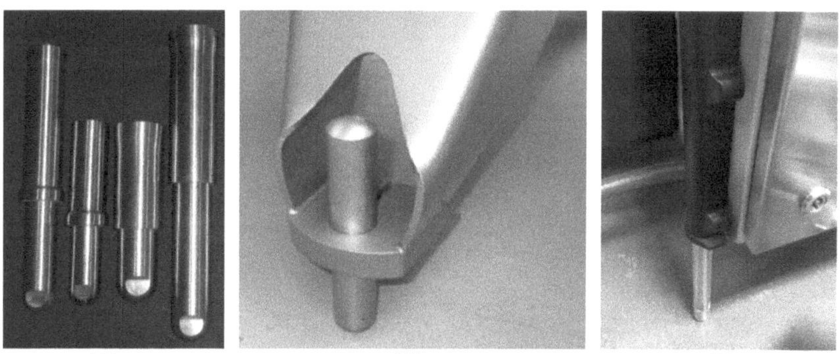

(A) (B) (C)

FIGURE 6.28 Ball feet with the threaded surface covered by an adjustable sleeve (A) and other foot designs without threads (B & C) provide easily cleanable designs. However, mechanically they will almost destroy the floor, because they exert a (very) high pressure locally, especially if they are used to support heavy equipment prone to vibration/oscillatory movement. Ball feet should only be used to support low-weight equipment. *Photo left, courtesy of Koss Industrial, Inc.*

Notice that ball feet (Fig. 6.28A & B) or other foot designs exerting high punctual loads on the floor (Fig. 6.28C) are not recommended. They leave uncleanable crevices between the floor and the foot. Moreover, mechanically they will almost destroy the floor, because—due to their very small contact surface with the floor—they exert a very high pressure locally. If the process equipment is heavy and prone to vibration, the floor will break up very quickly.

Foot ends with a ball-socket arrangement (Figs. 6.29 and 6.30) have a spindle with ball end that may freely swivel in a socket or internal cavity of

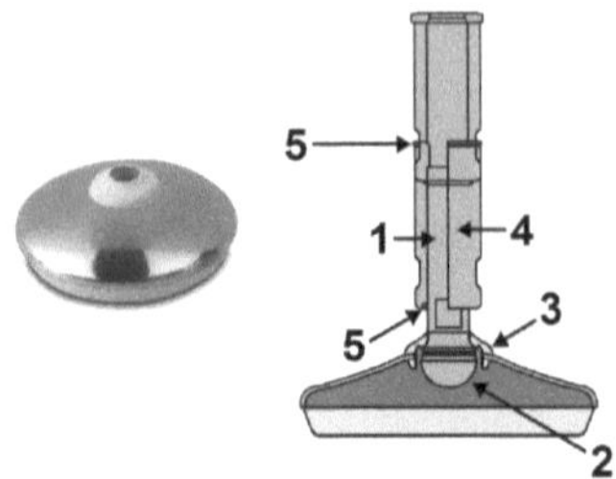

FIGURE 6.29 Hygienic adjustable equipment feet admitting slopes up to 10 degrees on floors. Footbase and spindle are two separate entities. The ball-end of the spindle (1) fits in the diametrically centered depression (socket or internal cavity) (2) in the footbase. The gasket (3) at the junction between the foot spindle and footbase plate (ball-socket joint) prevents any dirt from entering the socket or internal cavity. The adjustment sleeve (4) which covers the threaded surface also functions as a nut. O-rings (5) inserted inside the adjustment sleeve at both ends prevent dirt from entering the thread.

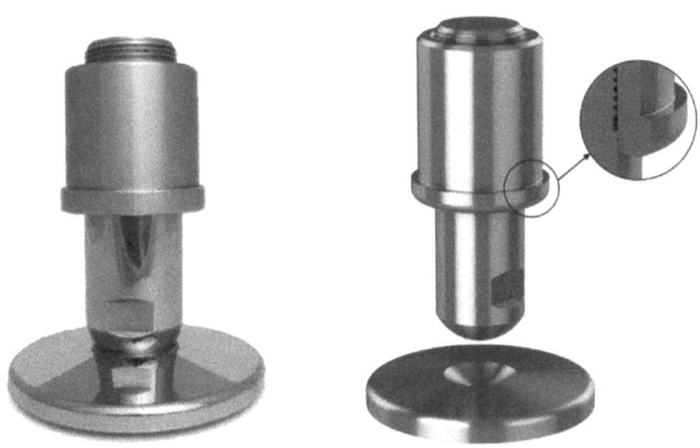

FIGURE 6.30 Heavy-duty adjustable feet with the threaded spindle completely covered by a sleeve provided with O-rings. By adhering to the smooth surface of the screw, this gasket prohibits any intrusion of dirt. Where required, the spindle can be welded to the footplate to avoid liquids or product residues from gathering in the hole in the footplate. *Photo left, courtesy of Martin S.R.L., photo right, courtesy of NGI A/S.*

a separate load-bearing footbase. Such a design provides stability under load (the load is evenly distributed about the entire spherical surface of the socket), as well as vibration/oscillatory movement absorption capacity and the ability to support the equipment on an uneven floor. To prevent any dirt from entering the socket or internal cavity of the load-bearing footbase, a rubber gasket at the junction of the footbase with the spindle (ball-socket joint) is essential. In other designs, the foot spindle has a concave end which can swivel over a diametrically centered convex elevation in the load-bearing footbase plate (Fig. 6.31). An O-ring fitted in the concave spindle end must prevent access of impurities, filth or bacteria in the joint. The load-bearing

FIGURE 6.31 Hygienic adjustable equipment feet admitting slopes up to 15 degrees on floors. Footbase plate and spindle are two separate entities. The concave end of the spindle fits in a diametrically centered elevation of the footbase plate, having the form of a spherical dome. An O-ring fitted in the concave spindle end must prevent access of impurities, filth or bacteria in the joint. The adjustment sleeve covers the threaded surface, and at the same time functions as a nut. The O-rings inserted inside the adjustment sleeve adhere tightly to the smooth surface of the screw, and prevent dirt from entering the thread section. *Courtesy of NGI A/S.*

foot may also include a rubber layer underneath or rubber can be embedded in the load-bearing foot. The elastomeric material may dampen the vibrations of the operating equipment and may prevent slipping of the foot on the support surface. The rubber pad may also prevent liquids and dirt from getting under the footbase due to the fact that the rubber compensates for the roughness and irregularities of the floor. A sufficiently low Durometer rubber provides a tight continuous seal with the flooring material.

For proper installation on uneven or inclined floors, it is not allowed to level food processing equipment with improvised shimming either with metal sheet (Fig. 6.32A) or a wooden plank (Fig. 6.32B). Equipment feet adjustable by min. ± 75 mm should be used. Adjustable feet may not leave (threads) (Fig. 6.33). When adjustable feet are used, the threaded spindle for leveling should be completely concealed in closed profiles/pipes (Fig. 6.34) or enclosed by a sleeve (Figs. 6.29–6.31), so as not to cause accumulation of dirt or contaminants in the thread and to facilitate the cleaning of the foot (Fig. 6.35). O-rings inserted inside the adjustment sleeve must prevent dirt from entering the thread section. USDA has imposed a new safety feature to prevent exposed thread parts due to overscrewing the sleeve (Fig. 6.36).

Also the fixation of feet to legs must be done in a hygienic way. Sometimes it is better to leave the leg end half open. In refrigeration

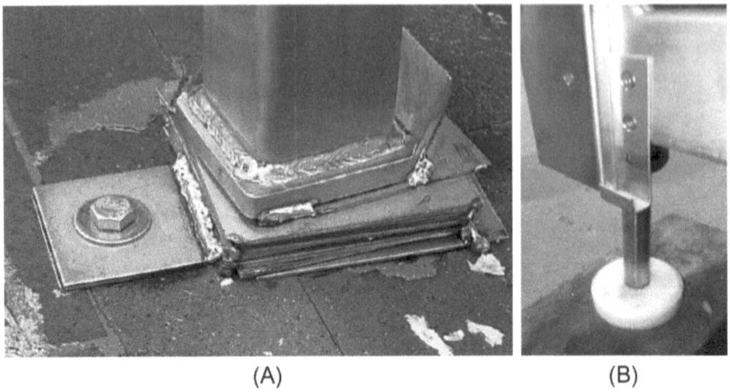

FIGURE 6.32 Leveling food processing and packaging equipment with improvised shimming creates an unhygienic mess. In case (A), several sheets of metal create plenty of metal-to-metal crevices in which dirt and liquids may collect (Don Graham, Graham Sanitary Design Consulting LCC, © 2010). In case (B), although wood may absorb moisture and food debris, a wooden plank is used to level the equipment. With time, this wood becomes prone to rot. The foot is fixed to the equipment subframe by means of rivets in an overlapping sheet construction. The horizontal section may accumulate dirt.

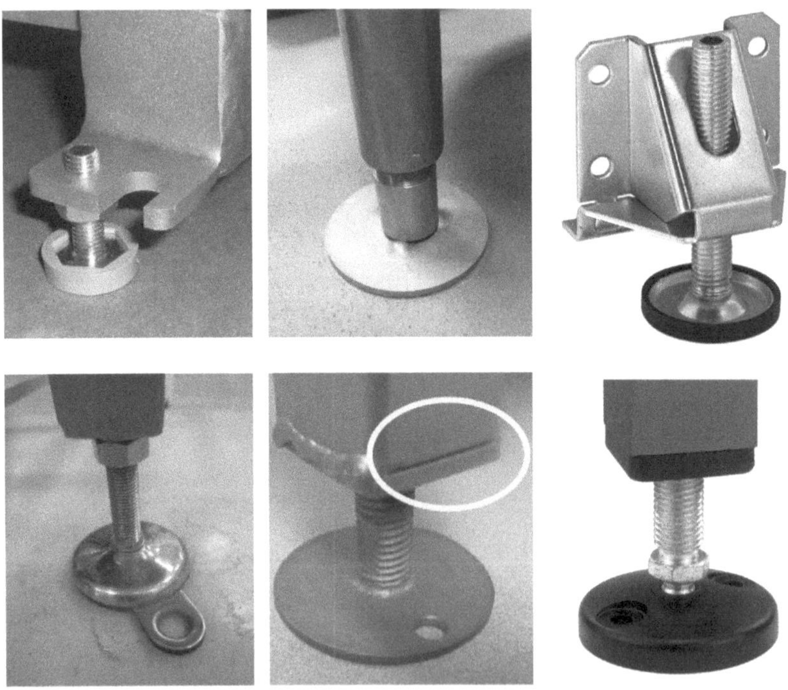

FIGURE 6.33 Adjustable feet with exposed threaded spindle easily become prone to contamination and are difficult to clean. Lower middle photo shows a hollow leg which is not hermetically closed. *Photo at lower center courtesy of Joe Stout, Commercial Food Sanitation LLC - Intralox, © 2016.*

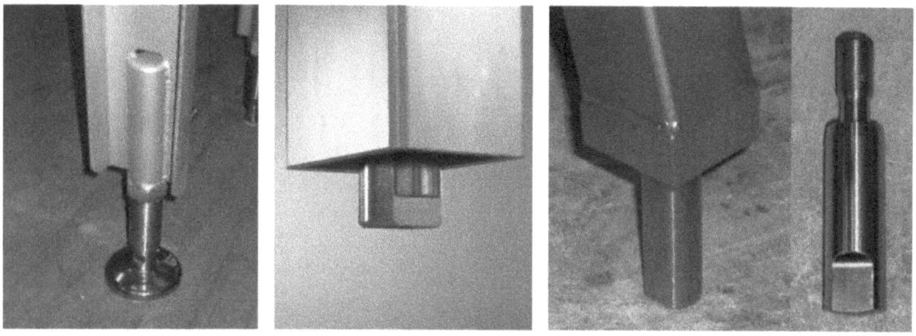

FIGURE 6.34 Adjustable feet with the threaded spindle completely concealed in closed profiles/pipes. In the first example (photo on the left), the thread of the foot is covered by the closed pipe, welded in a sheet metal leg *(courtesy of Den Rustfri Stålindustris Kompetencecenter)*. In the second example (photo in the middle), the foot can be screwed in and out at the bottom of the frame's legs *(courtesy of Alfa-Laval AB)*, while the thread still remains covered in the leg. In the third example (photo right), the leveling foot is screwed in leveling foot mount *(courtesy of Schenck Process LLC)*. In the second and third examples, the contact surface of the feet with the floor is rather small, which will locally exert a (very) high pressure, especially if used to support heavy equipment prone to vibration/oscillatory movement.

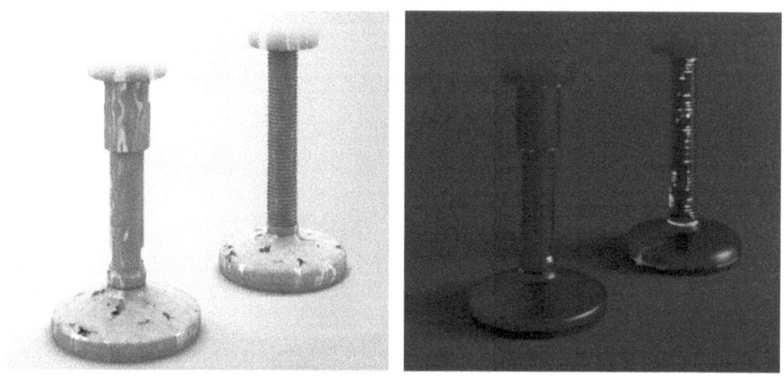

(A) (B)

FIGURE 6.35 Difference in cleaning results between (A) a hygienic leveling foot having the spindle covered by a sleeve and an arrangement of O-ring silicone seals in all the mobile joints (left) and a standard fully threaded foot (right). Both feet were soiled with sour yogurt containing fluorescents *(courtesy of NGI A/S)*. After rinsing to remove a large part of the soil, the feet were cleaned with detergent. In a postrinse step, the feet were flushed with water until visibly clean. (B) Both feet look visibly clean to the human eye. Opposite to the hygienic leveling foot (left), which was found to be free of residual soil and bacteria under UV light, the foot with exposed spindle thread (right) still contained residual soil and bacteria, visible as fluorescent spots under UV light.

equipment operating below 0°C, it is essential that any liquid entering the hollow leg can be drained to prevent cracking of the leg when ice forms at the inside (Fig. 6.37). But preferably, hollow legs must be hermetically closed, requiring adequate sealing of the points where the feet enter the

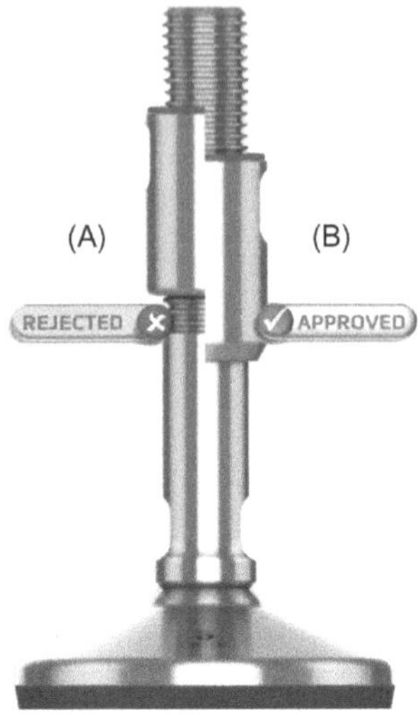

FIGURE 6.36 (left) Overscrewing the lock/sleeve may result in exposure of the thread in the area just below the sleeve. USDA therefore requires an additional feature to avoid overscrewing by blocking the sleeve or counter nut. *Courtesy of NGI A/S.*

FIGURE 6.37 The leg end is provided with a drain hole to allow draining of any liquids entering the hollow leg. However, in this design, dirt can get in the leg end and complete draining is not possible. The leg end may provide a niche where microorganisms can grow. Despite the drain hole, at below 0°C temperatures, the leg end can still crack under the impact of ice formed within the leg end. The footplate is also not hygienically fastened to the stainless steel floor plate. *Courtesy of Stephanie Olmsted, Safeway Inc.,© 2016.*

hollow leg (Fig. 6.38). Fixation of the leveling feet onto open profile legs may not leave thread parts exposed, requiring a top sleeve with O-ring seal and round top to cover the exposed thread parts (Fig. 6.39). Threadless adjustable feet are also available (Fig. 6.40).

(A) (B)

FIGURE 6.38 Fixation to the equipment or equipment legs must be done in a hygienic manner. (A) Stair riser leg, totally sealed, with sloped top and set off the riser *(courtesy of Joe Stout, Commercial Food Sanitation LLC - Intralox, © 2016)*. (B) The stringers of staircases make no floor contact due to the fixing of hygienically designed leveling feet on the stair riser legs welded onto the stringers (*courtesy NGI A/S*).

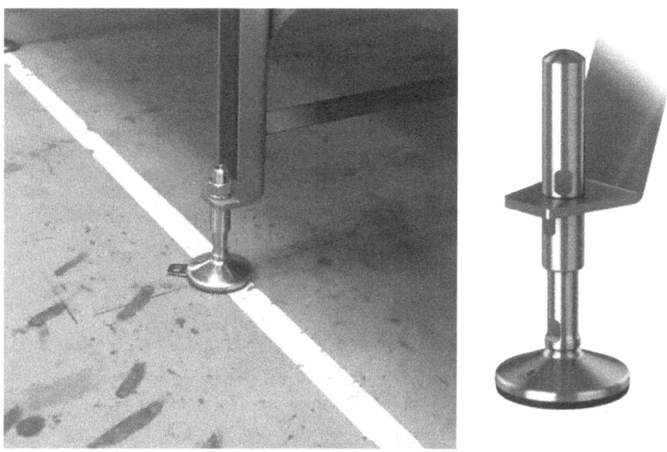

(A) (B)

FIGURE 6.39 (A) Fixation of the leveling feet onto open profile legs may not leave thread parts exposed. (B) A top sleeve with O-ring seal and round top must be used to cover the exposed thread parts. *Courtesy of NGI A/S.*

FIGURE 6.40 Threadless adjustable feet hygienically fixed onto open profile legs. *Courtesy of Rudi Groppe, Heinzen Manufacturing International, © 2016.*

6.9 CASTERS

Although preference should be given to feet rather than to casters, casters are used in those places where process equipment must be mobile in order to facilitate inspection and cleaning of the equipment and process rooms. Transportable equipment also allows changing of the layout of process lines in function of the food products that their customers want to be produced. Casters are excellent for increasing the flexibility of production processes (e.g., frozen vegetable industry).

6.9.1 Materials of Construction

6.9.1.1 Durable Materials of Construction Providing the Casters a Sufficient Load Rating

To prevent premature failure of wheels due to overloading, caster wheels should be constructed out of a material giving them a sufficiently safe load rating for the intended task. A safe load rating incorporates a safety factor for dynamic loading. On poor floor surfaces, casters experience dynamic loads well in excess of the static load. Hence, if the rating of the casters only just meets the minimum failure load, rapid deterioration of the casters will occur. Casters should thus be made of a material that suits the floor quality, the expected loading and the frequency of movement. If underspecified casters are used, the body of their wheels can break up due to being overloaded. To choose the best caster, the maximum load capacity that the casters are required to bear should be determined. That figure is commonly defined as the weight of the food processing equipment including the weight of the food product itself, divided by a number one less than the total number of casters needed. Casters shall further be made of a durable material that is corrosion resistant to food products, water, steam, cleaning agents, disinfectants, etc. They should be self-finish or dull-nickel plated, because the use of paint is not recommended. Based on these parameters and requirements, a caster made of a suitable material of construction that meets that specification can be selected.

6.9.1.2 Cast-Iron Casters

Although cast-iron wheels are virtually indestructible and are able to withstand the highest loads, their use in the food industry is not recommended (not acceptable), because they are prone to general corrosion of the caster components and because of the damage which they do to floor surfaces. Floor surfaces may break up, if wear is allowed to progress. Damaged floor surfaces then become an excellent harbor for contaminants and microorganisms.

6.9.1.3 Zinc-Plated Mild Steel Casters

Casters manufactured from zinc-plated mild steel should be avoided. When the zinc coating on the wheels wears away, corrosion will occur quickly, which may result in increased friction between the wheels and the caster forks (horns). Casters manufactured from zinc-plated mild steel require the swivel bearing to remain lubricated to prevent it from corroding. If corrosion of the swivel bearing occurs, it will prevent the caster from swiveling efficiently; as a consequence increased side loads will be applied to the caster components, which can result in premature failure of the caster. Zinc-plated mild steel casters with different types of swivel bearing seal arrangements are available that allow lubrication-in-place, and that prevent contamination of the bearing and loss of lubrication during the cleaning of the process equipment. But if lubrication is not regularly and properly done, corrosion and increased play still will be observed in these swivel bearings, with the result that, after a certain time, the caster will start to squeak until it stops swiveling again.

6.9.1.4 *Stainless Steel Casters*

Casters (body, mounting plate, etc.) manufactured from stainless steel with stainless steel swivel bearings need no lubrication to prevent corrosion. Although stainless steel axles in combination with an outer PTFE bushing provide self-lubrication of the wheel/axle surfaces, they still need to be replaced periodically. But PTFE bushes are relatively cheap and easy to replace, and are less time-consuming in maintenance than manual lubrication.

6.9.1.5 *Casters With Full Thermosetting Plastic Wheels*

Thermosetting plastics, particularly phenolics, are widely used in the food industry because they withstand high temperatures and can carry high loads. However, they can become damaged by poor quality flooring and by defects in floors, such as concrete joints and ridges. Phenolic wheeled caster types are often worn to a flatter profile or their tread is spalling. Some phenolic wheels may suffer from breakup due to impact damage, while others may absorb water when subject to damp or wet conditions for long periods. Hence, the correct type of phenolic must be selected if washing is carried out routinely.

6.9.1.6 *Casters With Full Thermoplastic Wheels*

Thermoplastic wheels have better impact resistance than phenolic wheeled casters, but they have poor resistance to higher temperatures. They are applicable in the food industry where high temperatures are not part of the environment.

6.9.1.7 *Rubber-Wheeled Casters*

An alternative to the phenolic wheel is the high-temperature rubber-wheeled caster. These wheels have a high-temperature thermoplastic center with a bonded high-temperature rubber tire. As with phenolic wheels, they will wear and may be damaged by poor or abrasive surfaces. They also have slightly lower load ratings than an equivalent diameter phenolic wheel because of their lower shear strength. A major drawback of rubber-wheeled casters is their poor resistance to acids, oils, chemicals and other substances harmful to rubber. These soft tread wheels, however, may ride more easily over bumps, level changes, joints, drainage gullies, etc., and are less destructive to tiled, linoleum, etc., floors. They also provide better shock absorption and traction.

6.9.2 Hygienic Design Requirements

Swivel casters (Fig. 6.41) only function well when they are securely mounted to a rigid frame so the swivel bearing kingpin axis remains vertical at all times. Rigid casters must be mounted (welded, sealed, or readily removable) in such a way that their axis and wheels are in alignment. All structural members (mounting plate and horn) shall have a minimum of horizontal flat surfaces. The plate mounting shall be constructed to have a flat top surface. The angle between the top surface and the edge of the plate shall be 90 degrees or less. Mounting holes and other devices provided for installation shall be so designed as to prevent the formation of pockets or areas difficult to clean. The horn assembly or fork shall be constructed so that the surface facing the wheel shall have no concave surface except that part joining the horn plate. Included angles between all surfaces should have a minimum radius of 6 mm. Kingpin assemblies, which have the nuts or rivets at the bottom, shall have suitable caps covering the ends. The minimum clearance

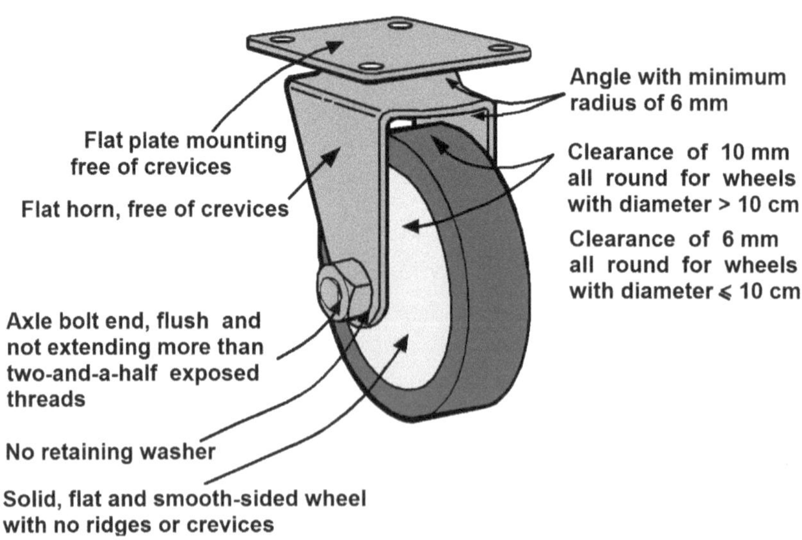

FIGURE 6.41 Hygiene design requirements that casters in the food industry must meet (APV Baker, 2001).

between horn assembly and wheel should be 6 mm all round, for wheels equal to or less than 10 cm in diameter, while the minimum clearance should be 10–12 mm all round for wheels with diameter larger than 10 cm. Brakes and locking devices should comply with the hygienic requirements mentioned previously.

6.9.3 Inspection and Maintenance of Casters

The frame and fasteners of casters should be regularly checked for distortions, loose bolts and nuts, and broken welds or deck boards. Where rigid casters are used, they should be inspected for bent caster horns. Wheel surfaces must be checked for chemical damage and tread wear. Flat spots may indicate foreign material, a loose caster or a frozen wheel, requiring the wheel to be replaced. Swivel assemblies should be regularly checked and replaced if they are loose. If they don't turn freely, bearing raceways should be checked for corrosion or dirt. Secure fastening of the king bolt nut should be verified

If no self-lubricating bearings (stainless steel with PTFE bushing) are used, wheel and swivel bearings should be lubricated every six months; while lubrication of bearings in corrosive environments is required once a month. In the food industry where the lubricant is washed away by daily cleaning, lubrication is sometimes even required after each washing. Single seals used to contain oil or grease in bearings will wear, ultimately allowing leakage. Therefore their integrity must be regularly checked, and their replacement at defined maintenance intervals is required.

6.10 BELT CONVEYOR

6.10.1 Conveyor Frame

Conveyor frames should have an open structure with a minimum of hidden areas/surfaces. But guards are required in places where a drive station, a pulley, rollers or the conveyor belt may cause injury. The guards, however, should be easy to dismount to allow for complete cleaning. Solid cross members as structural members are preferred over hollow section members. Open profile angle or channel sections must be installed in a manner such that horizontal ledges and crevices are absent. Where open profiles are used, the folding should be turned outward for easy cleaning (Fig. 6.42). Welding is preferred over fastening.

6.10.2 Conveyor Bed

Conveying surfaces shall be supported by a minimum amount of carrying surface or bed (Fig. 6.44) as required. The use of solid plate that expands the whole top surface of the conveyor table to provide support to a belt is likely

FIGURE 6.42 The conveyor frame and subframe should have an open structure without horizontal surfaces and with a minimum of hidden areas/surfaces. At the outside, the framework consists of open profile members turned outward for easy cleaning. The conveyor frame is an all-welded construction with flat cross-members welded at the outside framework. The cross-members not only act as structural frame members, but also as belt supports. The weld-on flat cross-members are provided with gaps to accommodate the freely located plastic wear strips that help to support the conveyor belt. No bolts, holes or nuts were used for fastening the ultra-high-molecular-weight (UHMW) polyethylene wear strips. The swivel-mounted roller permits release of the belt tension, hence providing improved access for cleaning the bottom side of the belt as well as the bearing strips. *Courtesy of Interroll.*

to increase contamination problems and cause excessive wear of the belt. Nonremovable bearing surfaces for belts cannot be cleaned easily. Rollers shall be used where practical, or line supports that are easily removable for cleaning. The conveyor belt should have minimal debris retention, and running under turned over sections of side cladding (overhanging belt edges) is not allowed because the whole surface of the belt cannot be cleaned, and the belt cannot be lifted up to allow cleaning and inspection of internal surfaces and support members (Fig. 6.2). But also pivoted covers can't be cleaned easily if continuous hinges are used. In continuous hinges, food debris and microbial slime are strongly retained in the hinge segments (Fig. 6.43). Side guides used to contain product should be capable of being removed. So, pivoting guides using pin hinges with removable pins are acceptable. But removable guides also may cause problems because of the possibility of the fastening system working loose. The conveyor frame must be designed so that the sides of the belt are turned up to form an integral guide to the belt. Besides this guide, cladding can be made removable allowing for effective cleaning (Fig. 6.44).

The most common design of a drive station is placing the drive pulley between two bearings, one at each side of the conveyor. Open bearings have low Ingress Protection (IP), making them sensitive for the removal of lubricant during cleaning operations (Fig. 6.45). Because they require more frequent lubrication, the risk of overlubrication increases. Self-aligning pillow block bearings or flange bearings with covers and constructed from materials

FIGURE 6.43 The conveyor belt running under turned over section of side cladding (overhanging belt edges) is not allowed because the whole surface of the belt cannot be cleaned, and the belt cannot be lifted up to allow cleaning and inspection of internal surfaces and support members. In this example, the side guides serve both as belt and product guide. They may pivot to the outside due to the hinge segments. Pin hinges with removable pins are used, which allows for minimal debris retention and excludes the likeliness of microbial growth and concomitant slime production. *Courtesy of PaxiomWeighPack Systems Inc.*

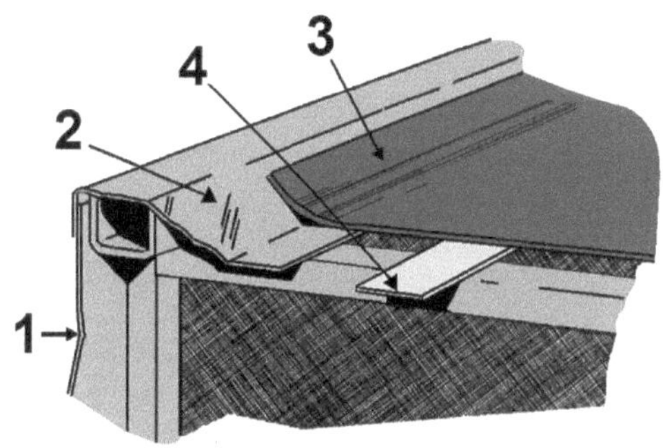

FIGURE 6.44 The conveyor frame (1) must be designed so that the sides of the belt are turned up to form an integral guide (2) to the belt (3). Besides this guide, cladding can be made removable allowing for effective cleaning. The conveyor belt shall be supported by a minimum amount of carrying surface or bed (4) as required. Rods, slats, rollers, or like supports shall be used where practical (CFPRA, 1983; Hauser et al., 2004b).

approved for food contact may provide waterproof and corrosion-free designs. The higher IP rating prevents lubricant inside bearings being removed during cleaning operations, making overlubrication less likely. Lubricated bearings, including permanent sealed types, should be located outside the direct product contact surface area with adequate clearance open for inspection between the bearing and any product contact surface (Kold et al., 2016).

If the specification of the gear motor allows it, the bearing on the gear motor side of the conveyor can be omitted, using the output shaft of the gear as a bearing. In this case the gear motor has to be fixed to the side of the

FIGURE 6.45 Lubricated bearings should be located outside the direct product contact surface area with adequate clearance between the bearing and any product contact surface. Open bearings have low Ingress Protection (IP), making them sensitive to the removal of lubricant during cleaning operations. Because they require more frequent lubrication, the risk of overlubrication increases. Excessive lubricant and grease should be removed to prevent them from coming into contact with the product. *Courtesy of John Butts, Land O'Frost.*

FIGURE 6.46 Although the drive motor has a drip pan, the difficult-to-clean motor is located too close to the product flow, because the fan may blow dust and dustborne microbes around and onto the food. *Frank Moerman, © 2016.*

conveyor (=direct drive motor). Drive motors should be located below the line of the product flow because the exposed motor may have a fan that will blow dust and dustborne microbes around (Fig. 6.46). It is also possible to place the gear motor away from the drive drum, driving it by means of a chain or toothed belt, for example. Such a design requires a guard around the chain or belt to avoid any contamination of food product, as well as for occupational safety reasons. However, a chain guard, when open, may provide a place where product may accumulate, allowing microbes to multiply in large numbers and so posing a contamination hazard for the food product on the belt (Fig. 6.47). Measures have to be taken to make the drive guard

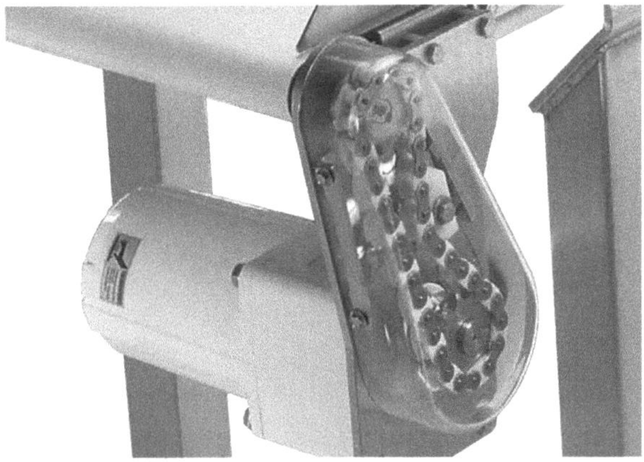

FIGURE 6.47 The drive motor is located below the line of the product flow. Gears, chains (stainless steel or polyacetal), and motors of belt drives must be covered to avoid any contamination of product. However, a chain guard (essential from an occupational safety point of view), when open, may provide a place where product may accumulate, allowing microbes to multiply to large numbers and so posing a contamination risk for the food product on the belt. A transparent chain guard may allow detection of any product debris inside the chain guard. *Courtesy of Dorner Mfg. Corp.*

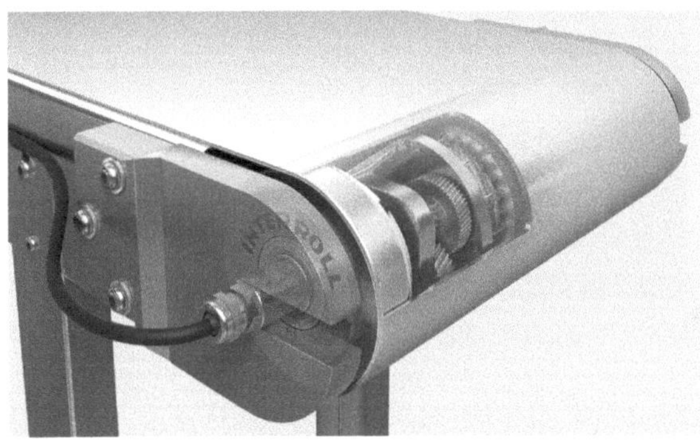

FIGURE 6.48 Drum motors make external gears and chains redundant. *Courtesy of Interroll.*

using a hygienic and easy-to-clean design (e.g., hermetically sealed housing). Furthermore, the guard must be designed in a way that the generated heat from the gear motor can be conducted away.

The drive motor is often of a type that cannot be washed with a high-pressure hose using water and cleaning agents. However, IP54/55/67 sealed wash-down or easy-clean motors which do not require ventilation or housings are available. There should be enough air space around the motor for cleaning and disinfection, maintenance and repair. Where possible, use drum motors (motorized pulleys) (Fig. 6.48) that are fully closed, nonventilated, conveyor belt drives where motor and gearwheels are at the inside, submerged in a bath of food-grade lubricant, providing at the same time lubrication and cooling. Drum motors make gears and chains redundant.

FIGURE 6.49 Rollers and pulleys shall not have hollow parts. *Don Graham, Graham Sanitary Design Consulting LCC, © 2010.*

FIGURE 6.50 Hollow rollers and pulleys shall not be used. *Courtesy of General Mills.*

Rollers and pulleys shall not have hollow parts (Fig. 6.49). The design of rollers, pulleys and sprockets shall be closed if hollow (Figs. 6.50 and 6.51) and shall be free of end recesses (Fig. 6.49). A welded construction should be preferred to a sealed design (Fig. 6.51).

Embedded reinforcements, as well as fabric backing materials in conveyor belts, must be covered to avoid contact with the product. Cut edges of belts which incorporate reinforcing materials must be sealed to prevent penetration by wicking (capillary action) of liquids into the interior (Fig. 6.52).

6.11 MOTORS

Motors made of aluminum or cast iron are prone to corrosion, while painted units create a hygienic hazard because paint may flake off. Motors with cooling ribs have a lot of hooks, corners, and dead spots, making them difficult to clean.

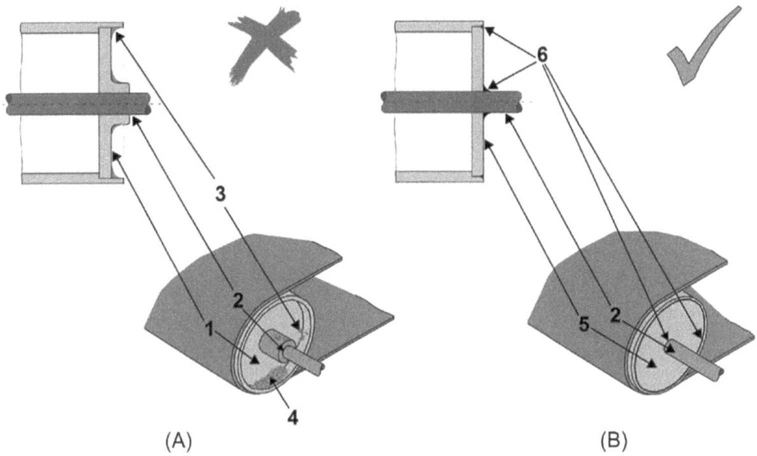

FIGURE 6.51 (A) Pressed-in roller ends (1) create dead areas and crevices (3), where residues of product and soil (4) may accumulate. (B) Flush roller ends (5) which are properly welded (6) to the roller and to the shaft (2) avoid any hazard and can be cleaned easily (CFPRA, 1983; Hauser et al., 2004b).

FIGURE 6.52 Cut edges of belts that incorporate reinforcing materials (e.g., fabric) are prone to penetration of liquids into the interior by wicking (capillary action). Therefore, embedded reinforcements, as well as fabric backing materials in conveyor belts, must be covered to avoid contact with the product. The edge should be suitably sealed and covered in a way that the covered edge is shaped with a round rim. *Courtesy of Ammeraal Beltech Holding BV.*

A cooling fan that ventilates the heat produced by the motor to the outside may act as a source of contamination, because—along with the cooling air—it concomitantly releases dust and dustborne microorganisms accumulated inside the electrical equipment in the process environment. Hence, such motors should not be installed in the product area (including the splash area). Installed over the product flow (Fig. 6.53), the motor also may contaminate the product with lubricants, condensate, or dirt discharged from the drive system.

FIGURE 6.53 Drive motor is installed above the manhole of the process vessel. When the manhole cover is open, the product may become contaminated with lubricants, condensate, or dirt discharged from the drive motor *(organization Sanitary Design Workshop, © 2016).*

FIGURE 6.54 Wash-down motor. However, it is preferable to avoid hosing of motors, outlets, and electrical cables. The electrical connection to the motor should be waterproof and the cable gland of the motor should be positioned in a direction pointing away from the food contact zone and away from direct spraying when cleaning. *Courtesy of CES NV.*

Stainless steel motors (Fig. 6.54) with smooth surfaces (e.g., milled, ground or polished) and without complex geometries or features (such as cooling ribs, screw heads, grooves in the gear assembly, unused or threaded holes) should be used. Hygienic motor assemblies ask for domed cap nuts and metal-backed gaskets under the nuts. A wash-down motor is a sealed electric motor that is designed to allow complete washing with a high-pressure hose, using water and cleaning agents, with no difference in operating characteristics at the end of the wash-down cycle.

As an alternative, easy-clean motors have been designed and built to reduce obstructions to cleaning, as far as operation and economics allow. However, easy-clean motors don't meet the standard completely. They can be washed down only if caustic solutions are not used (Moerman, 2011).

6.12 COVERS AND GUARDS

Chains, motors, gearboxes, sprocket wheels, etc. must be guarded or covered for the following reasons:

- To prohibit injury to operators during inspection, cleaning, disinfection, and maintenance.
- To protect difficult to clean machine components (e.g., drive parts) against contamination by food debris.
- To prevent liquids (e.g., cleaning solutions, condensate) from intruding into water-sensitive machine components.
- To prohibit damage to machinery components.
- To protect the food product from contact with drive parts such as lubricated chains and sprocket wheels.

However, the requirements of guarding or covering machinery to ensure safety in operation may easily conflict with the recommendations of EN 1672-2 as well as other hygiene requirements. By their nature, covers and guards may compromise food safety, unless considerable care is taken in their design, construction, installation, and maintenance.

6.12.1 Covers

Good hygienic design and a stringent cleaning/disinfection regime can solve many problems. Covers and panels must be designed so as to prevent entry and/or accumulation of soil and liquids. Upturned sections, that may provide a tray in which dirt and liquids may accumulate, are not allowed. To ensure good drainability, covers must be contoured to avoid horizontal surfaces and provided with an angle of 5 degrees away from food areas. Where horizontal surfaces can't be excluded, avoid panel joints and provide overlap where possible. Reduce exposed fixings to a minimum. Covers on their own—although practical—can often make cleaning more difficult. Wherever covers are used around drive parts such as chains and sprockets, they should be easily removable to provide access for cleaning, either by opening or unhinging them (Moerman et al., 2005).

Totally removable covers or cladding are not recommended as they may not be put back or they may be damaged during removal. Without covers, operators in the environment of the process equipment and exposed food products are at risk. Where possible, hinged covers that pivot outboard should be used. But use as few hinges as possible, and use concealed hinges with

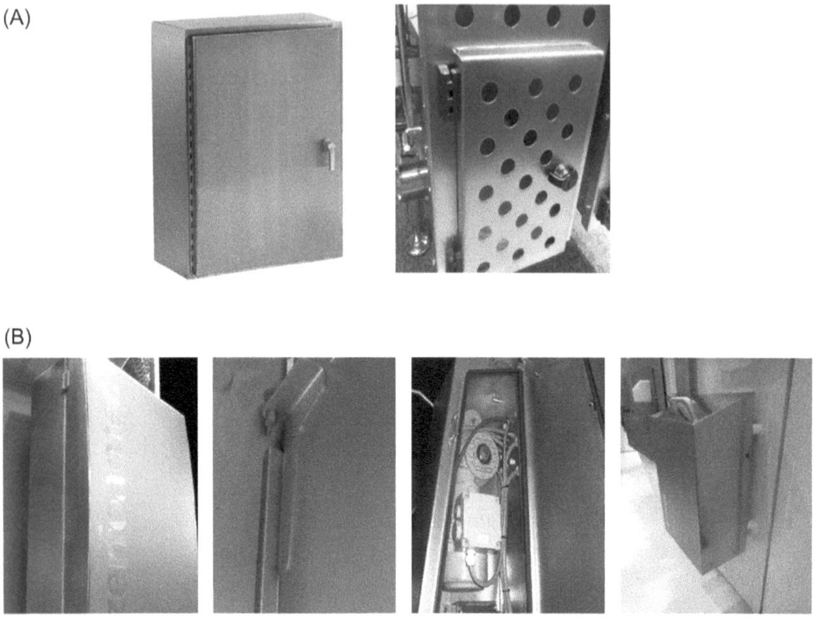

FIGURE 6.55 (A) Continuous and piano hinges are not allowed (Frank Moerman, © 2016). (B) From left to right: block hinges lift-off type, pin hinges with removable pins or concealed hinges should be used. An alternative could be a hingeless cover fixed by means of two small screws at both sides. *First and second photo, courtesy of Heinen Freezing GmbH & Co. KG; third photo, courtesy Den Rustfri Stålindustris Kompetencecenter; fourth photo, courtesy IQF Frost AB; Frank Moerman, ©2016.*

the least number of parts. In view of cleaning and disinfection, continuous and piano hinges (Fig. 6.55A) are not allowed. Block or pin hinges are a possible option but should have removable hinge pins or be of the lift-off type (Fig. 6.55B). Sliding covers are not recommended and if they are used, tracks and guides for covers must be designed to minimize retention of food particles, condensation water, spillage, and soil. As an example, the grooves shall be rounded. For small covers, a "D" handle mounted to the cover is acceptable (Fig. 6.56) (Moerman et al., 2005).

Clear plastic covers (Fig. 6.57) should be made of shatter-resistant material. Polycarbonate is especially prone to cracking if not installed properly (e.g., lack of room for expansion). To maintain the clarity of the transparent material over the life of the equipment, it must be resistant to cleaning agents and disinfectants, and—where necessary—hot water and steam. Acrylic and polycarbonate should be at least, respectively, 12 mm and 5 mm thick. Transparent covers preferably should be mounted unframed; however, for large covers a frame is often required. The frame should be mounted spaced away from that transparent cover and should be provided with a handle or rail to hinge open the cover without putting

FIGURE 6.56 Products transported by a conveyor belt are protected from the environment by means of covers provided with "D" handles at their sides. *Courtesy of Bay State Industrial Welding & Fabrication, Inc.*

FIGURE 6.57 Clear plastic covers/guard. *Frank Moerman, ©2016.*

stress on it. Plastic covers should not be used where frequent removal of covers is required (APV Baker, 2001).

6.12.2 Guards

As compared to covers, the main function of guards is providing a physical barrier to prevent access of personnel into potentially hazardous areas or equipment such as moving equipment components. Although casings, maintenance enclosures, covers and doors also may function as guards, real guards have a more open surface, allowing observation of objects located behind the guard. Although the guards must comply with current health and safety legislation, they must be quite open to permit access for cleaning and disinfection by spray nozzles or hosing down procedures. Typically, the equipment components to protect are less or not water sensitive and are less likely to contaminate the food product (Moerman & Fikiin, 2016).

Bars (Fig. 6.58), perforated/punched sheet (Fig. 6.59) and weld mesh stainless steel guards (Fig. 6.60) with an open area of 40–50% give good protection from moving equipment parts. Expamet expanded metal mesh should not be used. Bars must be made of full stock rod (never hollow sections), preferably with small cross section and positioned in a vertical direction (easily cleanable from all directions) (Fig. 6.58A). In a vertical position, bars should have a round, diamond, rectangular (with small side turned outward) or triangular (with flat face turned outward) cross section. Bars with square cross section or a triangular cross section with flat face turned inward are not recommended, because their backside—without opening the guard—is difficult to clean. Guards should have a minimum number of horizontal parts. If installed in the horizontal plane, bars should not form ledges where dirt may accumulate. Placed horizontally, full stock small round cross section rods, full stock square or rectangular section members mounted at 45 degrees and parts having a triangular cross section with flat face turned outward are

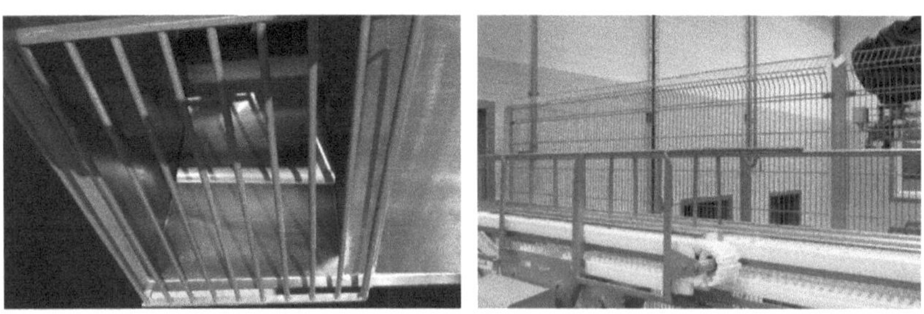

(A) (B)

FIGURE 6.58 (A) Bars made of full stock rod preferably with small cross section and positioned in a vertical direction (easily cleanable from all directions). (B) If weld mesh guards are used, they should have a minimum number of horizontally running wires. *Courtesy of NTF-Aalborg A/S.*

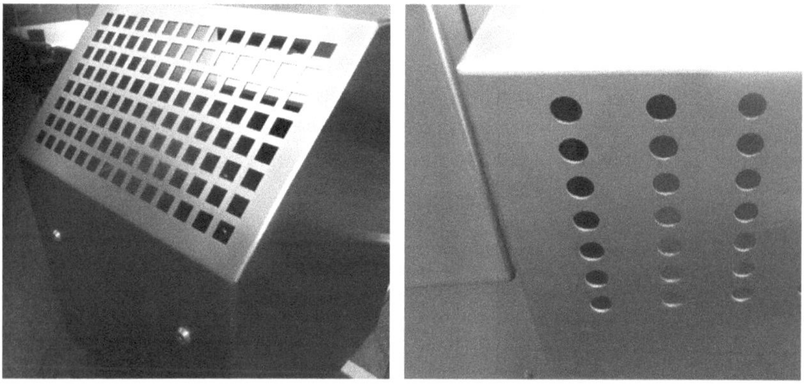

FIGURE 6.59 Punched stainless steel guards with 40/50% open area maximum. They must be opened for cleaning the inside. *Frank Moerman, © 2016.*

(A) (B)

FIGURE 6.60 Stainless steel guards: (A) Punched sheet (>95% open) gives good protection against the centrifugal fan, while allowing access for cleaning and disinfection. (B) Weld mesh guards positioned around the centrifugal fans have a sloped top surface. The fans themselves are positioned inclined on a sloped surface, allowing maximum drainage of condensate and cleaning liquids to the front side.

preferred, while bars with square or rectangular (with small side turned outward) cross section that provide a horizontal ledge are not recommended. Care should always be taken to pitch the bar according to safety requirements. If weld mesh guards are used, they should have a minimum number of horizontally running wires (Fig. 6.58B). Perforated/punched sheet guards with 40/50% open area (Fig. 6.59) must be opened to allow regular inspection and cleaning at the inside by a competent person. The fixing and removal of guards should reflect the frequency of necessary removal (e.g., for cleaning): interlocked guards should be used for high-frequency tasks (e.g., one shift a day) and fixed guards should be used for low-frequency tasks (e.g., monthly). Punched sheet guards (>95% open) and weld mesh stainless steel guards must not be removed for cleaning and disinfection.

6.13 ELECTRICAL EQUIPMENT, CABINETS, AND FIELD BOXES

Electrical and control installations should be limited to those that are necessary for the safe and correct operation of the process equipment. However, a significant part of both the electrical and control installations is still located within the process area, e.g., cabling to power motors of plant machinery, control cables connecting sensors via field boxes/cabinets to the plant control system, etc. But, little by little, wireless transfer of data between instruments and control equipment and battery-supplied low-energy sensors and actuators are finding their way in the industry.

6.13.1 Electrical Equipment

Electrical equipment should have an IP55 rating at a minimum. Preference is given to dust- and moisture-tight electrical equipment that can be hosed down with powerful water jets (IP66) or even better (IP67 & IP67K). IP69K rating, to German standard DIN 40050-9, is required for high-pressure, high-temperature, wash-down applications. Electrical equipment must be protected from things falling or product spilling on it. During cleaning, covering of electrical equipment with polyethylene or equivalent film is recommended.

Electrical equipment usually produces heat, and therefore must be in the possession of a dedicated cooling device. Fans ventilate (blow) heat out of the equipment in the environment by drawing in cooler outside air, circulating it throughout the case and expelling it to the outside. But along with the cooler air comes dust, microorganisms and eventually moisture, which easily accumulate in dry electrical equipment (Fig. 6.61). To avoid heat, dust and microorganisms being blown on the food during processing, the electrical equipment should be positioned away from the product zone. A more recommended alternative to ventilators is the use of the self-cooling capabilities of a casing by means of creating an internal air circulation and achieving temperature reduction through the casing surface. If this does not provide sufficient cooling, then additional cooling could be provided by fixing an air-to-water type heat exchanger to the casing (Moerman, 2011).

Direct or indirect incidental contact between the electrical installation and food cannot always be fully excluded and may possibly result in contamination of the food product. In all cases where product contact cannot be fully excluded, electrical installations have to be used that are suitable for these sensitive areas. Electronic devices positioned in the food contact area (direct and indirect) should be smooth, of a cleanable type and resistant against corrosive cleaning agents.

6.13.2 Electrical Cabinets and Field Boxes

Appropriate materials to construct field boxes and electrical cabinets are stainless steel AISI 304 or coated mild steel and plastic, provided they have

FIGURE 6.61 A cooling device that ventilates the heat produced by electrical equipment to the outside concomitantly accumulates dust and dustborne microorganisms inside the electrical equipment. Moisture- and dust-tight casings with high IP rating will reduce inside buildup of dirt, and therefore are especially recommended in dry material handling areas with high dust loads. Regular cleaning of the interior of electrical devices by vacuum is also a suitable option, but blowing with compressed air even in the presence of a dust extraction system is not recommended (Moerman, 2011).

a smooth finish. Enclosures (e.g., electric cabinets, junction and field boxes, as well as pneumatic/hydraulic enclosures) should have minimum ingress protection IP55. Any junction box expected to perform outside of an enclosure or cabinet and exposed directly to a hose-down needs an IP67 rating, at minimum. Although mounted in a moisture-proof housing, electrical distribution systems located in cold areas should be further protected against the formation of any condensate inside the cabinet by using an anticondensation heater within the cabinet. However, the heat generated by the electronic apparatus within the cabinet is often sufficient to avoid condensation (Moerman, 2011).

The design and installation of enclosures must prevent product, product liquid or water from accumulating onto the enclosure. As an example, in Figs. 6.62 and 6.63, the electric cabinet and field boxes are mounted at 30 to 45 degrees. However, a top roof with a minimum 30-degree inclination toward the front also allows water to run off and prevents tools being placed on the top of the cabinet. The front edge of the inclining cabinet top should reach beyond the front door and the seal, which preferably should be of a removable type to inspect for any potential dirt buildup under the seal (Figs. 6.64 and 6.65). Hinges must be located inside the sealing zone (Fig. 6.65) or shall be of the simple, take-apart type. When taken apart, hinges should be free of cracks or crevices.

The cabinets and field boxes must be mounted where they will be least exposed to splashes and hence they should not be placed in or above the

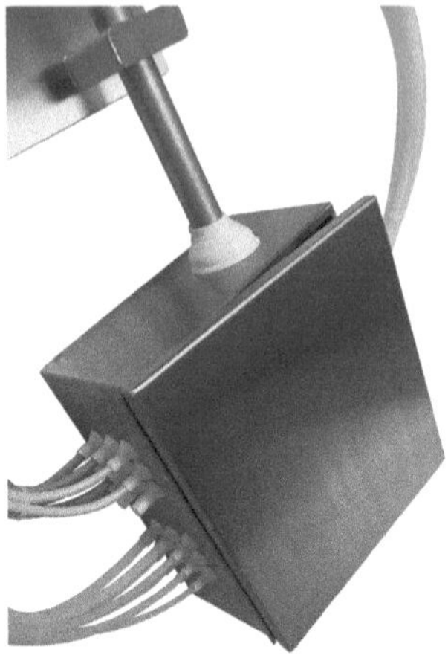

FIGURE 6.62 The electric cabinet is mounted at 30–45 degrees, preventing debris or liquids from accumulating on the top surface. The gaps formed by electric cables entering or leaving the enclosure can be closed by means of plastic caps. However, in this example, liquid tightness is not completely guaranteed. *Courtesy of Central States Industrial, www.pipetite.us.*

FIGURE 6.63 Sufficient distance is maintained between the cabinet base and the floor. The cabinet is positioned at a considerable distance from the wall of the enclosure. The top roof is sloped toward the front. The field boxes in smooth plastic are mounted at 45 degrees and sealed to the wall of the enclosure. Connections of cables and wires to field boxes are sealed. *Courtesy of GEA Group.*

food contact area. Placing in or above the food contact area would also result in condensate dripping from the field box of enclosure into the product. Field boxes and electrical enclosures also should be located such that easy access for maintenance and cleaning is practicable. It should be possible to open enclosure doors up to 90 degrees. Electrical control cabinets and field

FIGURE 6.64 Enclosure with a top roof having a minimum 30-degree inclination. *Courtesy of Rittal GmbH & Co. KG.*

FIGURE 6.65 Seal with folded lip along the top inside edge of the door should be of a removable type, and hinges installed inside the sealing zone create an easy-clean design on the outside. *Courtesy of Rittal GmbH & Co. KG.*

boxes mounted on the exterior of the equipment shall be sealed to the supporting member or equipment wall with food-standard silicon seal (Fig. 6.63) or spaced sufficiently away from the member/equipment wall to permit cleaning of all surfaces. In the latter case, a minimum of 20 mm between the enclosure and supporting member or equipment wall shall be provided. The distance between the cabinet base and the floor should be no less than 0.3 m. Dead spaces under cabinets or under false bottoms of electrical control cabinets, switching panels, or even computer closures should be regularly inspected for pest harborages, and treated with pesticide when necessary (Moerman, 2011).

All connections (e.g., cable ladders or wire trays, conduit, cable) to cabinets, field boxes, motors, and motor disconnects should be made via the bottom side. Occasionally, cable inlets can be made from the side. Connections must be made in a way that cables remain correctly separated (Fig. 6.66).

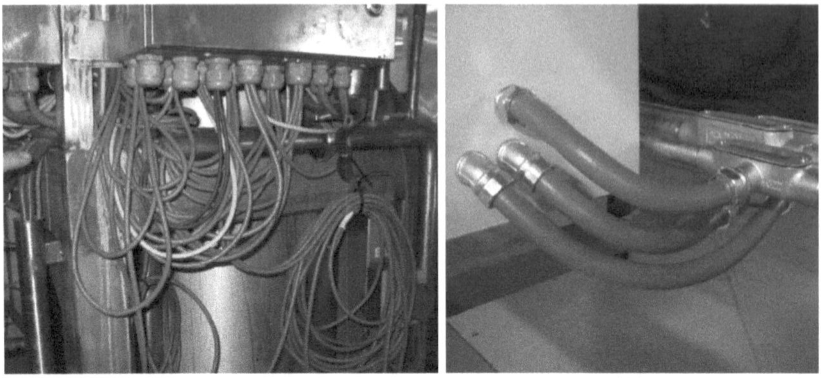

(A) (B)

FIGURE 6.66 (A) Unhygienic connection of cables to the enclosure. (B) Connections must be made in a way that cables remain correctly separated. Flexible silicone paste appropriately applied between the fitting and the enclosure provides a watertight acceptable connection. *Courtesy of Mondelēz International, © 2016.*

FIGURE 6.67 Hygiene compatible cable glands with seal (blue) directly at the cable entry. The outside of the gland has minimum radii on the hexagon. The connection does not leave an external thread. Due to the presence of a seal with square cross section, the termination between the cap nuts and the enclosure is flush. *Photo left: courtesy of Rittal GmbH & Co. KG; photo right: courtesy of Pflitsch GmbH & Co. KG.*

Watertight connections of cables and wires to housings are essential because production stoppages may occur by water seeping into electrical machine parts through the cable connection. For cable assemblies used in wet industrial applications, overmolded connectors—whose material chemically bonds to the cable's outer jacket during manufacture—provide a watertight seal. In older designs, food-standard flexible silicone paste appropriately applied between the fitting, coupling or gland and the enclosure provides a watertight acceptable connection. Nowadays, hygiene compatible cable glands could be used, as well as hygienic connectors (Fig. 6.67).

6.13.3 Electrical Cabling

As cables are exposed, they must withstand mechanical (abrasion, impact and other trauma) and chemical stress (cleaning and disinfection agents, water, and foodstuffs like vegetable oil, fat, acid, and salty food), as well as high temperatures (warm water and steam) and in some cases below 0°C temperatures (Fig. 6.68). With respect to the outer cable jacket, natural rubber or neoprene was for many years preferred for its superior abrasion resistance and flexibility. However, modern thermoplastic technology has produced a number of PUR and PVC compounds that are soft and flexible but also very tough. In a food processing plant where there is a significant amount of splashdown used daily to achieve hygienic standards, PVC is a better choice than PUR because it is more resistant to water and harsh cleaning chemicals. It is of utmost importance that the outer cable jacket can withstand corrosive cleaning agents and disinfectants to prevent it becoming porous. In areas at very low temperatures (e.g., cold-storage warehouse), cold-resistant jacket materials must be used such as PA, PUR, PTFE, neoprene, nitrile butadiene rubber, silicone rubber, and EPDM rubber. Neoprene, nitrile, and silicone rubber behave excellently in the presence of edible oils and fats; EPDM rubber cannot be applied in such an environment. The outer cable jacket should be of a smooth type without longitudinal crevices, and only electrical cables with a round cross section should be used. Corrugated cable housings and spiral wound power lines shall never be used in the food processing area, as they may accumulate a lot of dirt and be very difficult to clean.

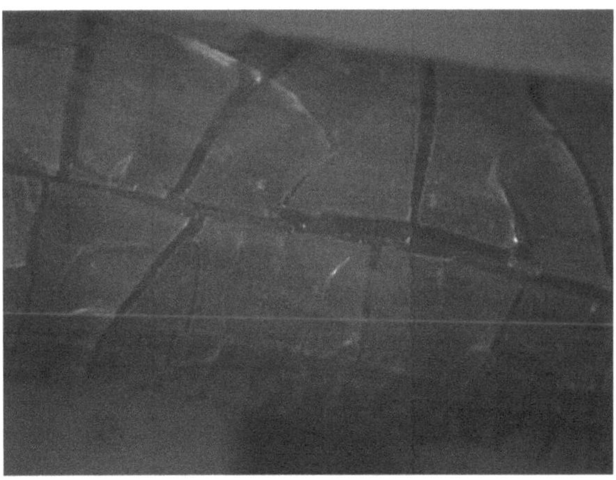

FIGURE 6.68 PVC-insulated wire after exposure to hot pressurized water of 130°C. The surface of the insulation has suffered from severe cracking. In an environment of high moisture content and high temperature, insulating materials are prone to hydrolysis. Hydrolytic attack is not concerned only with absorption of water, but with a chemical reaction in which the original insulating material is turned into a new material that no longer acts as an electrical insulator. If only water absorption occurs without hydrolysis, only a small decrease in its dielectric properties will occur, which can be tolerated.

FIGURE 6.69 Entangled cable arrangements may accumulate dirt, hamper inspection and appropriate cleaning, and become breeding grounds for vermin and pests. *Courtesy of Mondelēz International, © 2016.*

Tangled cable arrangements (Fig. 6.69), which may become breeding grounds for vermin and pests, should be avoided. These bundles of cables may also be the cause of accumulating product residue and may give rise to the development of microorganisms. Generally speaking, cables, hoses, etc. should be routed in a way that makes it possible to see dirt—e.g., the routings should be as open and visible as possible to facilitate cleaning around and between them. Therefore, electrical cables must be fastened individually at a distance from each other (no less than 25 mm) (Fig. 6.70) but cable binders should be avoided because they impede the effectiveness of cleaning operations. If strips are used, they should preferably be of a stainless steel type that can be detected by means of a metal detector. Alternatively, a plastic strip of a color that is not omnipresent in the food product and food factory could be used, which in most cases is blue. Nowadays, plastic strips are available with metal content dispersed throughout the head and strap, but it has been proven that cut-off sections cannot always be detected by means of metal detection.

In medium hygienic areas, cable ladders or cable trays should be used instead of conduit as a means to support current carriers over long distances. Furthermore, conduits should not be used in dry production areas; small wire trays should be used here because they allow dry cleaning. However, cable ladders or cable trays are less suitable in high hygienic areas because of the difficulties in cleaning them. In high hygienic areas, conduits can be used for short distances on the condition that they are suitably sealed at both ends by a proprietary cable gland/sealing gland where a cable does pass through. The index of protection for the conduit should not be less than IP55. The use of conduits with unsealed openings in medium hygienic areas is only

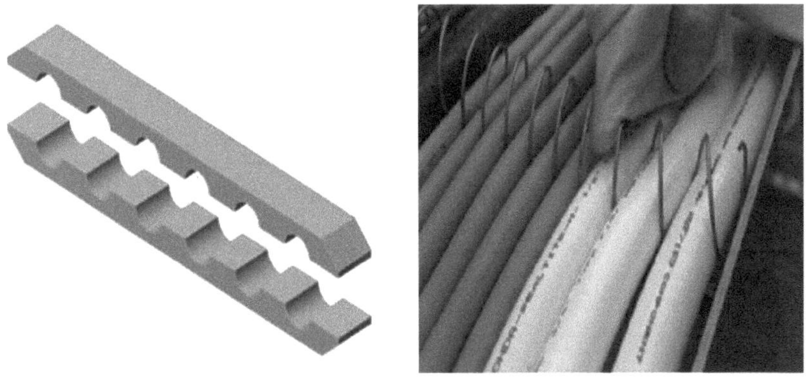
(A) (B)

FIGURE 6.70 (A) Cables should be separated by a distance of no less than 25 mm to prevent the buildup of soil and to ease cleaning. Cable separators of the type shown can be installed in wire trays (Moerman, 2011). (B) A hygienically designed cable support system with springs keeps the cables separated at a correct distance from each other. *Courtesy Anamet Europe B.V.*

FIGURE 6.71 Cables can be protected from dirt layering and damage by encapsulating them in hermetically closed cable housings such as stainless steel pipes, especially in the neighborhood of the food contact and splash area (Den Rustfri Stålindustris Kompetencecenter, 2006).

acceptable for small distances. When open conduits are used for multiple cables of small diameter sharing the same route, oversizing the conduit to allow for wet cleaning is common practice. Individual cables that do not share a common route with other cables are as hygienic as a single conduit run. However, a cable is usually more difficult to support in a hygienic manner than conduit. In the neighborhood of the food contact and splash area, cables also can be encapsulated in hermetically closed cable housings such as stainless steel, aluminum or hardened plastic pipes (Fig. 6.71) or flexible

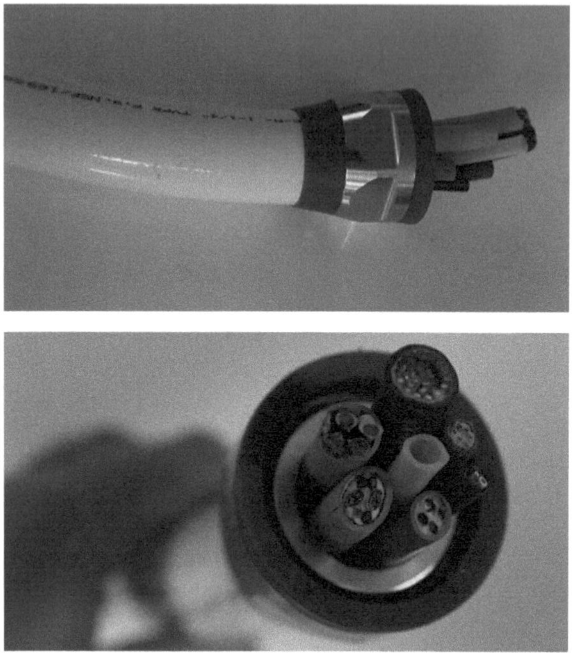

FIGURE 6.72 Multiple cables and wires can be combined in a flexible conduit with a smooth outer surface, further provided with a hygienic connector allowing a watertight connection to the maintenance enclosure or process equipment without visible thread in the screwed state. These hygienically designed conduits may reduce the number of cables that must be individually laid on cable ways, reducing the amount of contaminants that may build up on current carriers and enhancing the cleanability of the cable transport system. These flexible conduits are also suitable for connections to machinery subject to vibration (e.g., motors) (Moerman & Wouters, 2015). *Courtesy Anamet Europe B.V.*

conduits (Fig. 6.72). Cable mounting in pipes, however, still creates a hollow body and hence a hygienic risk when the integrity of the piping system is lost because of unsuitable fittings at cable entries and exits or because of damage to the thin-walled cable housing/pipe due to physical/mechanical impact. However, the flexible conduits (Fig. 6.72) provide a permanent and fluid-tight protection tube for multiple electrical cables and wires, as they are provided at their ends with stainless steel AISI 316 L fittings that allow ingress-protected seal connections with the equipment that must be powered (Moerman, 2011; Moerman & Wouters, 2015).

Cables may not be routed under machines and too close to the floor for the following reasons: (i) food residues may fall onto the cables, (ii) restricted access/visibility for inspection and cleaning, and (iii) liquids and dirt may splash from the floor onto the cables during cleaning operations. Mounting of cables also should be as far as possible away from the splash area. If there is no other choice, cables in the neighborhood of

FIGURE 6.73 Cable support system with a minimum number of members. The number of loosely fixed cables should be minimal, to avoid uncleanable entangled cable arrangements.

food processing equipment should be loosely mounted on cleanable open cable trays. To allow for cleaning, cable supports (trays, ladders, conduits, etc.) must be spaced at least 20 mm away from adjacent surfaces (Moerman, 2011).

Cable support systems are usually constructed of the same hygienic material as the equipment being supported (stainless steel AISI 304 L or AISI 316 L) and the exterior finish must be smooth. They should not have sharp edges, recessed corners, uneven surfaces, open hollow sections, unprotected bolt threads and screws. Brackets manufactured from angle or channel must be avoided or minimized. Cable support systems with a minimum number of members should be used where possible (example Fig. 6.73). Eventually horizontally mounted trays can be covered at the top with removable lids, so that dirt settles on the lid instead of between the cables. These lids should be wider than the trays and inclined to allow run off of liquids. It is essential that lids can be removed to allow cleaning of cable supports and cables (Moerman, 2011).

6.14 HUMAN INTERFACES

6.14.1 Hygienic Design and Installation of Switch Boxes

Control boxes should be preferably made of smooth, corrosion-resistant stainless steel plate with low surface roughness, and should be constructed with >6 mm radiused edges and without pits and crevices. Seams should be minimized and bolted connections should be avoided. With an IP67 to IP69K rating, they are protected against the penetration of water or damp during high-pressure hose-down cleaning operations. The switchbox should be mounted to equipment at least 6 cm from the equipment framework (Fig. 6.74), with suspending members constructed of solid steel round tubing.

6.14.2 Hygienic Design and Installation of Control Panels With Control and Indicator Devices

In noncomputer-based control panels, control and indicator devices are the machine components used as interfaces between man and machine. Adequate space should be provided between control and indicator devices for easy cleaning and disinfection (Fig. 6.75). Very often, control panels are provided with more holes than necessary for the installation of control and indicator devices. Unused holes in a control panel can be closed by means of blanking plugs. Installation of control and indicator devices in control panel bore holes that are larger than required can occur by means of adapter rings.

Push buttons, knobs, valve handles, switches and locks must be designed so as to ensure that food product, water or product liquid do not penetrate into the interface or accumulate onto the enclosure. Therefore push buttons,

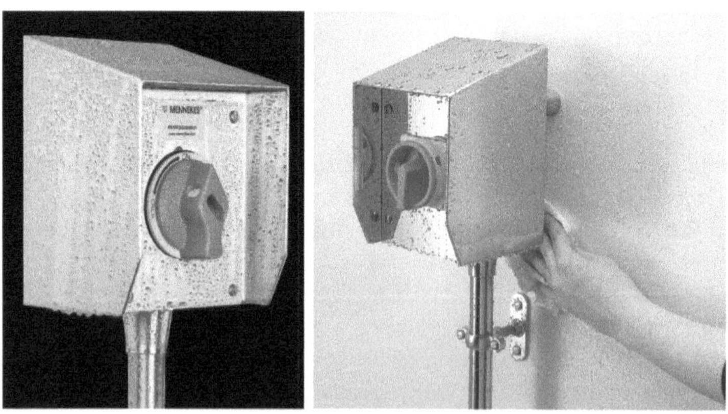

FIGURE 6.74 Control boxes either can stand apart *(courtesy of MENNEKES Electronics, Inc.)* or can be mounted remotely from the equipment framework/wall at a distance of about 6 cm. Suspending members are constructed of solid steel round tubing *(courtesy of Electrix International Ltd).*

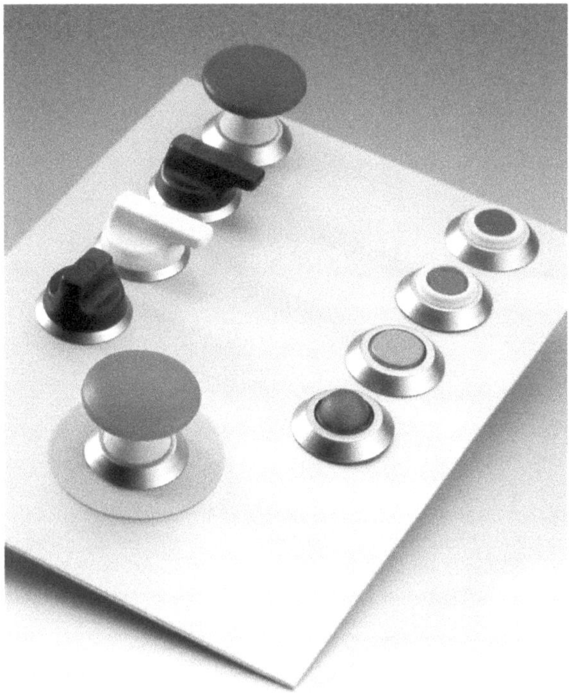

FIGURE 6.75 Control panel with hygienic control and indicator devices. An IP67 or IP67K ingress protection rating for control panel enclosures is highly recommended. *Courtesy of K.A. Schmersal GmbH & Co. KG.*

when touched, should not penetrate deeply in the front panel far beyond a protruding frame edge surrounding the button (Fig. 6.76).

Control and indicator devices must be constructed of durable and mechanically stable (unbreakable, resistant to steam, moisture, cleaning agents and disinfectants, abrasion and corrosion resistant) material, such as stainless steel or plastic (polyamide, polycarbonate, polyoxymethylene, silicone, and acrylonitrile butadiene styrene). Antimicrobial push buttons are commercially available. Knurling on hand grips should not be used, and device heads must have crevice-free and easily cleanable surfaces with smooth finish. Actuators of devices with grip or mushroom shape must have curvature radiuses ≥ 3.2 mm at all corners and edges. The device seals of the control devices shown in Fig. 6.77 make contact with the actuator and bezel (gray), hence preventing hygiene-critical gaps. The outer surfaces of the device seals all make a smooth, flush (in the case of push buttons and indicator lights) or continuous (in the case of other device versions) transition from the free outer surface of the actuator to the bezel. Front plate seals inside the control device help to avoid the penetration of pressurized water. Front plate and outer surface of the bezel are at an angle of approximately

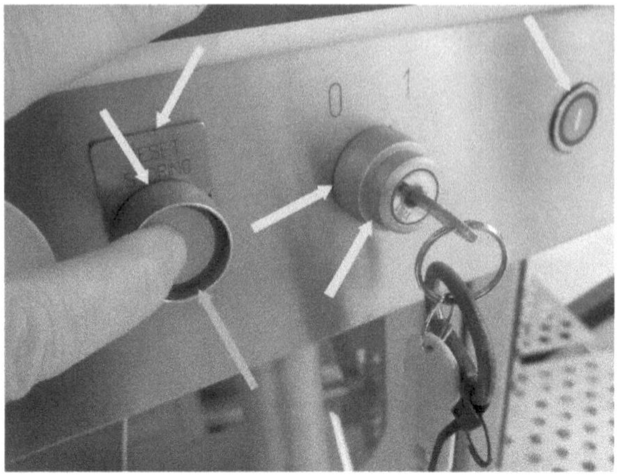

FIGURE 6.76 The push button, when touched, penetrates deeply in the front panel far beyond the protruding frame edge surrounding the button. Every time the button is pushed, food debris built up at the inside cylindrical frame moves deeper to the inside. As the inside of the actuator forms a niche in which microorganisms may grow, the inside of the protruding frame edge becomes heavily contaminated with microorganisms every time the button moves back. Thus, the actuator becomes a serious source of cross-contamination from operator to operator. The sharp corners and gaps, as indicated by the arrows in orange, are areas where, respectively, accumulation or ingress of product residues and cleaning solutions may occur. *Frank Moerman, ©2016.*

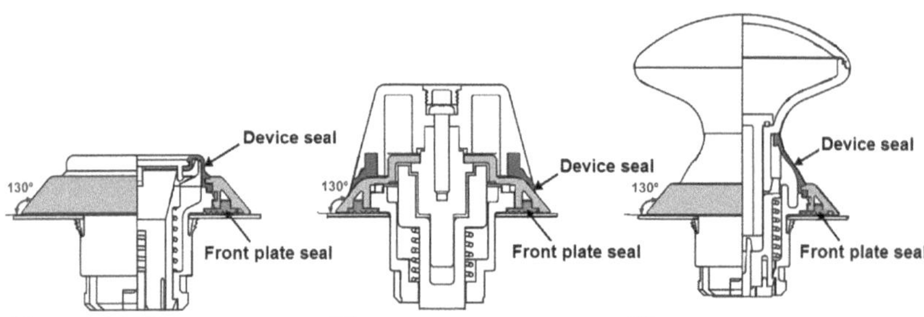

FIGURE 6.77 Control devices such as (A) push buttons, (B) position/selector switches, (C) mushroom buttons finally become integrated in a control panel. They must be shaped in such a manner that no accumulation of dirt and bacteria occurs and cleaning is facilitated. Also device head to front panel transitions must be smooth and without corners and edges. Perfect, hermetic device and front seals prevent the ingress of moisture, dust, and dirt within the control panel. Devices with damaged or destroyed seals should be replaced immediately. *Courtesy of K.A. Schmersal GmbH & Co. KG.*

135 degrees, thereby creating a surface without "sharp" transitions (Elan Schmersal, 2010; Moerman, 2011).

Control panels should be positioned away from the product zone, so that the operator does not have to lean over product to operate. The preferred installation positions for control and indicator devices are declining and

FIGURE 6.78 Touchscreen display installed inclined. *Courtesy of OctoFrost Group; Frank Moerman, ©2016.*

vertical surfaces, such that fluids (splashed food and cleaning solutions) are able to flow from the control panel, at least in the cleaning position.

6.14.3 Hygienic Design and Installation of Electronic Panels

More hygienic alternatives to control panels with push buttons and selection switches are membrane panels with a ≥2% inclination or touchscreen displays (Fig. 6.78). Vertically placed touchscreen displays are preferred over membrane panels, although the latter are more suitable when the input of huge amounts of information is needed. Touchscreen displays are often made movable, as they can pivot around a vertical axis fixed on the equipment. Thus, cleaning and maintenance of the equipment and its environment may proceed more easily.

Screens should be covered with an antistatic layer to avoid dust collection. Furthermore, they must be hermetically enclosed in a frame with IP67 or IP67K ingress protection rating and should be flush with the housing (no crevices due to protruding or intruding of screens in the screen housing). Screens, including the more fragile touchscreen displays, must be frequently wiped clean with a soft damp cloth and finally dried with a soft dry cloth. Disinfection may occur with wipes impregnated with 60% isopropyl alcohol and 200–400 ppm quaternary ammonium compound.

6.15 INSTALLATION OF THE FOOD PROCESSING EQUIPMENT IN THE FOOD FACTORY

6.15.1 Clearance With Respect to the Floor, Walls, and Adjacent Equipment

There should be enough clearance under the machine to allow for adequate cleaning and inspection to be carried out effectively. With that purpose, the process equipment should be installed as high off the ground as possible (Fig. 6.79). The minimum height should be a function of the depth of the bottom surface above the floor (indicative: 150–300 mm). For large-sized equipment, greater distances apply (at least 0.5 m from walls), as it is necessary to be able to walk around such equipment and have at least enough room to facilitate cleaning. If the equipment is sealed against the mounting surface, care must be taken to avoid gaps, cracks or crevices where insects or microorganisms can remain/survive after cleaning.

Installation of large equipment (e.g., freezing equipment, meat curing chambers) on feet is technically not always possible. An alternative is sealing the equipment onto the factory floor. Proper sealing of the perimeter between the equipment and the subfloor must prevent water from accidentally getting

FIGURE 6.79 Process equipment should be installed as high off the ground as possible to facilitate cleaning and reduce the risk of cross-contamination during cleaning operations. In this case, dirt released from the floor during high-pressure cleaning may splash on the process equipment. *Courtesy of Joe Stout, Commercial Food Sanitation LLC - Intralox, © 2016.*

into this space. But sealing, especially with silicone, has not always proven to be successful in excluding wet and unhygienic conditions.

Equipment must not be mounted beneath tanks or vessels so that maintenance and cleaning are impeded but must be easily accessible. Increased elevation of tanks and vessels facilitates cleaning and maintenance operations beneath them but water and condensation running down their sides may allow microbial growth and certainly must not fall onto exposed product.

6.15.2 Stairs, Raised Walkways, and Platforms

Stairs, walkways, and platforms comprising the secondary steelwork in the food factory must be constructed as follows:

- Painted or galvanized mild steel may be used in dry areas, while SS AISI 304, aluminum or galvanized mild steel should be used in wet areas depending on the type of cleaning agents used to clean the equipment in the area concerned. Where the cleaning agents corrode aluminum and galvanized steel, SS AISI 304 should be used. Stainless steel AISI 304 or better AISI 316 L always should be used for all fixed access equipment (e.g., decking, handrails) located in areas where foods are prepared and manufactured.
- Supporting and framing members should be designed to eliminate as many ledges, projections and pockets as possible. Open profiles are preferred over hollow sections, but consideration should be given regarding their orientation. Open profiles installed in the horizontal plane preferably should have their folding turned downward, although outward turned constructions are allowed as long as they are cleanable. In the horizontal plane, open profiles should never have their folding turned upwards. Installed in the vertical plane, open profiles must have their folding turned outward.
- If closed profiles have to be adopted (not recommended), frequent inspection for cracking should be carried out to prevent risk from contamination. Round or square section members turned through 45 degrees (Fig. 6.80A) that provide sloping surfaces are recommended
- If handrails are made from circular tube, they should be welded to the stanchions and any tube joints should also be welded and ground flush. All open ends of tubes must be sealed with a welded plate. All welded junctions must be smooth and continuous. In high hygiene areas, solid handrails could be considered.
- All framework parts of stairs, ladders, raised walkways, and platforms should be accessible for inspection, maintenance and cleaning.
- Connecting the framework of stairs, walkways, or platforms to a floor usually occurs by means of a base plate securely fastened to the floor by means of a fixing bolt (Fig. 6.81). However, to minimize the possibility for microbial presence and growth, a rubber seal between the floor and

(A) (B)

FIGURE 6.80 (A) The framework of the platform consists of closed square section members turned through 45 degrees (diamond configuration). The solid deck plates of the walkway are mounted a distance from these horizontal framework members. Kickplates are installed at the perimeter of the platform deck. Handrails and rungs of the ladder are made of round tubing *(courtesy of The Stellar Group)*. (B) The ladder is installed about 200 mm above ground level. Because the ladder makes no floor contact, the factory floor below the ladder can be cleaned very easily. Ladder rungs are covered with nonslip surfaces. Open metal grid is used as decking, although not recommended, especially if food is passing beneath the platform. However, solid deck plates provide fewer niches for food debris to accumulate and microorganisms to grow, and they are also much easier to clean. *Photo left, courtesy of The Stellar Group; photo right, courtesy of Bay State Industrial Welding & Fabrication, Inc.*

base plate is required to ensure a tight fit, especially because direct mountings onto floors are inevitably uneven. Base plates also can be fixed to a floor plinth (Fig. 6.82), thus prohibiting dirt and liquids from getting under the base plate during process and cleaning operations. Alternatively, the supporting framework members can be embedded in concrete. To facilitate the cleaning of the factory floor, the stringers of staircases should make minimal floor contact, and if possible no floor contact (Fig. 6.82).

- If ladders are used (Fig. 6.80B), provide handrails where possible. Ladder rungs should not be used as handholds, because they are a source of contamination (transfer of contaminants from feet to hands). Provide a safety cage where required (depending on the height of the ladder), and install the ladder in a manner that floor contact is minimized (Fig. 6.80B). Where possible, to facilitate the cleaning of the factory floor, the base of the ladder should not make floor contact, and preferably should be about 200 mm above ground floor level.
- Stairs, raised walkways (Fig. 6.81B and platforms over exposed product should be avoided as much as possible, because dirt may be transferred from clothing or footwear onto production lines beneath.
- The treads of stairs should be constructed from solid plates containing a raised antislip material (e.g., checker plate). Open grating is not allowed.

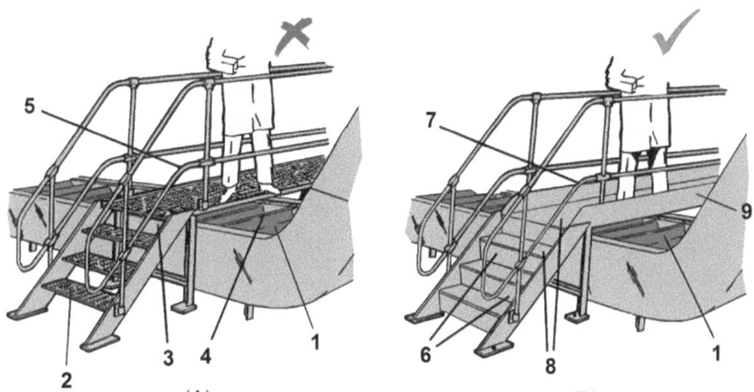

FIGURE 6.81 (A) If not appropriately designed, walkways and stairs over open product (1) may contaminate it. Open-mesh steps (2) that are not enclosed by closed vertical risers (3), the absence of a cover over the product area (4) and the handrail and its mountings hanging (5) over product area put the open food product at risk. (B) Now, the steps are enclosed (6), the handrail is mounted inside the walkway (7), solid antislip steps and floor plates are used (8), and fully welded, continuous kickplates are in place to prevent the open product from getting contaminated (Hauser et al., 2004b).

FIGURE 6.82 (A) Connecting the framework of stairs, walkways, or platforms to a floor usually occurs by means of a base plate securely fastened to the floor by means of a fixing bolt (*courtesy of Holland Applied Technologies*). (B) Base plates also can be fixed to a floor plinth, thus prohibiting dirt and liquids from getting under the base plate during process and cleaning operations. The stringers of staircases make no floor contact, hence facilitating the cleaning of the factory floor. *Photo left, courtesy of Holland Applied Technologies; photo right, courtesy of Bay State Industrial Welding & Fabrication, Inc.)*

The tread plates must have a small inclination (3–5 degrees) for improved drainability (Fig. 6.83).

- Where there is no risk for product contamination, the risers of the stair may be left open, but the stringers still must be of the closed type. Gaps

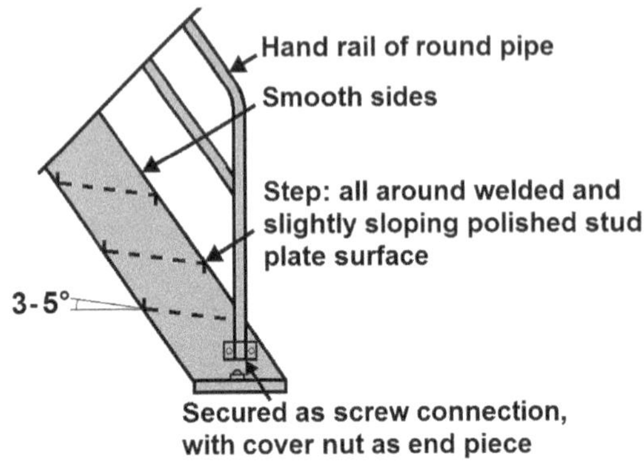

FIGURE 6.83 The tread plates must have an inclination of 3–5 degrees for improved drainability.

FIGURE 6.84 (A) Decking and adjacent kickplates give rise to difficult-to-clean square corners and crevices. Underneath the tread plate at the base of the stair, dirt and liquids may accumulate. (B) Kickplates and decking designed as a one-piece construction with generous radii in the corners of kickplates. *Courtesy of Mondelēz International,* © *2016.*

between the tread plates and stringers are allowed, as they may preclude buildup of contaminants in sharp corners and improve the cleanability of the stair.

- Where personnel movement is required in areas with exposed food products, the stairs must be totally encased (Fig. 6.81B). Completely closed risers now must form with the solid tread plates and stringers (closed type) an all-welded completely sealed staircase. Risers and tread plates of staircases should be constructed of the same impervious, noncorrodable, easy to clean, and impact-resistant antislip material as the deck. As sharp corners and crevices (Fig. 6.84A) allow food debris to accumulate, all corners must be rounded (Fig. 6.84B).
- When factory employees use walkways and platforms of poor hygienic design as crossovers of conveyor-systems, they may bring the food

beneath at risk. Although covers (Fig. 6.56) may protect the exposed food products, 150 mm high kickplates without any gaps between the kickplates and solid plate decking (containing a raised antislip material) may further preclude spillage of debris and liquids onto food products (Fig. 6.82). Kickplates and decking should be designed as a one-piece construction with generous radii in the corners of kickplates (Fig. 6.84B) to allow proper cleaning and disinfection. Kickplates are provided either by bending the perimeter of the solid deck plate or by full continuous welding. To protect the food processing activities below, it is essential that the kickplates be provided over the whole perimeter of the walkway or platform without leaving gaps and openings from where liquids may spill.

- Handrails should not overhang the walkway and must be attached to the inside of the walkway bridge or platform.
- Drainage of walkways and platforms, etc. is always difficult as the floors are rarely sloped to drains. The drains themselves have then to be piped and led to factory drains in the floor below in a way that does not endanger product safety. Untrapped drain lines shall be provided with an air gap at the discharge to the sewer and shall be removable for cleaning. If possible, the use of water on such structures should be avoided. If unavoidable, safe mechanisms of disposing surface water should be planned at the time of construction.

6.16 CONCLUSIONS

If open food processing equipment is not designed for easy cleaning and disinfection, small amounts of liquid and dirt, as well as microorganisms may become large enough to unroll both chemical/physical contamination and microbial growth. Moreover, the small bits of food debris from a previous product batch can compromise the quality of product (e.g., taste, texture, etc.) produced in a subsequent batch, make food unfit for consumption on religious grounds, affect its authenticity or induce allergic reactions in consumers (if allergens are present). Good hygienic engineering and design practices have proven to successfully reduce these problems, with a final result that costly product recalls can be avoided. Moreover, apart from being effective in preventing one batch cross-contaminating the next one, good equipment design also allows considerable savings during cleaning operations. The time required to clean and disinfect can be minimized and thus prolong the time to produce, and the consumption of water, cleaning chemicals and energy to heat the cleaning solutions can be reduced. These features make hygienically designed, open food processing equipment much more sustainable and cost-effective in a long-term perspective, although it is initially more expensive than poorly designed equipment.

REFERENCES

APV Baker (2001), Hygienic Design Handbook, 2nd edition, Peterborough, United Kingdom, 53 p.

Campden Food Preservation Research Association (CFPRA) (1983), Hygienic Design of Food Processing Equipment, in Dudley, K. (Ed.), Report prepared by the Working Party on Hygienic Design of the Heat Preserved Foods Panel in conjunction with the Research Association, Technical Manual No. 7, Chipping Campden, Gloucestershire, United Kingdom, 93 p.

Den Rustfri Stålindustris Kompetencecenter (2006). Conveyors, with a focus on hygiene, *Guideline N° 3*, version 1.0, Danish Technological Institute, Kolding, Denmark, 60 p.

EDQM (European Directorate for the Quality of Medicines and Healthcare of the Council of Europe) (2013), Technical guide on metals and alloys used in food contact materials and articles, A practical guide for manufacturers and regulators, ISBN 978-92-871-7703-2, Committee of Experts on Packaging Materials for Food and Pharmaceutical Products (P-SC-EMB), 1st edition. Strasbourg, France, 216 p.

Elan Schmersal (2010). Control Devices and Indicator Lights for Food Processing Machines and Heavy-Duty Applications, 22.3 mm Diameter Installation. Catalogue N/09 - Type series N, 34 p.

Hauser, G., Curiel, G.J., Bellin, H.-W., Cnossen, H.J., Hofmann, J., Kastelein, J., et al., 2004a. Hygienic Equipment Design Criteria, EHEDG Guideline N°8, 2nd edition. EHEDG subgroup "Design Principles", EHEDG, Frankfurt, Germany, pp. 1–16.

Hauser, G., Curiel, G.J., Bellin, H.-W., Cnossen, H.J., Hofmann, J., Kastelein, J., et al., 2004b. Hygienic Design of Open Equipment for Processing of Food, *EHEDG Guideline N°13*, 2nd ed. EHEDG subgroup Design Principles, EHEDG, Frankfurt, Germany, pp. 1–24.

Kold, J., Tomsett, R., Hopma Zijlema, P.K., Silverman, C., Lucchi, M., Scheffler, R., et al., 2016. Hygienic design of belod industry. EHEDG guideline N° 43. EHEDG subgroup Conveyors, EHEDG, Frankfurt, Germany, pp. 1–76.

Lelieveld, H.L.M., Mostert, M.A., Curiel, G.J., 2003. Hygienic Equipment Design. In: Lelieveld, H.L.M., Mostert, M.A., Holah, J., White, B. (Eds.), Hygiene in Food Processing, 1st edition. Woodhead Publishing, Cambridge, England, pp. 122–166. Book N°88. (Chapter 8).

Moerman, F., 2011. Hygienic Supply of electricity in food factories. In: Lelieveld, H.L.M., Holah, J. (Eds.), Hygienic Design of Food Factories, N° 216. Woodhead Publishing, Cambridge, United Kingdom, pp. 369–411. (Chapter 19).

Moerman, F., Fikiin, K., 2016. Hygienic design of air nlast freezers. In: Lelieveld, H.L.M., Holah, J., Gabrić, D. (Eds.), Handbook of Hygiene Control in the Food Industry, 2nd ed. Elsevier, Cambridge, United Kingdom, pp. 271–316. (Chapter 22).

Moerman, F., Kastelein, J., 2014. Hygiene design and maintenance of equipment. In: Motarjemi, Y., Lelieveld, H.L.M. (Eds.), Food Safety Management: A Practical Guide for the Industry. Academic Press, San Diego, United States, pp. 673–739. (Chapter 26).

Moerman, F., Partington, E., 2014. Materials of construction for food processing equipment and services: requirements, strengths and weaknesses. J. Hygienic Eng. Design 6, 1–37.

Moerman, F., Wouters, P., 2015. Hygiene concepts in food factory design. In: Leadley, C., Sykes, R. (Eds.), Innovation and Future Trends in Food Manufacturing and Supply Chain Technologies, N° 293, 1st ed. Woodhead Publishing, Cambridge, United Kingdom, pp. 81–133. (Chapter 4).

Moerman, F., Perryment, A., Cain, T., Fröderberg, I., Declerck, D., Berghoff, R., et al., 2005. Hygienic Design of Freezing Equipment, 1st guideline draft. EHEDG subgroup Refrigeration, EHEDG, Frankfurt, Germany, 41 p.

Partington, E., Besuchet, P., Godwin, A., Hall, K., Holah, J., Holland, P., et al., 2005. Materials of construction for equipment in contact with food. EHEDG Guideline N° 32. EHEDG subgroup Materials of Construction, EHEDG, Frankfurt, Germany, pp. 1–48.

Chapter 7

Hygienic Design of Closed Equipment for the Processing of Liquid Food

F. Moerman
Catholic University of Leuven – KU Leuven, Leuven, Belgium

7.1 INTRODUCTION

There is a global trend in the food industry toward minimal food processing and preservation. But the general tendency to apply mild processing and preservation techniques often shortens the shelf life of food, may put foods at risk and may compromise consumer health. It forces food manufacturers to pay more attention to hygiene during the manufacturing of food. Poorly designed or maintained closed equipment for the processing of liquid foods may compromise the product's hygienic condition. Cross-contamination may occur even after applying rigorous cleaning and disinfection practices. Typical contaminants are spoilage microorganisms and pathogens, as well as allergens. Additionally, broken equipment parts may find their way in as foreign body contaminants, while poor installation and lubrication practices may result in food contaminated with lubricants.

Closed equipment used for the processing of liquid food falls into two categories: (1) process equipment that can be cleaned in-place and can be freed from relevant microorganisms without dismantling and (2) process equipment that in addition is sterilizable and impermeable to microorganisms so as to maintain its aseptic status. Examples of closed process equipment (components) are closed vessels (reactors) often provided with insulation and cladding, piping and pipe joints, pumps, valves, measurement devices, etc. Good hygienic design of this equipment is essential to ensure that the required level of food safety is maintained, in compliance with compelling national and international food safety legislation, as well as food safety management systems built on the well-known concepts of Good Manufacturing Practices (GMPs), Hazard Analysis and Critical Control Point (HACCP) and prerequisite food safety and quality programs. On request of their customers, many

Food Protection and Security. DOI: http://dx.doi.org/10.1016/B978-1-78242-251-8.00007-2

food manufacturers also need to certify their food processing operations against the SQFI (Safe Quality Food Institute), BRC (British Retail Consortium), FSSC 22000 (Food Safety System Certification 22000) or GFSI (Global Food Safety Initiative) standards and/or certification schemes. Interest in correct equipment design has also grown in the light of rising production costs, one reason being that hygienically designed process equipment is much more sustainable and cost-effective from a long-term perspective. Although sometimes initially more expensive, hygienic design may (1) reduce the risk of costly product recalls, (2) reduce labor costs by minimizing the effort and time needed to clean and disinfect, and (3) allow cost savings in the consumption of water, cleaning agents/disinfectants, as well as the energy required to heat the cleaning/disinfection solutions.

Therefore, this chapter aims to familiarize users of closed equipment for the processing of liquid food with food safety hazards related to the design, construction and application of these systems. Furthermore, process equipment manufacturers are made more aware of state-of-the-art engineering solutions to improve the hygiene friendliness of their process equipment. In Section 7.2, an overview is given of the current legislation and standards dealing with the hygienic design of food processing equipment. Section 7.3 lists the basic hygienic design requirements that food processing equipment must meet to produce microbiologically safe food products. Section 7.4 describes the hygienic and food grade materials that can be used in the manufacturing of food processing equipment, followed by a section that outlines the requirements for the food contact surface finish (Section 7.5). In the next sections, we discuss the hygienic design of closed equipment (components) for the processing of liquid food such as closed vessels and reactors, including agitators (Section 7.6), processing and utility piping, including hoses (Section 7.7), pumps (Section 7.8), valves (Section 7.9), pressure measurement devices (Section 7.10), and temperature measurement devices (Section 7.11).

7.2 LEGISLATION, STANDARDS AND GUIDELINES COVERING HYGIENIC DESIGN

7.2.1 Legislation

Food processing equipment intended to be sold in European countries and designing operations in food factories must comply with the European Machinery Legislation, consisting of the Machine Directives 2006/42/EC and 98/37/EC, and an endorsing guidance document published by the Industry and Enterprise Department of the European Commission with the title: "Guide to application of the Machine Directive 2006/42/EC" (European Commission, 2010). Annex I of the Machine Directive 98/37/EC (formerly 89/392/EEC and its amendments 91/368/EEC and 93/44/EEC), and Annex V of Council

Directive 93/43/EEC on the Hygiene of Foodstuffs require that all equipment used to handle food should be hygienically designed.

7.2.2 European Standards and Guidelines for Liquid Food Processing Equipment

7.2.2.1 EN Standards

The Comitée Européen de Normalization technical committee CEN/TC 153 has developed two Harmonized European standards with respect to food machinery: prEN1672-1 and EN1672-2. With respect to hygienic design, the most important is EN 1672-2, which sets design principles for both safety and hygiene objectives.

7.2.2.2 EHEDG Guidelines

In Europe, the European Hygienic Engineering & Design Group (EHEDG) is the most experienced organization in the field of hygienic design. EHEDG is a European-based nongovernmental organization, more specifically a consortium of process equipment manufacturers, food industries, universities, research institutes and public health authorities, founded in 1989, with the aim to promote hygiene during the processing and packing of food products. It has developed 45 guidelines. In Table 7.1, an overview is given of those EHEDG guidelines, dealing with the hygienic design of closed equipment for the manufacturing of liquid food.

7.2.2.3 Other European Standards

Other specifications used in the food industry are the requirements of the International Standardization Organization (ISO) and the German Standardization Authority (DIN) (more specifically for fittings), the bulletins of the International Dairy Foundation (IDF) and the British Standards BS 5750. Note that many food and equipment manufacturers have developed their own hygiene standards for internal use.

7.2.3 US Standards and Guidelines

7.2.3.1 3-A Sanitary Standards

The International Association of Milk, Food, and Environmental Sanitarians, Inc. (IAMFES) and the committee on sanitary procedures 3-A (an independent organization of equipment manufacturers, food processors and regulatory agencies) introduced the first industry hygienic standards for equipment, which relate to the cleanability of dairy equipment. 3-A Sanitary Standards provide material specifications, design criteria and other necessary information for several types of equipment. It also provides a third-party evaluation for some food producing equipment. With respect to US sanitary standards,

TABLE 7.1 Available EHEDG Guidelines for the "Hygienic Design of Closed Equipment for the Manufacturing of Liquid Foods"

Doc. No.	Title
2	A method for assessing the in-place cleanability of food processing equipment
8	Hygienic equipment design criteria
9	Welding stainless steel to meet hygienic requirements
10	Hygienic design of closed equipment for the processing of liquid food
14	Hygienic design of valves for food processing
15	A method for the assessment of in-place cleanability of moderately sized food processing equipment
16	Hygienic pipe couplings
17	Hygienic design of pumps, homogenizers and dampening devices
18	Chemical treatment of stainless steel surfaces
20	Hygienic design and safe use of double-seat mixproof valves
25	Design of mechanical seals for hygienic and aseptic applications
32	Materials of construction for equipment in contact with food
35	Hygienic welding of stainless steel tubing in the food processing industry
37	Hygienic design and application of sensors
42	Disc stack centrifuges—design and cleanability
45	General principles of cleaning validation in the food industry

both NSF and 3-A cooperate with EHEDG, so as to achieve global harmonization of guidelines and standards.

7.2.3.2 Other US Sanitary Standards

In the United States, the following government agencies and private organizations also have published sanitary standards for food processing equipment:

- US Public Health Service: Food and Drug Administration (FDA) and GMPs
- American Society of Mechanical Engineers (ASME): ANSI-ASME F2-1: "Food, Drug and Beverage Equipment"

- American Society of Mechanical Engineers (ASME): Bioprocessing Equipment guideline, ASME BPE-2014
- Baking Industry Sanitation Standards Committee: BISSC Sanitation Standards
- Association of Food and Drug Officials of the United States: "AFDOUS Frozen Food Code"

7.3 BASIC HYGIENIC DESIGN REQUIREMENTS

In all stages of design, construction, installation and maintenance of food processing equipment, hygienic design aims to reduce the buildup of food material or microorganisms in individual items of equipment and the complete line, and to ensure that all detectable soil is removed after cleaning and disinfection. According to European Standard EN1672-2, soil is "any matter, including product residues, microorganisms, residual detergents or disinfecting agents." Food processing equipment should at least meet the following basic hygienic requirements (Moerman & Kastelein, 2014):

- Piping and components should be constructed from the same materials, so as to prevent contact corrosion (bimetallic corrosion) between dissimilar metals (Fig. 7.1). However, this is not always possible. The tendency for corrosion increases as the difference in electrode potentials between different metals increases. The user has to check whether any dissimilar materials will or will not corrode.
- Smooth product contact surfaces must minimize the adhesion and colonization of microorganisms, so as to prevent the formation of biofilms.

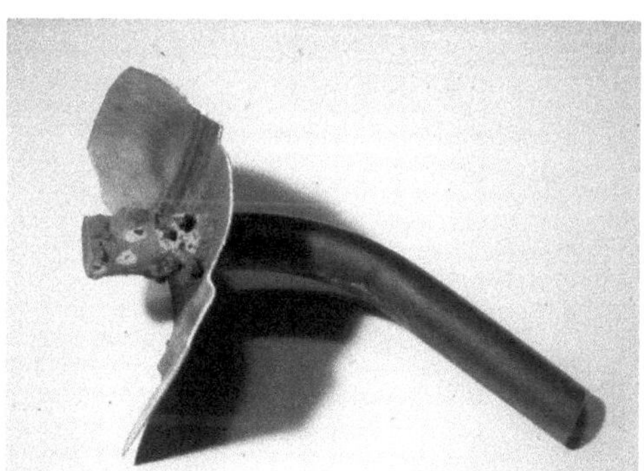

FIGURE 7.1 A badly corroded copper pipe is entering a large stainless steel hot water tank. Stainless steel is cathodic to copper in fresh water. Where the area ratio of stainless steel to copper is large, severe galvanic corrosion can occur.

- Food processing equipment design and construction may not allow bacterial ingress, survival, growth and reproduction on both product and nonproduct contact surfaces. Effective and efficient cleaning over the whole lifetime of the equipment must be guaranteed.
- The Machine Directives 2006/42/EC & 98/37/EC states that assemblies, e.g., joining of two or more parts, preferably should be made by welding or continuous bonding so as to reduce projections, edges and recesses, as well as fastenings, to a minimum. The joint (e.g., welds) must be smooth and have neither ridges nor crevices, pits or cracks that could harbor organic materials. Welds must be ground and polished to a standard of finish equal to that of the surrounding material.
- Overlapping sheets of metal must be avoided. Wherever possible, the two pieces should be butt welded. If the weld is also exposed to the product, the weld must be ground and polished to the same finish as that of the adjacent surfaces.
- In some cases, continuous welding may result in unacceptable distortion. In this case, although not recommended from a hygienic point of view, exceptionally intermittent welds may be employed (Fig. 7.2).
- Where components are fixed together by means of fasteners, these joints are often of a semipermanent nature. It means that they are not routinely broken, e.g., on a daily or even weekly basis, for cleaning. All such joints must be sealed by means of a gasket. The gasket material must be of food quality, nonabsorbent, nontainting and trimmed flush both internally and externally. Their physical and mechanical properties must be suitable to prevent the ingress of product, liquids (from cleaning) and microorganisms. The condition of gaskets should be checked periodically because some materials such as rubber eventually harden and crack in service.

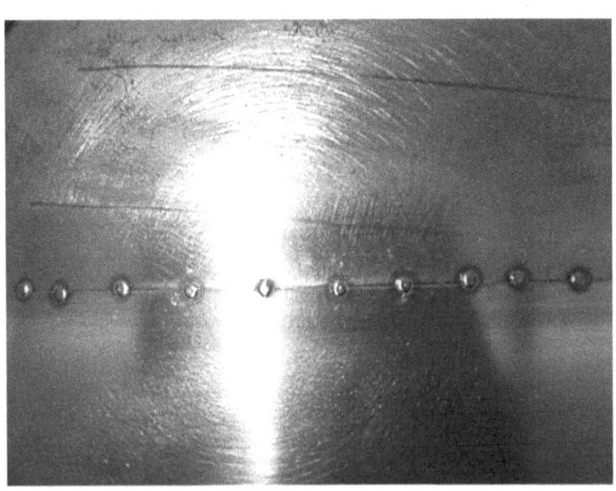

FIGURE 7.2 Spot welds create crevices and may retain product residues, as well as harbor microorganisms.

- Where fasteners are removed on a daily basis, to remove components or assemblies for cleaning purposes, for example, they are unlikely to give rise to hazardous conditions. However, fasteners that are cleaned infrequently may give rise to hygiene problems.
- Avoid exposed screw threads, nuts, bolts and rivets whenever possible, certainly in product contact areas. The slots and sockets in screw or bolt heads can retain product. Preference should be given to domed hexagonal-headed screws or bolts. Sometimes, nuts and screws also come loose, a problem that is often solved by using spring washers, which create a gap between the ends where dirt may accumulate and microorganisms may find a niche to grow. A thread locking compound or component (e.g., washer with a compressible rubber insert) may form a bacteria-tight seal.
- Niches such as pits, cracks, crevices, open seams, gaps, lap seams, inside threads, holes that may accumulate dirt and hamper the cleanability of the food processing equipment are not allowed.
- All inaccessible horizontal flat areas, ledges, projections, protrusions, recesses, edges, etc., where product residues can accumulate, should be eliminated.
- The Machine Directives 2006/42/EC and 98/37/EC state that equipment must be designed and constructed so as to prevent the ingress of liquids and living creatures (e.g., insects) into any areas that cannot be cleaned. In addition, organic matter must not be permitted to accumulate in such areas (Fig. 7.3). Retained product residues or cleaning fluids may subsequently contaminate the product on rejoining the product stream.

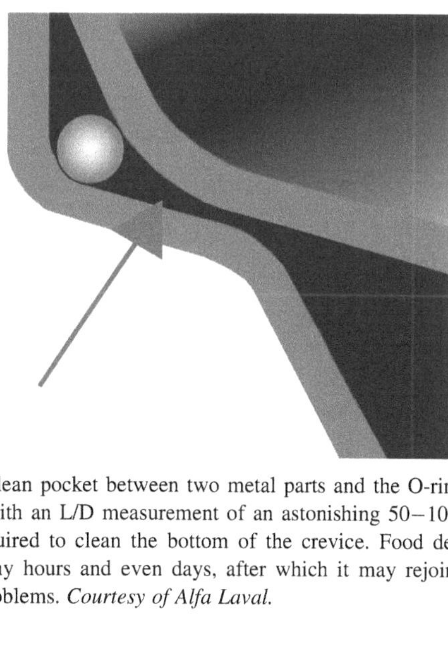

FIGURE 7.3 Hard-to-clean pocket between two metal parts and the O-ring. This crevice/dead zone forms a dead leg with an L/D measurement of an astonishing 50–100. It is impossible to achieve the velocity required to clean the bottom of the crevice. Food debris can be retained in these pockets for many hours and even days, after which it may rejoin the product stream causing food spoilage problems. *Courtesy of Alfa Laval.*

There is also significantly reduced transfer of energy to the food residues (soil) in dead areas in process equipment that is placed outside of the main flow of cleaning liquids than there is to the soil in the main flow. Such areas are difficult to clean and therefore should be avoided or sealed. If unavoidable, their presence should be taken into account when devising the cleaning procedures. Where possible, any space must be designed for regular dismantling to permit the space to be cleaned and disinfected. Typical shadow zones, for example, can be found in the legs of T-pieces in pipelines (Fig. 7.4). In many food factories T-pieces are used to mount sensors such as pressure gauges.

- External surfaces that are not in contact with foods should be corrosion resistant, smooth, easily cleanable and free of protruding parts and crevices where debris may accumulate.
- The exterior of nonproduct contact surfaces should be so arranged that harboring of contamination in and on the equipment itself, as well as in its contact with other equipment, floors, walls or hanging supports, is prevented.
- All parts of the equipment shall be readily accessible for inspection, so as to facilitate the detection of all potential contaminants on representative surfaces throughout the product contact zone. So, all surfaces in the product zone must be immediately visible for inspection, or the design of the equipment shall allow dismantling readily without use of any tools.
- Disassembly and reassembly must be as simple as possible, so that the surfaces and parts exposed to bacterial contamination can be cleaned within 15 min. For that purpose, the number of working parts must be minimal, while their weight and dimensions may not be too large. Ideally components that require frequent cleaning should be easily manageable by one person, and suitable racks should be provided to hold dismantled components off the floor. Heavier components may be more suitably removed by some form of a hoist. Quick release devices, e.g., captive bolts with coarse threads or

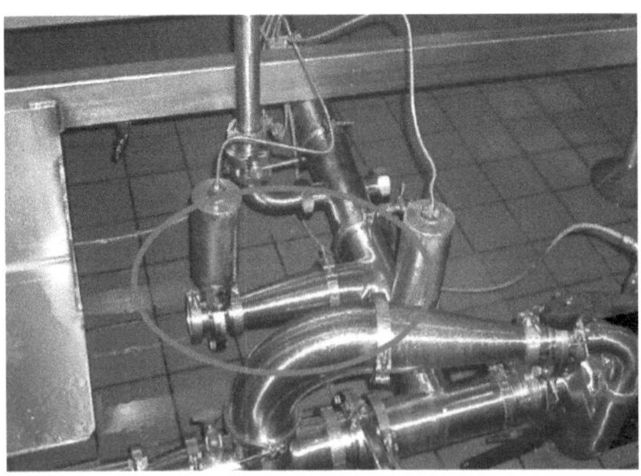

FIGURE 7.4 The too-long T-piece functions as a dead leg, where fluid movement is reduced. *Courtesy of Mondelēz International, ©2016.*

clamped joints, allow fast and easily dismantling and have the advantage of overcoming the microbiological hazards associated with screw threads as mentioned earlier. Whatever quick release devices are used, it is important to ensure that only the simplest tools are required in the dismantling and reassembly of the equipment. Equipment designed for simple and speedy dismantling and cleaning is going to be cleaned more enthusiastically and efficiently than equipment that proves tedious and difficult to handle.

- Equipment surfaces must be readily accessible for manual cleaning and disinfection, unless it can be demonstrated that the result of in-place cleaning and disinfection procedures without dismantling is equivalent to the result of dismantled and manual cleaning procedures. All potential obstructions to cleaning, disinfection and maintenance should be avoided or minimized.
- For the same reason and to facilitate cleaning, sharp corners in the product area should be avoided. A radius can be obtained when metal is bent or machined. When components are bolted or screwed together, a sharp corner is usually unavoidable. Constructions where the sharp corners are continually swept, such as in lobe pumps, are acceptable.
- Where a joint is necessarily close to an internal angle, a butt-welded joint should be made away from the corner on the flat surfaces, and must be smooth. Preferably the weld should be made at the non-product side.
- Microbes can flourish in stagnant pools of water, especially when nutrients are trapped in the internal pockets. Hence, accumulated and pooling cleaning and disinfection solutions may contaminate food products. Moreover, when liquid food, cleaning and disinfection solutions, and rinsing water are retained during idle periods, corrosion may occur. Therefore all equipment surfaces in the product zone and piping must be so arranged that they are self-draining, which means that they should be sloped toward drain points. There should be no ridges that may hamper draining. Where it is not possible to build equipment in such a way that proper draining is possible, procedures must be developed to ensure that residues of cleaning and disinfection liquids can be removed in another way.
- Equipment design, therefore, should not permit the formation of condensate that may enter the food zone and contaminate product or product-contact surfaces.
- Bearings should be either of the "sealed-for-life" or "double-sealed" variety and should be located external to the equipment in order to minimize contact with product. When bearings are mounted outside the product area, contamination of food products by lubricants is avoided, as well as the ingress of bacteria. There is also less chance that the bearing gets damaged due to ingress of product. Where possible the bearings should be self-lubricating. Bearing covers should be fitted where possible. Note, however, that seals ultimately wear and leak. Their condition must therefore be monitored regularly and they must be repaired periodically as part of planned maintenance programs.

- When the bearing is within the product area (e.g., slide bearings, such as bottom bearings of top-driven stirrers or bearings in scraped-surface heat exchangers), it should be self- or product-lubricated. In the case of a foot bearing for an agitator shaft, the design of the shaft and bush should allow the passage of cleaning fluid to all surfaces. Grooves must be provided in the shaft through the whole length of the bush.
- It is advisable to avoid steam cleaning of lubricated bearings. After contact with fluids or steam (where essential), bearings should be allowed to dry out before they are relubricated.
- Food grade lubricant should be used, and leaking of lubricant onto food product must be excluded. To protect the product zone a drip pan should be used, or motors driving equipment components such as belt drives, etc. should be placed outside the product area. If they are within the splash area, they should be protected by a removable cover. Where possible, food grade grease such as acetylated monoglycerides should be employed. Any excess of lubricant should be removed.
- Equipment design must ensure hygienic compatibility with other equipment and systems, such as hydraulics and electrical, steam, air and water systems.
- Shaft passages and seals may leak product to the outside. Microorganisms may then multiply in the product and grow back to the product side. In the case of dynamic seals, such as those for shafts of valves, pumps and mixers, the movements of the shaft will assist the transfer of product to the outside and the transfer of microorganisms to the product side. This applies to reciprocating shafts, and to a lesser extent to rotating shafts. However, rotating shafts also display some axial movement. Reciprocating shafts can be sealed by means of flexible diaphragms or bellows. To prevent the ingress of microorganisms in rotating shafts, double seals with microbicidal barrier liquids should be used. If not replaced in a timely manner, however, such barriers may become a growth medium for microorganisms. In order to prevent damage to the seals, they should be designed to minimize the ingress of product particles. In some cases, however, the total exclusion of particles is not economically possible. Under these circumstances, the seals should be subject to frequent cleaning.

7.4 SELECTION OF THE CORRECT MATERIALS OF CONSTRUCTION

Food contact materials must meet specific requirements. They must be inert to the product under all in-use conditions, such as temperature and pressure, as well as to any chemicals used, such as detergents or biocides. They may also need to be resistant to pressurized hot water or steam sterilization. They must be corrosion resistant, nontainting, mechanically stable, smooth, and nonporous. Materials of construction that are used for surfaces in contact with food should allow the original finish to be maintained and no porosity should develop. In addition, materials should be resistant to deformation, denting, chipping, flaking, and delamination. Materials that may release unacceptable concentrations of harmful components to the food may not be used in contact with food. Particular care must be taken when elastomers, plastics, adhesives, and signal transfer liquids are used, as these may contain toxic components that could be leached out into product. The suppliers of such components must provide clear evidence that the materials meet all legislative requirements. With the above requirements in mind, it is not surprising that the range of construction materials available is rather limited.

7.5 SURFACE FINISH

Product contact surfaces must be finished to a degree of surface roughness that is smooth enough to enable them to be easily cleaned and disinfected. The surface finish must have a roughness area R_a as low as practicable and without cracks, pits or cavities where water or soil might remain. Surface roughness R_a can be defined as the arithmetic average value of the departure of the profile above and below the mean line throughout the specified sampling length. A surface finish of roughness $R_a \leq 0.8\ \mu m$ is considered as acceptable for closed equipment used for handling liquid food and normally cleaned in-place. A roughness R_a exceeding 0.8 μm may be acceptable if test results have demonstrated that the required cleanability can be achieved through other design features or more intensive cleaning methods.

7.6 HYGIENIC DESIGN OF CLOSED VESSELS

A wide range of tanks is used in the food industry, which can be classified either as storage tanks or process vessels. The function of storage tanks is purely storage of raw materials and intermediate or final products at different stages of the production process. Process vessels can include mixing, blending, heating, cooling, separation, and fermentation operations. Process vessels are more complex in design than plain storage tanks because of the presence of various internal components. However, the design criteria are the same for all types of tank.

7.6.1 Interior and Exterior Design of Closed Vessels

Vessel and appurtenances must be designed to facilitate the tank cleaning process (ASME BPE committee, 2014):

- Vessels need to be designed with smooth, straight walls and curved corners that can be cleaned easily by liquid spray produced by stationary (e.g., static spray balls) or rotary cleaning devices. Corners shall be well-rounded, with a radius equal to or larger than 3 mm (Fig. 7.5).

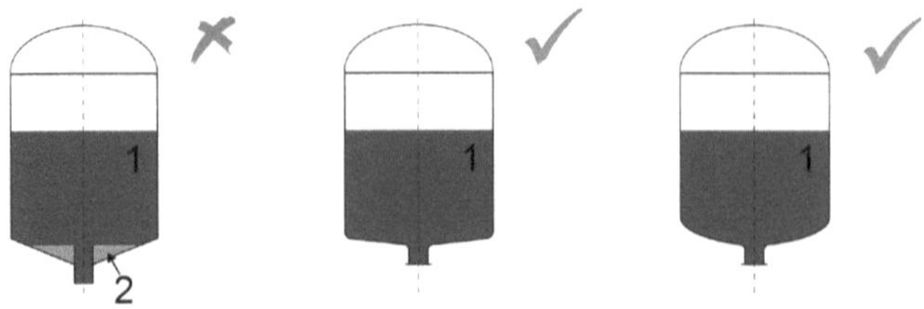

FIGURE 7.5 Full drainage of products and cleaning solutions is required. If discharge outlets are above the lowest level of the closed vessel, self-draining will be hampered, and residual products and cleaning solutions (2) will be left in the vessel. Closed vessels must be self-draining with discharge outlets at the lowest level. Bottoms must be sloped and all corners should be well-rounded (Hauser et al., 2004b).

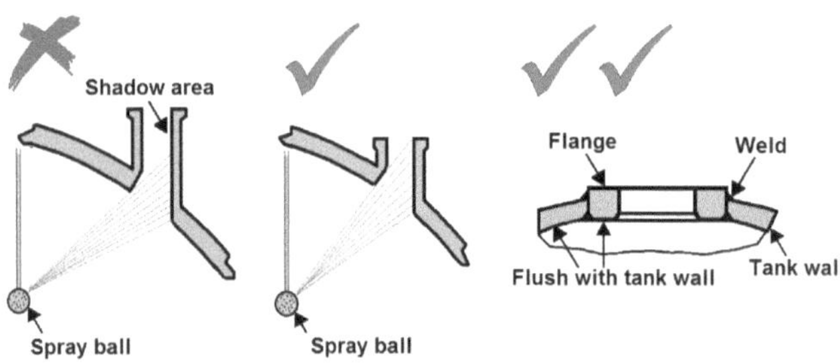

FIGURE 7.6 The tank-cleaning process by means of tank-cleaning devices can be facilitated by applying short-neck nozzles, which means tank head ports with reduced L/D ratios, short in length and large in diameter. During the cleaning action of the tank-cleaning devices, internal shadows in the top nozzles can be reduced to a certain extent if the lower part of these top nozzles is sloped toward the center of the vessel.

- Junctions between vessel and piping must be smooth, flush and without crevices.
- Closed vessels with bottom outlets must have their discharge outlet at the lowest level and their bottom shall be sloped (Fig. 7.5). Tank outlets should be flush with interior surfaces and self-draining.
- Flat top surfaces should pitch 4% from center to sidewalls to encourage the continuous flow of cleaning solutions sprayed on these surfaces toward the side walls.
- Death corners in the top of the vessel or tank should be eliminated. Difficult to clean areas are the annular space between the neck of the top nozzles in the tank head and agitator shafts, as well as down pipes, installed in the tank by means of an exterior tank connection. The ratio of nozzle neck length to annular space gap width should be $\leq 2:1$.
- Short-neck ports must be used, which means tank head ports with reduced L/D ratios. To avoid a dead leg, the maximum recommended length to tank head port diameter ratio shall be two-to-one. Top nozzles should preferably be flush with the tank wall (Fig. 7.6).

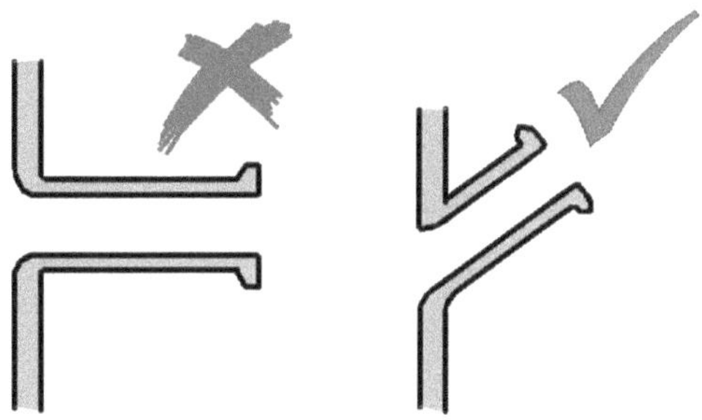

FIGURE 7.7 Use sloped side ports, rather than ports perpendicular to the vessel wall.

- The depth of manways must be reduced to avoid interior shadows, especially because they are harder to clean and a source of possible contamination.
- Sloping the lower part of the top nozzles toward the center of the vessel may eliminate shadows and provide the tank cleaning devices good "sight" angles into the top nozzles (Fig. 7.6).
- For maximum cleanability, side-wall sensor ports must be provided with a slope (5-degree angle) (Fig. 7.7), rather than ports perpendicular to the vessel wall.
- Eliminate dead corners in lower tank parts.
- To provide a reasonable flow across the tank bottom surfaces for moving suspended solids, the bottom of flat vessels should pitch no less than 2% from rear to front outlet, and 4% from side to center outlet for round bottom vessels.
- A probe (e.g., pH sensor) in the reactor wall shall be inserted in a sloped side port, with an O-ring seal to prevent the ingress of soil into the sensor port and the probe. An elastomeric O-ring seal should be placed as close as possible to the vessel wall so that only a short crevice is formed. When this seal is placed at the entrance of the port (end opposite to the tank wall), then a long and uncleanable large crevice is formed. Where cleaning relies on a free falling film, protrusion of stationary parts like sensor probes in a vessel wall should be avoided. They may form a shadow area during cleaning (Fig. 7.8).
- Baffles only partially fastened onto the side wall of the tank should be used instead of full-length fastened baffles. The internal support members to fasten the baffles to the tank wall must be made from solid round bar stock having a downward slope of 5 degrees (Fig. 7.9). When gaps are left between the baffle and tank wall, the flow allows the baffles and the tank wall to be cleaned more easily. Recommended gaps between the baffles and the vessel wall are equal to 1/72 of the internal vessel diameter, and 1/4 to 1 full baffle width between the bottom of the baffles and the vessel base. Instead of full-length baffles, the use of baffles can be limited to the lower part of the tank or the tank may be provided with intermittent

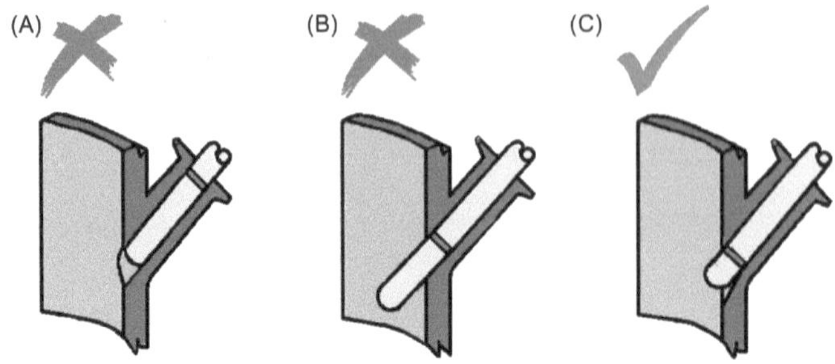

FIGURE 7.8 Probes shall be inserted in sloped, welded-in side ports, the weld being polished to obtain a surface finish comparable to that of the original finished metal. (A) An elastomeric seal at the entrance of the port (end opposite to the tank wall) gives rise to an uncleanable long and large crevice between the interior surface of the sensor port and the outside probe surface. (B) Protrusion of probes in a vessel wall should be avoided, as they may form a shadow area during cleaning. (C) When the elastomeric O-ring seal is placed close to the vessel wall, only a short crevice is formed.

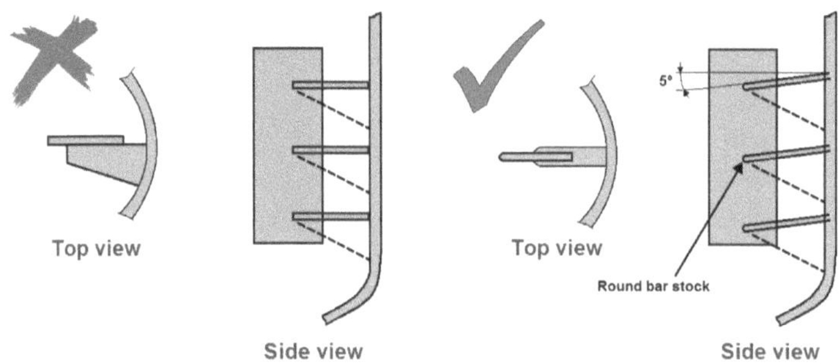

FIGURE 7.9 The internal support members to fasten the baffles to the tank wall must be made from solid round bar stock having a downward slope of 5 degrees. When gaps are left between baffle and tank wall, the flow allows the baffles and the tank wall to be cleaned more easily (ASME BPE committee, © 2014).

baffles (underbroken baffle, resulting in two shorter baffles, one below the other), without loss in agitation efficiency. Baffles can be omitted in small tanks (<500 L), and in designs where the agitator is mounted off-center and angled at the same time. Where material can hang up or becomes trapped in stagnant regions around the baffles during drainage, profiled baffles instead of flat-plate baffles are recommended (Myers et al., 2002; ASME BPE committee, 2014).

- Suitable covers or lids must be provided and should be made of the same material as the vessel. Covers on smaller vessels should be close fitting and easily removable for cleaning. Covers on such vessels should preferably be unhinged since the hinge, especially piano hinges (Fig. 7.10), can collect dust and food debris which might fall into the product when the cover is opened. Where hinges are used they should pivot sufficiently

FIGURE 7.10 The piano hinge prohibits removal of the cover, and allows food residues to accumulate in the hinge. In the piano, microorganisms may find a niche to grow (Don Graham, Graham Sanitary Design Consulting LCC, ©2010).

outwards to avoid product contamination (Fig. 7.11). For maximum cleanability, hinges should be removable and, where possible, consist of a simple hook-on type without bolts. Larger vessels are usually fitted with manholes, either in the top or side, which preferably should be ca. 90 cm in diameter for easy access. Mandoor covers intended to protect the food products may accumulate dirt or liquids on top of the lid while in a closed/horizontal position. If these mandoor covers are not correctly designed and mounted, these liquids and dirt may slide into the vessel opening and, respectively, spill and fall into the product when the lid is opened. Policy should specify that no tank is opened during production unless absolutely necessary. Correct design and mounting of covers must prevent soil and/or liquids from dripping/falling into the product during the opening of the cover. In Fig. 7.11A, the cover protecting the vessel opening is at the back side provided with a sloped edge, draining any dirt or liquids away from the vessel opening. In Fig. 7.11B, drip of soil and liquids into the food product is prevented by the curved edges at the left and right side of the bolted flat cover plate. Seals must be of a removable type (Fig. 7.12) to allow for inspection, cleaning, and replacement.

- Instead of a weld-on top surface, vessels, bins, etc. also can be closed with a (detachable) lid. Flat lids provide a horizontal surface (Fig. 7.13A) where dirt may accumulate. Hence, where nonremovable lids are used, preference should be given to domed lids with sloped tops that collect less dirt and allow for proper drainage of liquids (Fig. 7.13B). When the vessel is permanently covered by a lid, no sharp top corner (junction vessel wall–lid), which may hamper the cleanability, should be created.
- Conventionally designed right-angled grooves containing O-rings invariably create gaps and crevices that are impossible to clean in-place and/or to sterilize in-line (Fig. 7.14). One cause is that the elastomer material of the O-ring has a significantly higher thermal expansion coefficient than

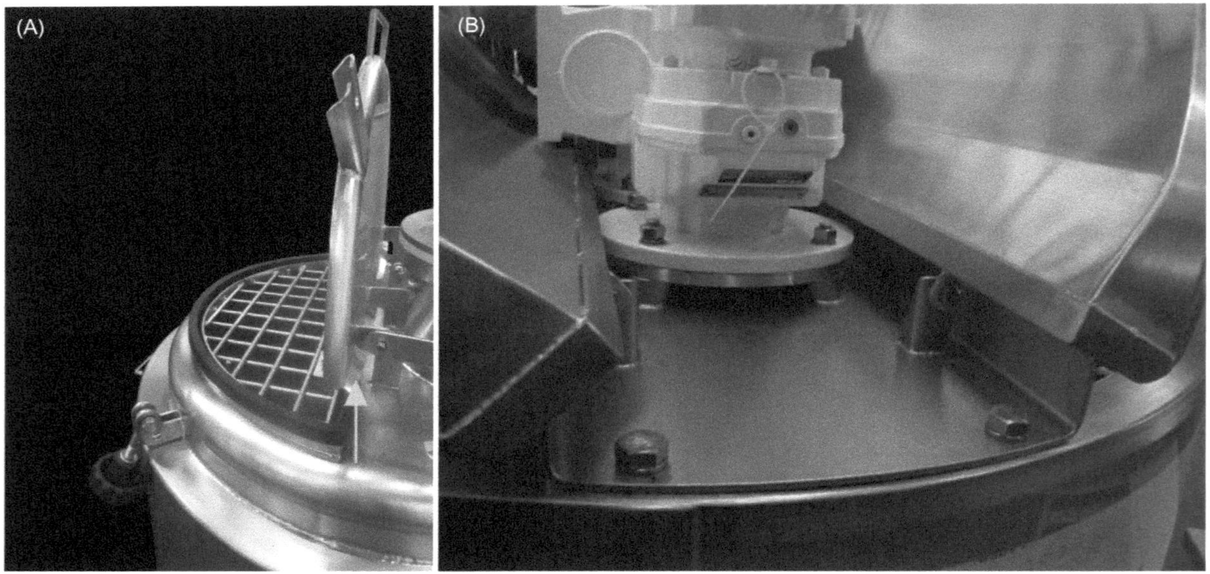

FIGURE 7.11 Correct design and mounting of covers must prevent any soil and/or liquids from accumulating on the cover while in a horizontal position, as they may fall into the product during the opening of the cover. (A) The cover is at the back side provided with a sloped edge (*arrow*), draining any dirt or liquids away from the vessel opening (Frank Moerman, © 2016). The pin hinges are less sensitive to the accumulation of debris. (B) In this example, drip of soil and liquids into the food product is prevented by the curved edges at the left and right side of the bolted flat cover plate. Pin hinges prevent the buildup of dirt. *Courtesy of Fineweld Stainless Steel Pty Ltd.*

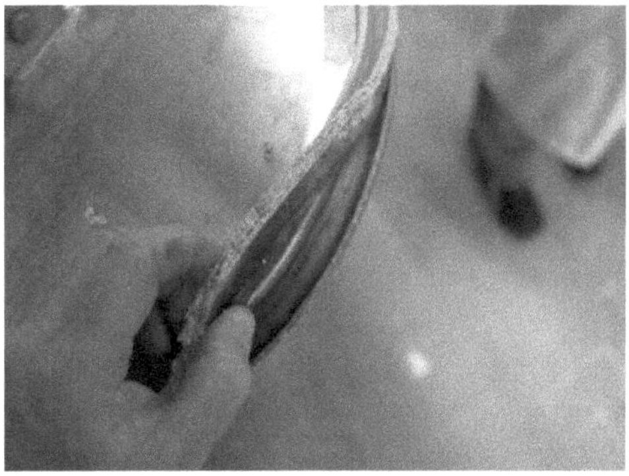

FIGURE 7.12 Seals for mandoor covers must be of a removable type to allow for inspection, cleaning and replacement. Food debris may accumulate under the seal, providing nutrients for microorganisms. *Courtesy of Burggraaf & Partners B.V.*

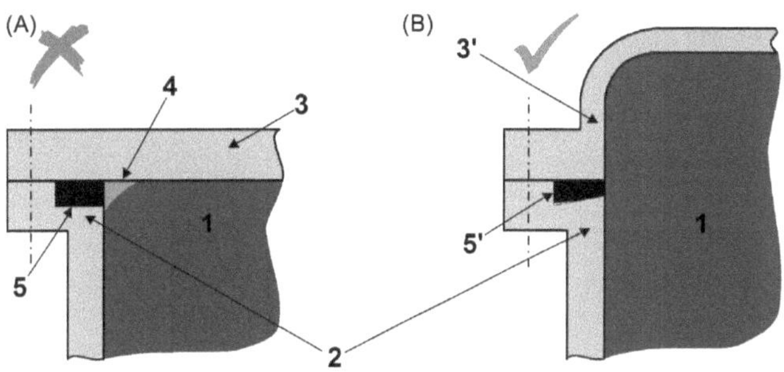

FIGURE 7.13 (A) Covers are used (e.g., for process vessels, tanks, bins, etc.) to avoid contamination of food product (1) from the environment during processing or storage. When the vessel (2) is covered with a flat lid (3), a horizontal surface is provided where dirt may accumulate. Moreover, a sharp corner (4) is created at the top near the seal. This seal (5) is not very appropriate because overcompression may lead to protrusion of the seal in the product area, thereby impeding cleaning, while undercompression may lead to both indentations and crevices and failure to provide a reliable seal. Even when it is not visibly leaking, the seal may permit the ingress of microorganisms. (B) Preference should be given to domed lids (3′) with a sloped top that collect less dirt and allow for proper drainage of liquids. The present sloped gasket groove allows for controlled compression of the gasket (5′) at the product side while providing space for expansion at the nonproduct side (Lelieveld et al., 2003; Hauser et al., 2007).

steel. During heating the seal will expand to cover an increasingly larger surface of steel, protecting microorganisms trapped between the O-ring and the steel surface against contact with hot water, chemical solution or steam. Although the seal contact surface will usually reach the correct temperature during treatment with hot water or steam, the water activity in the grooves will be too low for the destruction of many microorganisms at the temperature and time applied. After cooling down and shrinkage of the seal, the surviving microorganisms may be released and will multiply and

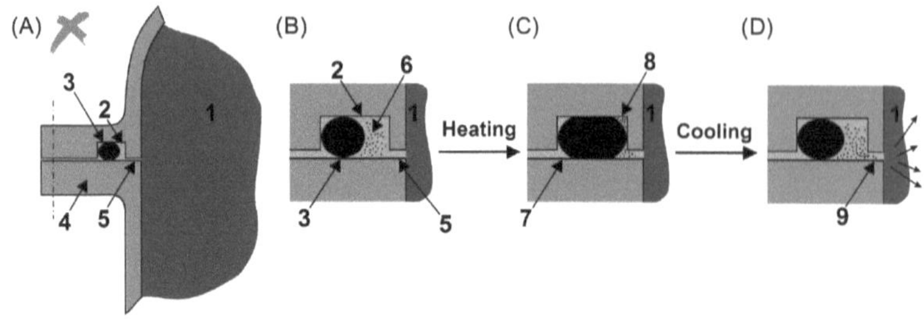

FIGURE 7.14 (A) A conventionally designed right-angled groove (2) contains an O-ring (3) that is compressed between the sealing faces of two stainless steel surfaces (4) to separate the product area (1) from the outside. (B) Such a rectangular groove–O-ring design invariably create gaps and crevices (5) that are impossible to clean in-place and/or to sterilize in-place. The groove provides sufficient space for microorganisms (6) to enter via the crevice. (C) During heating, due to the difference in thermal expansion between metals and elastomers, the O-ring will expand (7) to cover an increasingly larger surface of steel, protecting microorganisms (8) trapped between the O-ring and the steel surface against contact with hot water, chemical solution or steam. (D) After cooling down and shrinkage of the seal, the surviving microorganisms may be released (9) and will multiply and contaminate the product (Lelieveld et al., 2003; Hauser et al., 2007).

contaminate the product. Additionally, repeated thermal expansion of the seal into the product flow may result in damage which will not only contaminate the product but may also progressively reduce its ability to seal again upon recooling.

7.6.2 Hygienic Design and Installation of Agitators in Closed Vessels

The traditional arrangement of agitators is a driveshaft with an overhead drive unit and impeller blades mounted on the shaft. A wide variety of blade designs are used and typically the blades cover about two-thirds of the diameter of the vessel.

7.6.2.1 Hygienic Design of Permanently Installed Agitators

Top entering agitators with shaft seals are typically mounted to a vessel using a flanged or hygienic clamp connection, with hygienic O-rings or gaskets to seal between the mating surfaces. Agitators and agitator shaft assemblies passing through the seals shall be designed and constructed to be smooth, with all surfaces meeting all the hygienic design criteria applicable to a product contact area. Agitator ends shall have surfaces of minimum area immediately adjacent to the recipient ends and no longer than necessary to ensure proper incorporation of ingredients into a mix. The design of agitator product contact parts should minimize the occurrence of crevices and sharp corners, and be free of pockets, screw threads, void spaces, and dead spaces in grooves (Fig. 7.15).

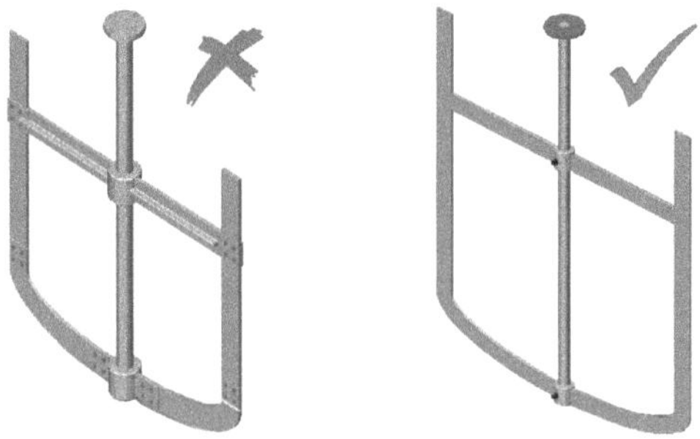

FIGURE 7.15 Install hygienically designed agitators, free of pockets, sharp corners, crevices, screw threads, etc. *Courtesy of Post Mixing Optimizations and Solutions, LLC.*

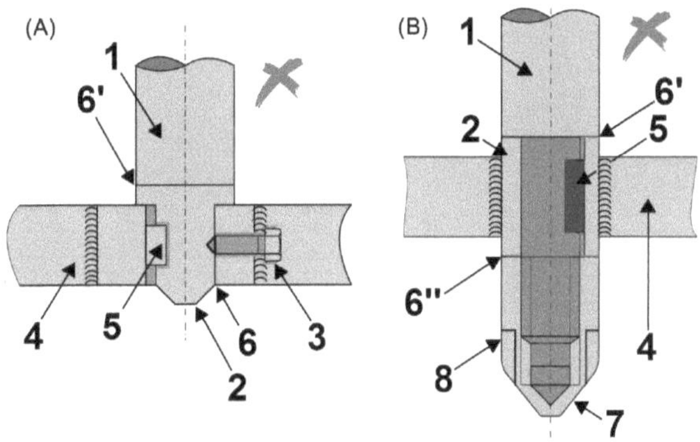

FIGURE 7.16 (A) The hub (2) is secured to the shaft (1) by means of a screw (3), which is exposed to product that may collect in and around the screw head. The hub-to-shaft connection gives rise to a metal-to-metal joint (6′) that may permit the ingress of product and bacteria. Agitator blades (4) should be welded to the hub, although screw connections are sometimes observed. These exposed screw heads (even bolts with dome head nuts and washers of suitable food grade material) again will create a food safety hazard, and the blade-to-hub connection gives rise to a new metal-to-metal joint (6). To avoid the latter problem, the joint between the blade and the lug on the hub can be sealed by a thin gasket. Keyways (5) exposed to product are not recommended, because product and microorganisms may be retained in the keyway. Keyways may require additional design and/or cleaning practice to ensure drainage and cleanability, e.g., spray ball and wand additions, increased CIP flow and adjusted spray coverage. (B) Once the hub (2) is secured to the shaft (1), an end cap (impeller nut, 7) is screwed on the interior male thread end of the shaft. The nonwelded impeller hub-to-shaft and hub-to-end cap connections give rise to crevices and metal-to-metal joints (respectively 6′ and 6″) that may allow the ingress of product and bacteria. In that way, the keyway (5) may retain product and microorganisms. The sharp corners of the spanner flats (8) on the end cap may be difficult to clean (Hauser et al., 2004b; Moerman and Kastelein, 2014).

Less suitable, not recommended designs are those in which hub, impeller blades and end nut are assembled together by screw joints (bolting) (Figs. 7.16A and 7.17A). Debris may collect on exposed screw threads, even if bolts with dome head nuts and washers of suitable food grade material are

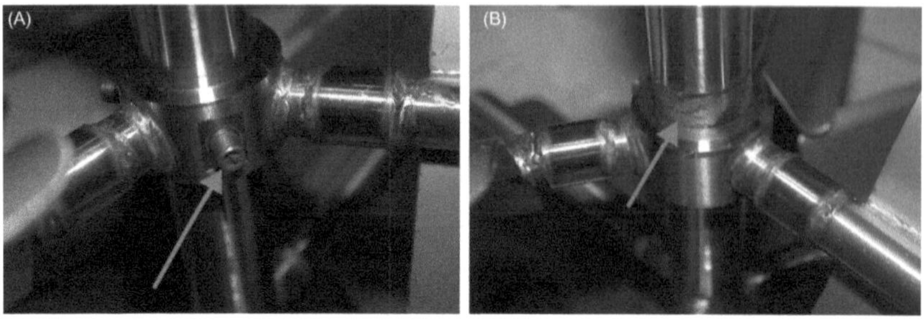

FIGURE 7.17 (A) The hub is secured to the shaft by means of bolts with dome head nuts. However, product may collect in and around the screw head (*red arrow*). (B) This nonwelded hub-to-shaft joint also lacks a food grade gasket that could seal the dead spaces in the groove and avoid crevices at points of metal-to-metal contact. Ingress and accumulation of product and/or microorganisms at the inside (*red arrow*) are shown. Welds also have a high degree of roughness. *Courtesy of Burggraaf & Partners B.V.*

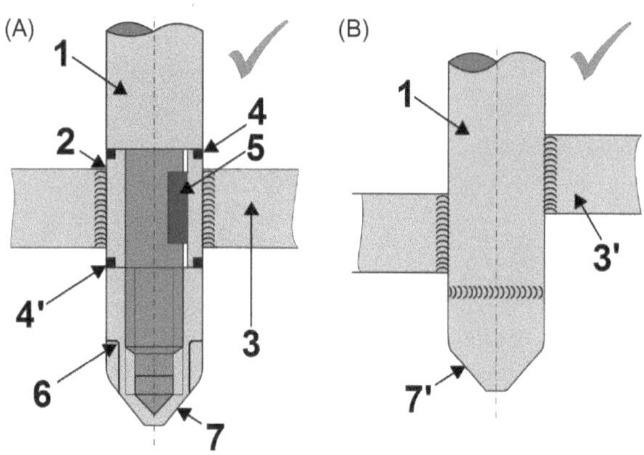

FIGURE 7.18 The hub (2) with blades (3) is secured to the shaft (1). All voids should be closed by either fabrication (welding) or approved sealing techniques (O-rings, seals, etc.) to give surfaces ground flush and free of crevices at points of metal-to-metal contact. (A) Food quality gaskets under controlled compression respectively may seal the propeller hub to the shaft (4) and end cap (4′). Keyways (5), where employed due to mechanical design considerations, shall have edge radii not less than 3 mm. The corners of the spanner flats (6) on the end cap have been radiused. (B) An all-welded impeller assembly (e.g., hub, blades, end cap) is still preferred. Impeller hubs welded to the shaft are preferred over removable hubs. Although the designer may omit the hub, and immediately can attach the blades to the shaft by welding (3′). Finally, also the end cap (7′) can be welded to the shaft (Hauser et al., 2004b; Moerman and Kastelein, 2014).

used. Metal-to-metal joints (e.g., keyways, hub-to-shaft joints, hub-to-end cap joints, etc.) may allow ingress and accumulation of product and/or microorganisms (Figs. 7.16B and 7.17B). Food quality gaskets under controlled compression may seal the propeller hub to the shaft and to the impeller nut (end cap) that secures the end of the agitator shaft (Fig. 7.18A). Alternatively, the hub should be welded to the shaft and the end cap (Fig. 7.18B), with the blades of appendages (stirrers, homogenizers, mixers, etc.) welded to the hub. As an alternative to impeller blade-to-hub attachment, blades can be immediately attached to shafts by welding (without hub) (Fig. 7.19). In this all-welded one-piece design, all welds must be ground flush and polished.

Permanently joined metal surfaces with a total included internal angle less than 135 degrees on agitators (e.g., at hubs and nuts) shall have a radius of not less than 3 mm tangential to both adjacent surfaces. Corners (e.g., at hubs, nuts, spanner flats, etc.) must be radiused to facilitate cleaning, and horizontal areas must be sloped to prevent debris from becoming lodged on the surfaces and to allow for maximum drainabilty. Machined transitions such as shaft steps, coupling surfaces, spanner flats, etc. should have 15- to 45-degree sloped surfaces. Impellers with flat, horizontal surfaces (e.g., flat-blade disc turbines, concave-blade disc turbines) may require additional design and/or cleaning practice to ensure drainage and cleanability, e.g., drain holes, spray ball and/or wand additions, increased CIP flow, adjusted spray coverage and faster impeller rotation. Where permanently installed agitators are equipped with an outer frame to which rubber, plastic or other similar scraping edges (Fig. 7.20) are attached, these scrapers shall be readily removable from the agitator. They should be regularly checked for integrity. Cases are known where plastic scrapers were broken and pieces lost in the product as a foreign-body contaminant.

Welded in-tank shaft connections are preferred, although in-tank shaft couplings (Figs. 7.21A–C and 7.22A–B) and in-tank threaded shaft connections (Fig. 7.22C) are allowed if they are of acceptable hygienic design. Threaded shaft connections are preferred over in-tank shaft couplings, although shaft rotation of the first is limited to a single direction to avoid the shaft

FIGURE 7.19 All-welded impeller assembly (e.g., hubs, blades, end cap). The agitator paddle blades being attached to the shaft by welding must have their welds ground and polished. *Photo right, courtesy of Intechwell/Alfa Laval AB.*

FIGURE 7.20 This agitator may cause hygiene problems because of the bolts and plastic scrapers. Cases are known where plastic scrappers were broken and pieces lost in the product as a foreign body contaminant.

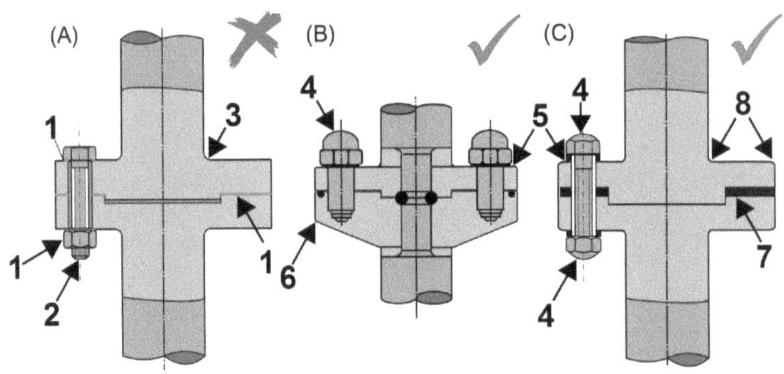

FIGURE 7.21 (A) Bolted agitator couplings with flat hexagon head screws without elastomer gasket under the bolt head and the nut give rise to metal-to-metal crevices (1) that may allow the ingress of food product and bacteria. Moreover, debris may lodge in and around the bolt thread (2). The absence of a circumferential O-ring or flat gasket gives rise to another metal-to-metal crevice, and product and microorganisms may be retained in the cavity (3). (B, C) Agitator couplings made by means of domed hexagon bolt heads and nuts (4) provided with an elastomer gasket (5) under the bolt head and the nut allow for a crevice free joint without metal-to-metal contact. Due to the presence of a circumferential O-ring (6) or flat gasket (7), no product and microorganisms can enter inside the agitator coupling. Corners are radiused (8). However, there is still a horizontal flat surface at the upper side of the agitator coupling where debris may lodge (Hauser et al., 2004b; ASME BPE committee, 2014; Moerman and Kastelein, 2014).

sections separating. The designer must ensure that the use of a threaded shaft connection is appropriate for the selected shaft diameter and design loads. To avoid exposure of the threads to the product, O-rings or flat gaskets (preference for the first mentioned) should be used to seal mating surfaces (Fig. 7.22C). Hygienic bolted coupling construction may be used where appropriate for the

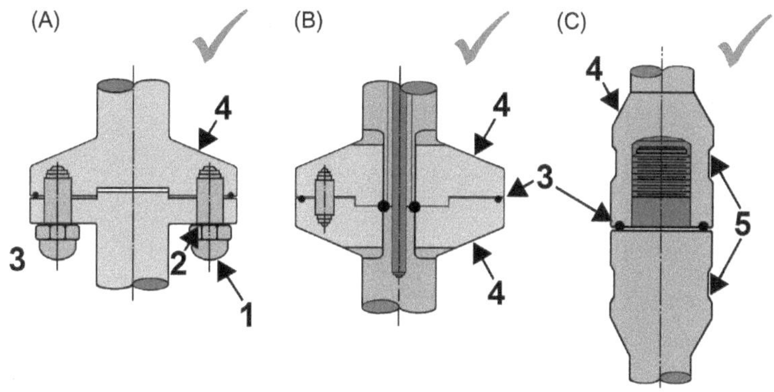

FIGURE 7.22 (A, B) Aseptic applications require the fastening of hardware at the bottom side of the agitator coupling. Agitator couplings made by means of domed hexagon bolt heads and nuts (1) provided with an elastomer gasket (2) under the bolt head and the nut allow for a crevice-free joint without metal-to-metal contact. Due to the presence of a circumferential O-ring (3), no product and microorganisms can enter the inside of the agitator coupling. The upper parts of the coupling should be sloped to a minimum 15–45 degrees (4) to prevent debris from collecting at these places and to allow for maximum drainability. (c) The optimal agitator coupling in an aseptic environment is a threaded shaft connection with O-rings or flat gasket (preference for the first mentioned) (3) to seal the mating surfaces to avoid exposure of the interior thread. The corners of the spanner flats on the end cap have been radiused (5) (ASME BPE committee, 2014; Moerman and Kastelein, 2014).

particular application. The preferred location for fastening hardware is on the underside of couplings, and the fasteners typically used should be hex-head cap screws, acorn-head cap screws and threaded studs with acorn nuts (Fig. 7.22A). These fastener heads shall be free of raised or engraved markings that might inhibit cleanability. Again, O-rings or flat gaskets (preference for the first mentioned) should be used to seal coupling mating surfaces. Elastomer seal washers (Figs. 7.21B–C and 7.22A) must avoid metal-to-metal contact.

7.6.2.2 Top Mounted Installation of Agitators

Agitators permanently mounted are not required to be removable if they are readily accessible to be effectively cleaned via spray, directed flow, immersion or cleaning-in-place (CIP) and if they do not interfere with drainage from the tank. Top entering agitators with shaft seals are typically mounted to a vessel using a flanged or hygienic clamp connection, with hygienic O-rings or gaskets to seal between the mating surfaces. The selected mounting arrangement must support the agitator mounting design loads while achieving an appropriate seal. The upstand for the top mounting of the agitator should have limited length L because of the difficulty of cleaning of the annular space in-place. The annular space between the agitator shaft and agitator nozzle shall, for cleaning purposes, have the target maximum L/A ratio of 2:1. At least a 25 mm gap is required to

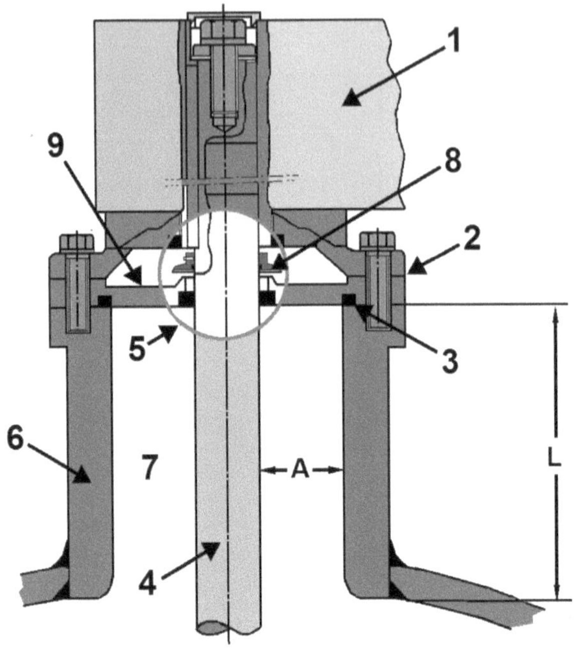

FIGURE 7.23 The top entering agitator with motor (1) is mounted to a vessel using a flanged or hygienic clamp connection (2), with hygienic O-rings or gaskets (3) to seal between the mating surfaces. A retained gasket having limited compression is more hygienic than an O-ring in the face for sealing the joint. The agitator shaft (4) passes through the mounting flange via a seal (5). The upstand (6) for the top mounting of the agitator should have limited length L because of the difficulty of cleaning the annular space (7) in-place. The annular space between the agitator shaft (4) and agitator nozzle (6) shall, for cleaning purposes, have the target maximum L/A ratio of 2:1. Agitator motors (1) should be equipped with permanently lubricated bearings. Where lubrication is required, the design and construction shall be such that lubrication cannot leak, drip, or be forced into the product zone. Self-lubricating agitator shaft (packing) seals (8) shall be provided with convenient means for adjustment to prevent leakage and to allow for complete drainage to the exterior. In that way, accumulations of foreign material in the event that leakage does occur can be avoided. Further, a drip protection plate (9) can be provided to prevent lubricant from entering the product zone (Moerman and Kastelein, 2014).

facilitate CIP spray coverage (Fig. 7.23) (CFCRA, 1997; BISSC, 2003; ASME BPE committee, 2014).

Agitator motors should be equipped with permanently lubricated bearings. Where lubrication is required, the design and construction shall be such that lubrication cannot leak, drip, or be forced into the product zone. Self-lubricating agitator shaft (packing) seals shall be provided with convenient means for adjustment to prevent leakage and to allow for complete drainage to the exterior (Fig. 7.23). In that way, accumulations of foreign material in the event that leakage does occur can be avoided. Further, drip protection is commonly provided to prevent lubrication from entering the product zone.

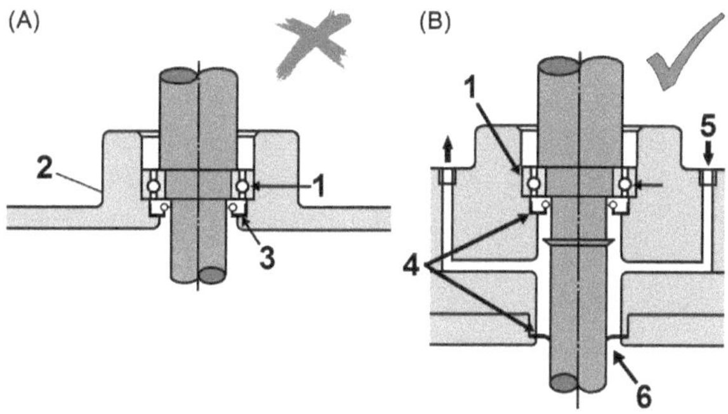

FIGURE 7.24 Rotary shafts running at a high number of revolutions are held in-place in an adaptor sleeve with a radial roller bearing (1). (A) Single dynamic seals (2) are lubricated by a lubricant (top-mounted agitator) or the product (bottom-mounted agitator) which may be transported past the seal and back again, contaminating the product further. They may be easy to clean if properly designed but they will not prevent the passage of microorganisms, and hence they are not suitable in aseptic process equipment. There is also a narrow annular space (3) at the product side in the proximity of the seal, which makes cleaning very difficult. (B) A double seal arrangement (4) allows the use of a barrier medium (5), such as steam, hot water, condensate, or a disinfectant solution that makes it well-suited from a microbiological standpoint. The volume of the annular gap around the shaft is increased (6), improving the cleanability of the seal and its proximity (Holah, 2000).

All surfaces of shaft seal ring assemblies passing through a bowl or cover shall be accessible, removable or retractable to permit cleaning of all product zone surfaces.

Rotary shafts running at a high number of revolutions are held in-place in an adaptor sleeve with a radial roller bearing. Single dynamic seals (Fig. 7.24A) will not prevent the passage of microorganisms. If properly designed, they may be easy to clean but not bacteria tight because rotating shafts always exhibit some axial mobility. This makes single dynamic seals unsuitable for aseptic equipment. A narrow annular space at the product side in the proximity of the seal, such as that shown in Fig. 7.24A, must be avoided because it is difficult to clean. The space around the seal should be as wide as possible. Rotary shafts with double seal arrangement allow the use of a barrier medium, and have been shown to be well-suited from a microbiological standpoint. In Fig. 7.24B, one seal is seated rigidly in the housing (longitudinal shading), while the other moves with the shaft. The sealing surface between the two seals must be lubricated. If the shaft opening has product flowing through it, which could be the case with agitators having a shaft entry from the bottom of vessels, the product itself can be directly used as lubricant. The product flowing through can be carried away by the barrier medium, which could be steam, hot water, condensate or a disinfectant solution (e.g., alcohol). The sterile fluid may scavenge the microorganisms that enter the space between the seals, maintaining sterile conditions. Which flushing fluid should be used will depend on the product and the process but both the barrier medium and lubricant chosen must be product-compatible. To avoid transfer of microorganisms from the outside of the equipment to the inside the distance between the two seals must always be sufficiently large (Lelieveld et al., 2003; Hauser et al., 2007).

Bearings in the product area should be avoided but an application may mandate the use of foot bearings. In the example given, if the shaft of a top entry agitator is very long, a foot bearing may be required at the bottom of the vessel to steady it. It shall be of a packless bearing type. The foot bearing must be mounted well clear of the base so as not to impede free draining of product and to allow easy cleaning of their supports. Design features and/or procedures required to ensure cleanability are: drain holes, spray ball, and/or wand additions, increased CIP flow, operating the steady bearing immersed in CIP fluid. The arrangement of wear surfaces (bushing, shaft, or shaft sleeve) shall facilitate drainage. A longitudinal or helical groove may be cut in either the bush or the shaft. It should be deep enough to allow access into the bearing of either the product as a lubricant or the detergent for cleaning (Fig. 7.25). Sealed bearings should not be used in the product area because they can cause hygiene risks at their seals. If, however, their use is unavoidable, their lubricants should be specified as being allowed in contact with food.

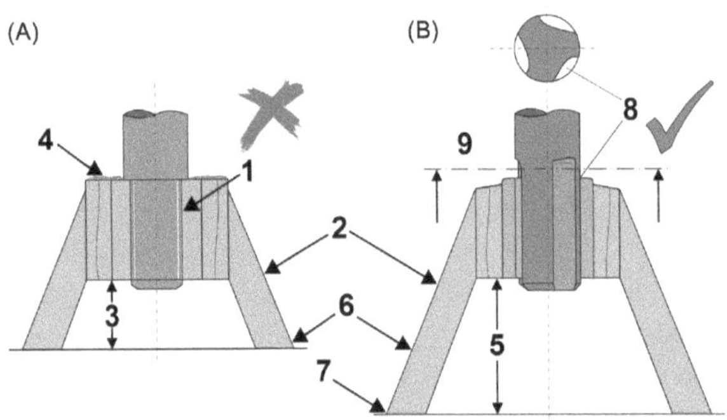

FIGURE 7.25 (A) Cleaning may be impeded due to too-tight clearance (1) in the foot bearing itself (2), and due to too little clearance between it and the base (3). Horizontal ledges (4) where product may accumulate or where liquids are not allowed to drain must be avoided. (B) The foot bearing is now mounted clear of the bottom of the vessel (5), allowing free flow of product and cleaning solution around it. Bearing pedestal support members (6) should preferably be made of solid construction. Hollow constructions are not recommended, but if used, they shall be of sealed (welded) construction and inspected for integrity. Round legs are preferred over flat members, even if the latter are radiused. The legs should be flush welded in-place to the tank bottom (7). All welds must be ground and polished to blend smoothly with the adjacent surfaces. The agitator shaft is provided with grooves (8) in the bearing area to facilitate both lubrication by fluid products and cleaning. Sloped and radiused surfaces (9) reduce the probability of debris getting lodged on the top of the foot bearing and allow for proper drainage of liquids (e.g., cleaning solution) (CFCRA, 1997; Lelieveld et al., 2003; Hauser et al., 2004b; ASME BPE committee, 2014). *Courtesy of Campden BRI.*

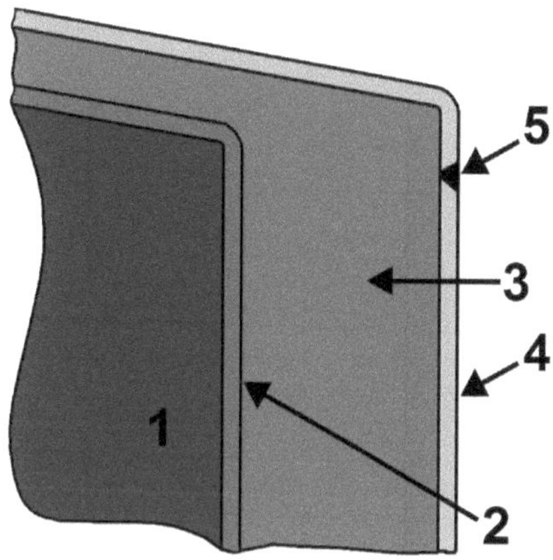

FIGURE 7.26 The inner tank wall (2) is provided all around with thermal insulation to keep food products (1) at the correct temperature. Rockwool (3) doesn't give rise to unhygienic conditions provided it is kept vapor tight by installing aluminum or stainless steel cladding (4) of appropriate thickness. Joints facing downwards must be continuously welded (5) to avoid any ingress of dust, liquor, air, and moisture.

7.6.3 Good Insulation Practices

Non-chloride-releasing insulation material should be used. For thermal insulation of vessels, appropriate qualities of rock wool are acceptable. It is highly recommended to install fully welded, vapor-tight, aluminum, or stainless steel cladding of appropriate thickness that resists tear and abrasion (Fig. 7.26). The exterior of the insulation protection should be smooth, properly sealed to avoid ingress of dust, liquor, air and moisture, and should be installed in a correct way with joints facing downwards. Such ingress could promote corrosion between the walls, assisted by possible microbial growth.

7.7 HYGIENIC DESIGN OF PROCESS AND UTILITY PIPING

7.7.1 Drainable Process and Utility Lines Without Dead Ends

Pipes must ensure minimum resistance to flow, and therefore there should be no sudden changes in cross-sectional area or obstructions that are likely to hinder the flow. However, valves and flowmeters may restrict the flow. Pipes also must be hygienically designed and meet standards that are required nationally and internationally. Consideration of CIP should be integrated into the mechanical and process design at an early stage, rather then being incorporated into an already fully specified plant. Hence, pipework must be so designed that as far as possible a minimum velocity of 1.5 m/s is achieved over the whole trajection of pipes, unless data for the specific soil indicate otherwise. As a consequence, substantial flowrates for larger-diameter pipework are required. Failure to achieve 1.5 m/s does not necessarily mean that effective cleaning cannot be achieved, but the process is likely to be nonoptimal.

To avoid the formation of standing "pools" of liquid that can support the growth of microorganisms, process and utility piping runs should be sloped to at least 3% in the direction of flow and should be properly supported to prevent sagging (Fig. 7.27). Nondrainable pipe sections are not allowed (Fig. 7.28). Where valves are fitted, additional support must be given. Where plastic piping is installed, special care should be taken to avoid sagging by increasing the frequency of support. Pipelines and valves should be supported independently of other equipment to reduce the chance of strain and damage to the equipment, pipework, and joints.

A properly designed food processing line must not have dead legs, as blanked-off tees constitute a hazard. A dead space, being an area outside the product flow where liquid or gas can become stagnant and where water is not exchanged during flushing, is formed. An air pocket may be present if the branch of a blanked-off tee is pointing vertically upwards (Fig. 7.29A). Hence it will prevent liquids (cleaning solutions, disinfectant solutions, or hot water) from reaching all surfaces to be treated, with the result that CIP and

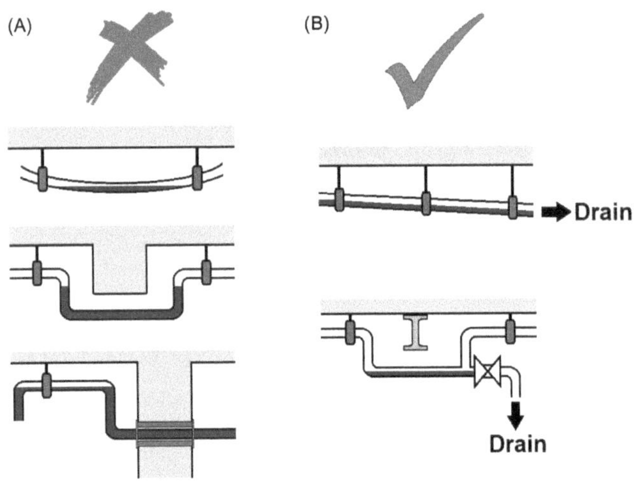

FIGURE 7.27 Pipes must be completely self-draining. (A) Sagging of piping must be avoided because standing "pools" of liquid can support the growth of microorganisms. Changes in the level of horizontal runs of pipelines should be avoided; otherwise, there will be an undrainable section. Horizontal runs of pipe which are routed vertically up and then down to bypass beams, doorways or other obstructions will allow air to collect in the raised section. (B) Process and utility piping runs should be pitched at least 3% in the direction of flow. Piping must be installed in a way that air doesn't collect in the raised section. While automatic air release valves can be installed (on top of elevated horizontal pipe sections) to remove trapped air, the resulting dead leg may cause contamination and/or cleaning problems. Where liquid collects in a lower horizontal pipe section, fitting a valve in a shortened tee allows liquid to be drained (CFCRA, 1997). *Courtesy of Campden BRI.*

FIGURE 7.28 Nondrainable pipe section. *Courtesy of Mondelēz International, © 2016.*

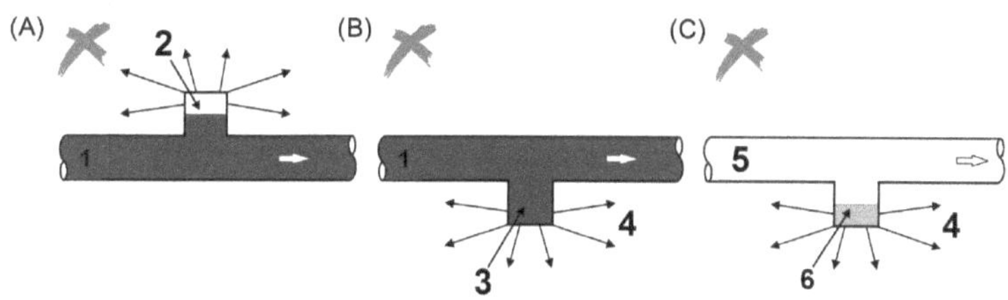

FIGURE 7.29 (A) When cleaning and disinfection solutions (1) flow through the piping, an air pocket (2) will be formed if the branch of a blanked-off tee is pointing vertically upwards. This will prevent the solutions from wetting the surface in the dead leg. (B) Drain points pointing downwards (3) again act as a dead leg, providing an area of entrapment that may not be reached by cleaning or sterilizing procedures, and hence they may lead to contamination of the product. Moreover, during a hot water treatment, the hot water also will stagnate in the downwards pointing pocket, so that the temperature of the surfaces in the dead area may be lower than required as the consequence of heat loss (4). (C) A downwards pointing dead area also will collect condensate (6) due to heat loss (4) during steam sterilization (5), with the result that again the temperature of the surfaces in the dead area may be lower than required. Blanked-off tees should be positioned such that they are a few degrees above the horizontal (Lelieveld et al., 2003; Hauser et al., 2007).

decontamination processes will be unsatisfactory. Drain points pointing downwards act as a dead leg (Figs. 7.29B and 7.30) and are not acceptable because they provide an area of entrapment that may not be reached by cleaning or sterilization procedures. During a hot water treatment, the hot water also will stagnate in the downwards pointing pocket, so that the temperature of the surfaces in the dead area as a consequence of heat loss may be lower than required. A downwards pointing dead area also will collect condensate during steam sterilization (Fig. 7.29C), with the result that again the temperature of the surfaces in the dead area may be lower than required. As a consequence, the thermal disinfection or sterilization of the dead space is compromised.

FIGURE 7.30 This drain point pointing downwards acts as a dead leg (organization Sanitary Design Workshop, © 2016).

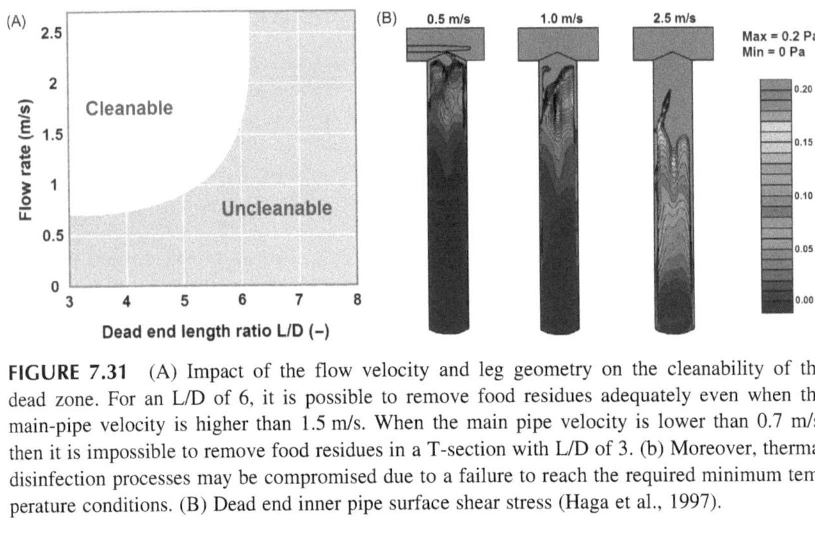

FIGURE 7.31 (A) Impact of the flow velocity and leg geometry on the cleanability of the dead zone. For an L/D of 6, it is possible to remove food residues adequately even when the main-pipe velocity is higher than 1.5 m/s. When the main pipe velocity is lower than 0.7 m/s, then it is impossible to remove food residues in a T-section with L/D of 3. (b) Moreover, thermal disinfection processes may be compromised due to a failure to reach the required minimum temperature conditions. (B) Dead end inner pipe surface shear stress (Haga et al., 1997).

Even with turbulent flow within the horizontal pipeline, the shear rate and temperature within the dead leg will fall rapidly with distance from the junction between the horizontal and vertical sections (Fig. 7.31).

The direction of the flow of food product has a significant influence on the residence time in the dead leg. When the food product flows in the direction as indicated in Fig. 7.32A, B and C, part of the product will stand still in the dead leg, especially if the length or depth of the T-section is too long. If the length of the T-section is equivalent to the diameter of the main pipe, a flow velocity of 2 m/s in the main pipe will already result in a reduced velocity of

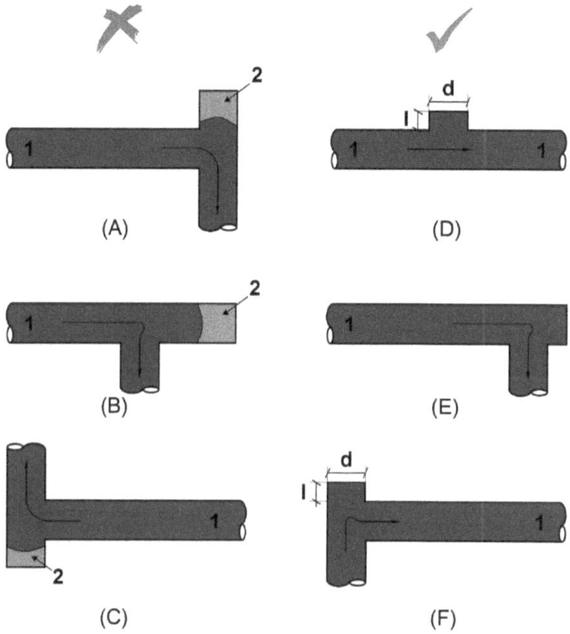

FIGURE 7.32 When the food product flows in the direction as indicated in (A, B, and C), part of the product will stand still in the dead leg, especially if the length or depth of the T-section is too long. Long T-sections outside of the main flow of cleaning solutions are also very difficult to clean. For most liquids, the dead leg should be positioned as shown in (D, E, and F). In particular, the configuration in (F) is quite acceptable if $l \leq d$, because the flow directed into the short dead leg provides sufficiently high velocities for proper cleaning. If the dead leg is very short ($l \leq d$), configuration (D) is acceptable, although flow across a dead leg results in much lower velocities within it and thus only provides moderate cleaning. Configuration (E) may not be suitable, if products contain any particulate matter that may accumulate in the dead leg (CFCRA, 1997; Lelieveld et al., 2003; Hauser et al., 2007).

0.3 m/s in the T-section. This decrease in flow velocity provides a relatively stable pocket or dead leg in which product residues can accumulate and microorganisms begin to multiply. Long T-sections outside of the main flow of cleaning solutions are also very difficult to clean. During cleaning there is much less transfer of thermal (heat), chemical (detergent and disinfectant chemicals) and mechanical energy (action of turbulent flow) to the food residues in the T-sections (outside the main flow of cleaning liquids) than to the soil in the main flow. Notice that flow away from the dead leg such as in Fig. 7.32A and C further gives rise to more contamination problems and worse cleaning, as velocities in these dead legs are even much lower.

For most liquids, the dead leg should be positioned as shown in Fig. 7.32D, E and F. If the dead leg is very short, configuration Fig. 7.32D is acceptable, although flow across a dead leg results in much lower velocities within it and thus prolongs the time for cleaning. Configuration Fig. 7.32E may not be suitable, if products contain any particulate matter, which may accumulate in the dead leg. The configuration in Fig. 7.32F is the most acceptable, because the flow directed into the short dead leg provides sufficiently high velocities for proper cleaning

For pipe diameters of 25 mm or larger, T-sections should have a depth/length of preferably under 28 mm, while for smaller pipe diameters this length should be smaller than the diameter. Blanked-off tees should be positioned such that they are a few degrees above the horizontal. The dead leg will then be drainable but not necessarily cleanable even if made as short as possible. If a sensor must be installed in a process line, it should be installed in a bend on a shortened tee in a position that the flow of cleaning fluid should be directed into the tee (Fig. 7.32E and F). Where an angle valve is installed in the process piping circuit, this valve also must be mounted in a shortened tee so that no or a minimum of annular space above the side branch is formed. Again, the flow of cleaning solution must be directed into the tee. In all cases, the cleaning procedure must take the presence of the dead leg into account.

Flow diversion should not be done in a way that would cause part of the product to stand still in a dead leg. The two-valve system for flow diversion (Fig. 7.33A) creates a dead leg toward the closed valve. The correct type of valve is shown in Fig. 7.33B.

For horizontal piping, eccentric reducers should be used instead of concentric reducers, because the latter provide a dead spot where condensate and dirt may collect (Fig. 7.34).

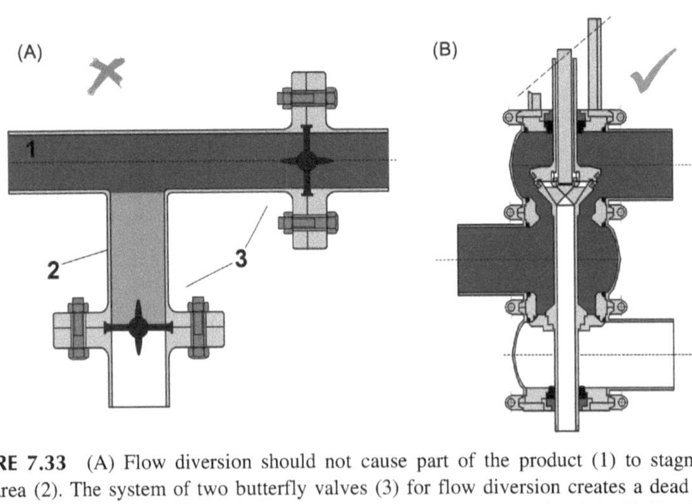

FIGURE 7.33 (A) Flow diversion should not cause part of the product (1) to stagnate in a dead area (2). The system of two butterfly valves (3) for flow diversion creates a dead area (2) toward the closed valve. (B) The correct type of valve is shown on the right (Lelieveld et al., 2003; Hauser et al., 2007).

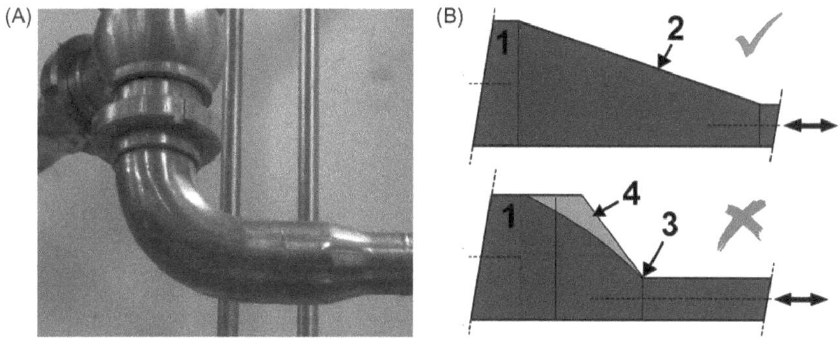

FIGURE 7.34 Changes in pipe diameter should be made by the use of reducers to ensure a smooth transition of the product flow. (A) In vertical piping, a concentric reducer is fully acceptable for food product to flow. However, this is not the case for horizontal piping, where the concentric reducer prevents full drainage if product flow is in the wrong direction. A dead spot is created where condensate and dirt may collect. (B) For horizontal piping, eccentric reducers are preferred. The reducers should be long enough (2) to avoid shadow zones during product flow (1). If a short eccentric reducer (3) is applied, a potential shadow zone (4) will be created (Lelieveld et al., 2003; Hauser et al., 2007).

7.7.2 Pipe Joints

Pipe work may be designed either for rapid regular dismantling to permit cleaning (sterilization takes place after reassembly), or for CIP without dismantling the plant. In the first case, couplings can be used, while in the second case preference is given to welding. It is strongly recommended to minimize the number of joints, whether welded or detachable. Cold bending of pipes is highly preferable to the use of prefabricated bends that have to be installed using couplings. Pipes may be joined together either by welding or by couplings, which again should meet the highest standards, both technically and hygienically. It is important to avoid crevices and gaps where product residues can accumulate and potentially begin to decompose. Although more hygienic, welded joints may be the weaker places in a process system, depending on the quality of the weld.

7.7.2.1 Permanent Pipe Joints (Welded Joints)

Welding is the preferred method of joining, provided that it is done correctly. Stainless steel hygienic tubing joints should be made by automatic orbital welding (Fig. 7.35) where possible and hand welding in those places that are difficult to access. However, those welds that are difficult to access should, wherever possible, be completed in the workshop prior to installation on the plant. The applied materials should be easily weldable, and higher alloyed filler metal in comparison to the welded material should be used to improve the corrosion resistance. Piping with the correct interior diameters should be applied because any mismatch in diameters or thickness may result in misalignment,

FIGURE 7.35 Stainless steel hygienic tubing joints should be made by automatic orbital welding where possible. Orbital welding provides smooth and crevice-free junctions (Kopitzke et al., 2006).

introducing a step in the wall or bore. If the diameters of the pipes to be joined are not the same, then the smaller pipe should be expanded to match the larger one. Misalignment also can be due to incorrect fitting up (missed coincidence between the axes of the two coupled components) prior to welding. Alignment and clamping tools are available to ensure accurate alignment. Misalignment tolerance must be limited to less than 20% of the wall thickness.

For proper welding, the parts to be welded should be adequately prepared. Cutting should be done with a mechanical mill or saw to ensure that the cut face is exactly at right angles to the longitudinal axis of the pipe. Any burrs must be removed with either a file or emery paper. Care must be taken not to remove the corner edges of the pipe, as this can give rise to problems with fusion of the root of the weld. The pipe surface 25 mm either side of the weld should be roughened up with a stainless steel wire brush, or emery paper. Then both pipe ends and the roughened surface area should be degreased with a solvent and cleaned from contaminants, because any organic substances remaining on the metal surface are vaporized during the welding process and form bubbles (porosity) in the weld metal. Pores in welds may trap product residues.

After two deburred pipe ends are aligned and butted together to a gap of less than 0.25 mm between both pipe faces, a butt weld joint is made by fusing together the two stainless steel edges with the aid of filler material. If the gap during the joint preparation is too wide, a crack running along the weld metal itself may be the result (center line cracking). Full penetration welds should be used whenever possible to avoid pockets where volumes of gas or contaminants can be trapped. Single pass welds should be utilized instead of multipass welds to avoid trapped volumes. The weld metal should exactly fill the joint and remain flush with the surface. Underpenetration leaves a crevice at the joint, while excessive overpenetration can give rise to hold up of product in pipework once taken into service. The weld metal in the joint must be fully fused to the parent metal; otherwise a crevice will form at the interface between weld and plate. Weld zones should be continuous, smooth, and flush with this parent metal. The welding process always should occur with sufficient weld seam protection, because insufficient inert gas shielding or no internal purge will result in roughened welds of lower corrosion resistance, which are prone to increased adhesion of soiling and are difficult to clean. Typically, where inert gas shielding was inadequate, significant discoloration or carbonization in the heat-affected zone is observed.

Weld slag and debris generated within the pipe must be removed from the inside and outside of the weld by proper maintenance and cleaning practice with an alkaline detergent solution prior to the start of the production process. This is followed by rinsing with water of good microbiological quality, usually chlorinated water to 2 ppm available chlorine maximum. After draining, the access points should be covered and sealed. In some circumstances there is an additional requirement to passivate the weld area on the product contact side. The welds may be mechanically polished (outside) or electro-polished (inside and outside), but air leakage should be monitored after the polishing procedure.

Weld seams finally should be visually inspected for any discoloration and surface breaking defects, usually by endoscopy and aided by dye penetrant tests that highlight these defects. Inspection personnel should be trained and act with caution to avoid internal surface damage while handling endoscopic tools (Hauser et al., 1993; Kopitzke et al., 2006).

7.7.2.2 Dismountable Pipe Joints

From a hygienic point of view, the use of threaded piping is not recommended, because it provides crevices and areas where bacteria can adhere and proliferate. Couplings must be crevice-free and only external threads are allowed. Several well-established pipe couplings and seal arrangements (Table 7.2) have been assessed for applications in the food industry.

TABLE 7.2 Overview of Market Available Pipe Couplings and Seal Arrangements (CFCRA, 1997; Lelieveld et al., 2003; Hauser, 2008b)

Type	Hygiene Characteristics	Application
3-A coupling—ground seat 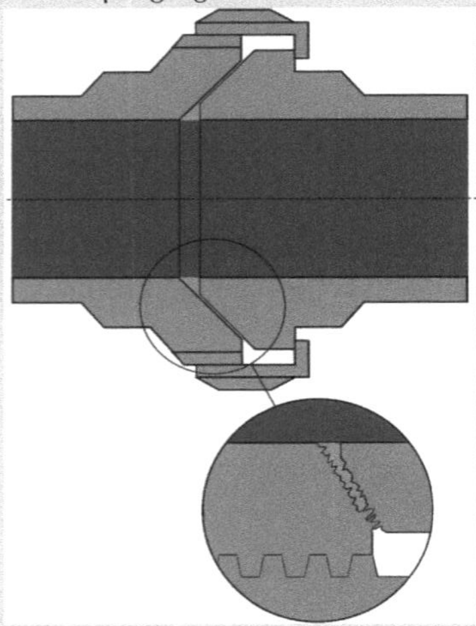	When these surfaces become permanently damaged, it becomes more difficult to obtain a tight seal after every disconnection. The metal-to-metal seat does not prevent the partial penetration of low viscosity liquids nor the ingress of microorganisms. Even if the joints are not visibly leaking, the ingress of microorganisms is possible. Furthermore, the seal obtained is very unlikely to be continuous at the interface with the product. More likely, the actual seal follows an irregular line between the inside and outside. The resulting annular crevice will trap product.	Not recommended for use in hygienic plant pipelines and CIP installations, because the internal annular crevice may retain product during production and/or after CIP. It is widely used in situations where a gasket is unacceptable.
3-A coupling—gasket seat 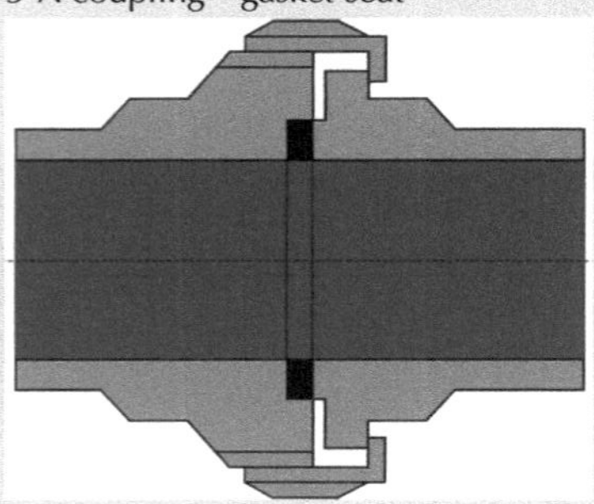	When correctly fitted and assembled a smooth, crevice-free internal surface is obtained.	Suitable for handling most products and for CIP.

Dairy coupling DIN 11851—standard gasket 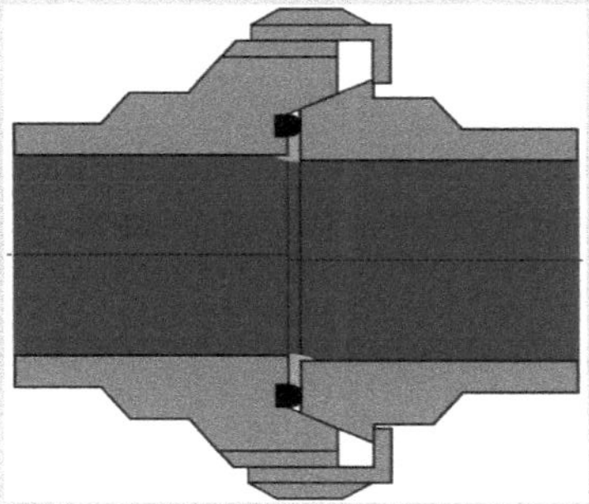	There is an internal annular crevice between the ends of the coupling parts and the bore of the gasket. Product may be retained during production and/or after CIP. An additional potential problem with this design of fitting is that it has a clearance on the cone fitting, with the consequence being that the two pipes are not automatically aligned. This could give rise to a potential step in the pipe joint. The coupling does not comply with 3-A or EHEDG hygienic design criteria.	Often found in the food industry (pipes and tanks) due to the fact that it is reasonably priced. Not considered as suitable for CIP, which means that the fitting should only be used where the pipework is manually cleaned.
Dairy coupling DIN 11851—nonstandard collared gasket 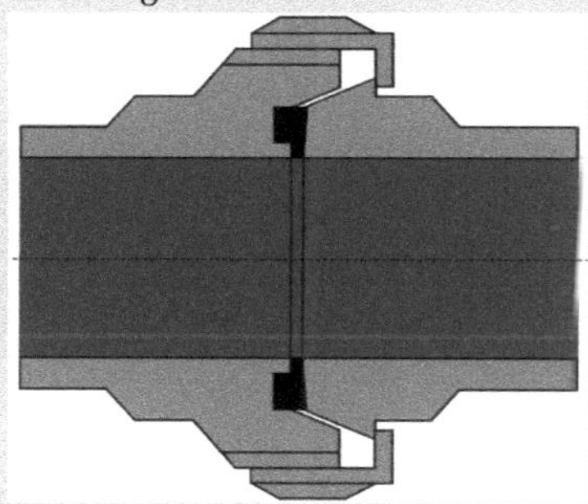	It provides a smooth, crevice-free internal surface when correctly fitted and assembled. However, because of the mobility of this type of coupling and the alternate expansion and contraction of the gasket, this gasket may be damaged by shear. Does not comply with 3-A or EHEDG hygienic design criteria.	Not recommended for use in hygienic plant process lines and CIP installations. Expensive and does not fulfill standard hygienic design criteria.

(Continued)

TABLE 7.2 (Continued)

Type	Hygiene Characteristics	Application
Dairy coupling DIN 11851—alternative gasket with SKS ring 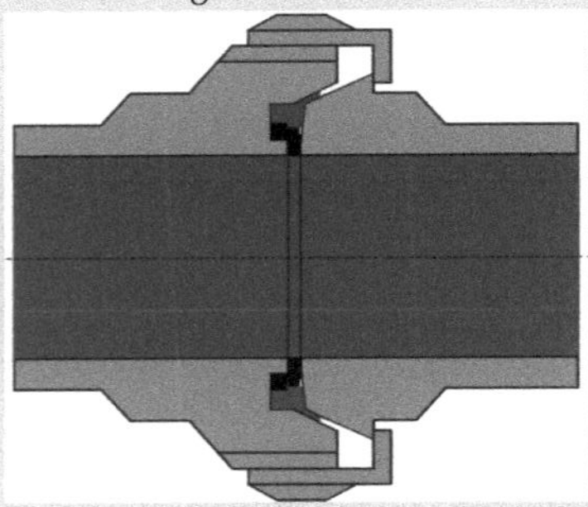	With support of the steel ring the gasket remains flush with the surface. The special designed gasket fills all dead areas in the coupling and will expand to the outside in cases of high temperature. At elevated temperatures, expansion of the seal to the inside is limited. This solution takes all critical points of a DIN 11851 coupling away. Complies with 3-A or EHEDG hygienic design criteria.	A stainless steel center ring and gasket are an easy solution to upgrade a DIN 11851 coupling to a hygienic status. The smooth surface gives excellent cleanability.
IDF coupling ISO 2853 with L-gasket 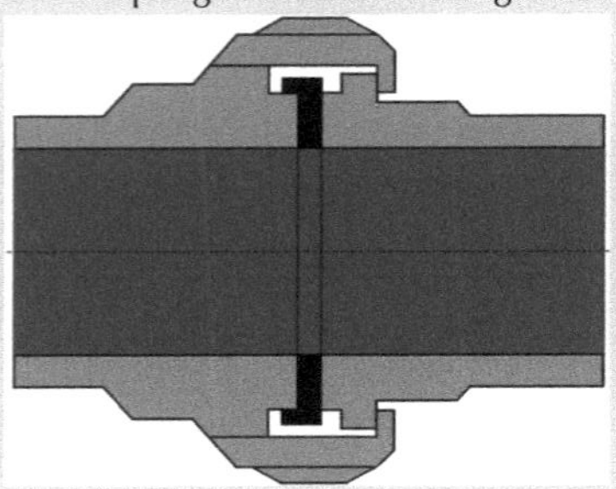	When the coupling is correctly fitted and assembled, a smooth continuous bore and internal surface without crevice is obtained, so that cleaning may be performed without any problems.	This coupling is recommended for applications where CIP is normally applied. Widely used for pasteurized circuits where dismantling is infrequent.

IDF coupling ISO 2853 with nonstandard T-shaped gasket 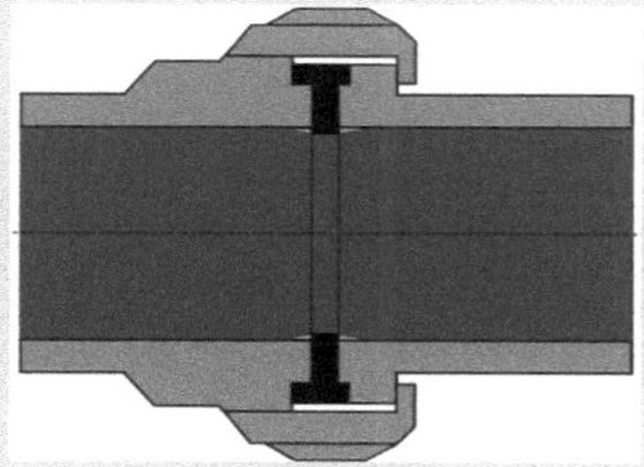	When properly made up, the joint is crevice-free and has a smooth bore, flush with the pipe walls. If overtightened, the gasket may expand into the bore of the pipe, which creates a step where product can become trapped. Unless the nut is tightened correctly, the coupling will not be bacteria-tight.	Most suitable for permanent or semipermanent installations that are going to be cleaned in-place. If the seal material is suitable, then it can be sterilized.
IDF coupling ISO 2853 with metal-backed T-shaped gasket 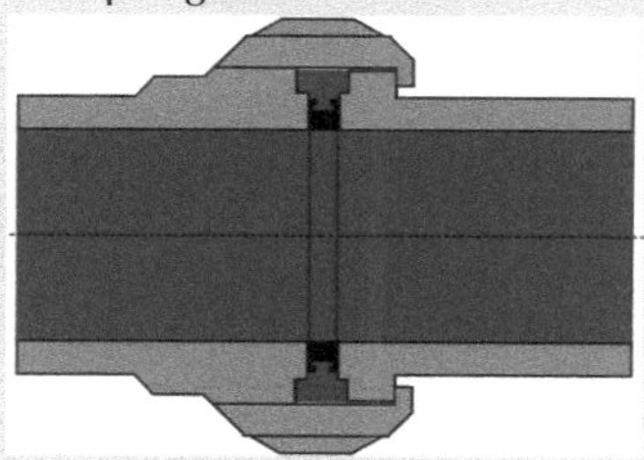	By supporting the seal with a stainless steel ring, both axial stop and centering can be achieved, allowing the connection to meet the requirements of hygienic design. The rubber is specifically shaped to give a flush interior joint when the union is tightened.	Most suitable for permanent or semipermanent installations that are going to be cleaned in-place. If the seal material is suitable, then it can be sterilized.
Recessed ring joint type (RJT) screwed coupling 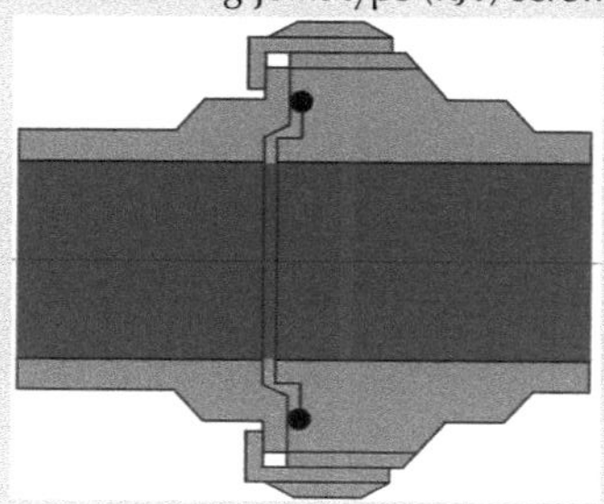	There is an internal annular crevice between the liner and the male part and the bore of the joint ring. Hence, product may be trapped and retained between the two metal components during production and could cause problems if certain products are handled. Does not comply with 3-A or EHEDG hygienic design criteria.	This type of coupling is recommended for use where piping systems are frequently dismantled, but is not suitable for CIP. It is used in the brewing and dairy industry in applications where pipework is manually cleaned. Excellent for flow-plates, owing to the wide dimensional tolerance on mating bends.

(*Continued*)

TABLE 7.2 (Continued)

Type	Hygiene Characteristics	Application
Coupling DIN 11864 form A 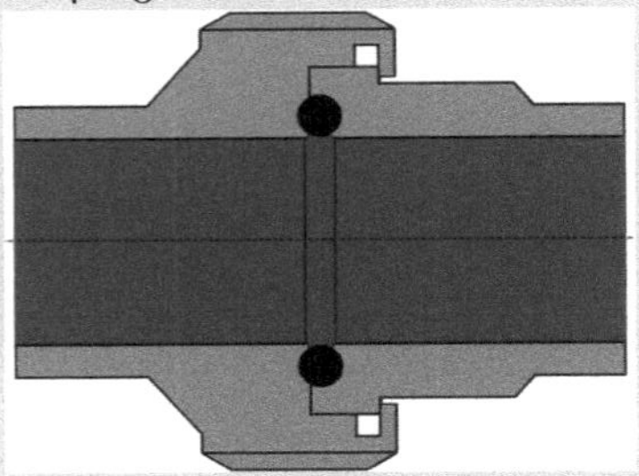	A smooth interface within the pipe work while simultaneously achieving a metal-to-metal seat behind the joint. A sufficient gap is created between the seal and the product space to facilitate rinsing in cleaning processes. This gap also serves as an expansion space that can accommodate volume expansions in the material as a result of heat or the influence of media without forces that can result in shearing. The groove is designed to minimize protrusion of the O-ring into the pipe bore. Complies with EHEDG and 3-A design criteria.	Optimal for aseptic operations because they are successfully tested for CIP-ability, steam sterilizability and bacteria tightness.
Coupling DIN 11864 form B 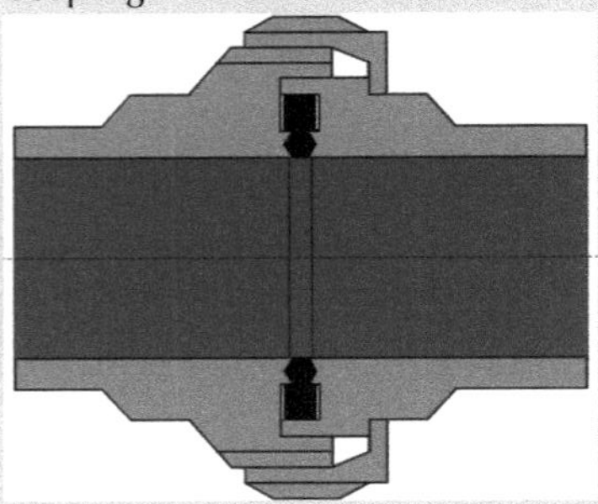	The volume of the functional part of the gasket (diamond section) is minimal to limit the effects of thermal expansion. A small area of the gasket is exposed to the product. The width of the exposed part of the gasket is only 1 mm. The block of elastomer behind the seal will accommodate the thermal expansion, relieve stress buildup on the sealing faces and limit expansion into the product stream to a minimum. The small functional part of the gasket may expand into two directions. To prevent air from being trapped between the gasket shoulder and the male part groove, small slits are provided on the outside, acting as vents.	Optimal for aseptic operations because they are successfully tested for CIP-ability, steam sterilizability and bacteria tightness.

Standard SMS 1145 coupling 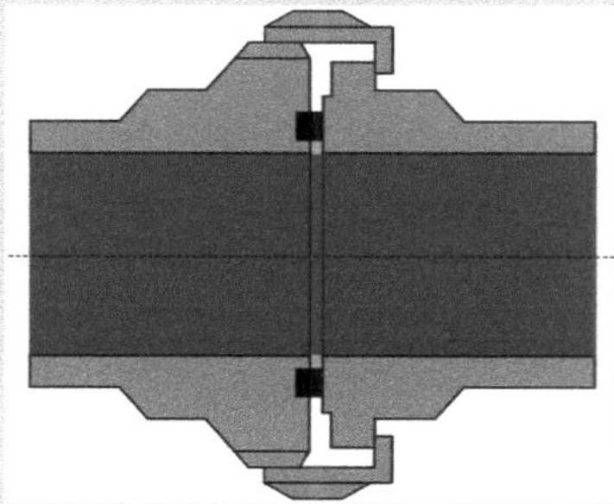 (DS coupling is similar to this coupling)	Standard SMS couplings are not hygienic because an internal annular crevice is formed in which product may be retained during production and/or after CIP. The bore of the gasket may retain product. L-profile gaskets are available but do not provide self-centering. A later version, when correctly fitted and assembled, provides a smooth, crevice-free internal surface. Does not comply with 3-A or EHEDG hygienic design criteria.	Only the latter version is suitable for handling viscous products and for in-place cleaning.
SMS 1145 coupling—Alternative gasket 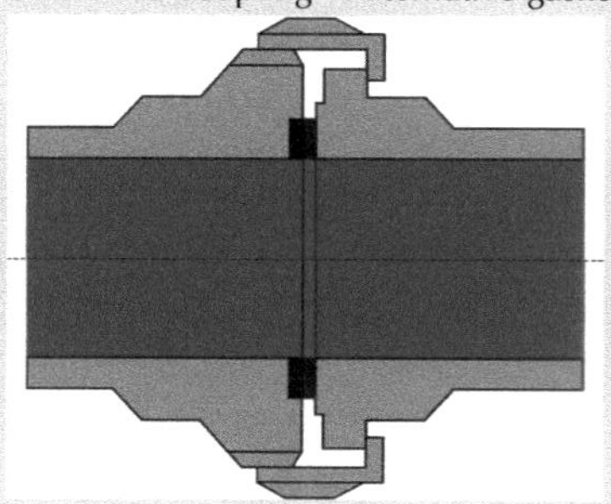	When correctly fitted and assembled, provides a smooth, crevice-free internal surface.	Suitable for handling viscous products and for in-place cleaning.

(Continued)

TABLE 7.2 (Continued)

Type	Hygiene Characteristics	Application
Clamp coupling ISO 2852 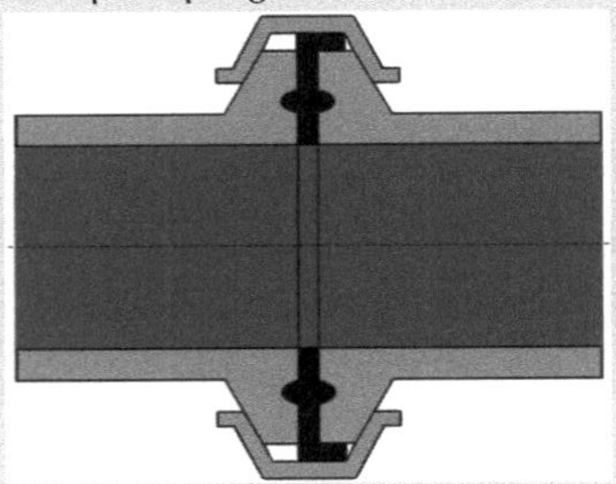	The seal is considered to form a smooth, crevice-free joint between the liners, which makes clamp type couplings suitable for CIP duties. Some users have indicated a preference for clamp fittings rather than screw-type couplings because in the event of a spill, screw threads cannot be decontaminated effectively. Clamp-type couplings are perceived to have the advantage that in the event of a product spillage at the fitting there is no thread to become filled with product that may be difficult to clean.	Often found in the food and pharmaceutical industry, in pipes and tanks. Not considered suitable for CIP.
VARIVENT flange coupling 	VARIVENT flange coupling ensures a smooth transition, free of dead space. It complies with EHEDG and 3-A design criteria.	Successfully tested for CIP-ability. Suitable for aseptic processes.
NEUMO BioConnect 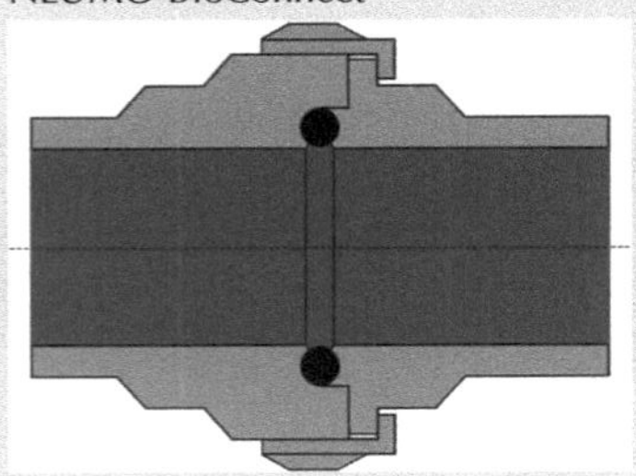	The seal is almost completely encapsulated. The highest press-on power is found at the transitions to wetted areas, preventing dirt and germs from penetrating into the sealing space behind the sealing element. Dead volume is minimized. Complies with EHEDG and 3-A design criteria. Successfully tested for CIP-ability.	Optimal for aseptic operations because they are successfully tested for CIP-ability, steam sterilizability and bacteria-tightness.

They are often covered by national, international, or internal company standards. Many of these couplings have been in use for some considerable time and are not considered to be compatible with current requirements in some areas of the food and drink industry. In most examples, a joint ring, usually made of rubber, is clamped between two stainless steel ferrules drawn together by a nut or split clamp. The ferrules must be of a similar grade to that of the tube and securely fitted to the pipe by welding. The inside diameter of the ferrules and tubes must be the same.

To make detachable joints, the use of conventional O-ring grooves is not recommended, because these groove designs leave a considerable free space in the groove. With O-rings, product and bacteria can accumulate in the crevice on the product side, up to the O-ring. If O-rings are to be used, they must be fitted as close to the product area as possible to minimize the depth of the crevice formed. So properly located gaskets are preferred, but then the seal material must be compatible with both the system product and the cleaning fluids, which may be at a much higher temperature.

When making bolted flange fittings, a lot of care should be taken to avoid offsets, gaps, penetrations, and voids. Careful tightening is necessary since, if overstressed, the seal ring may become extruded. Therefore, an axial stop for controlled compression of the seal is essential. However, if too little torque is applied, bacteria can gain access from the outside into the product area even though no leakage of product occurs from the inside to the outside. The seal also must be retained in the correct position at all times, flush with the internal bore of the pipe. Depressions and steps of more than 0.2 mm in the pipe work may prohibit cleaning fluids from thoroughly washing the surface and proper drainability of the piping will be hampered. Therefore, coaxial alignment of the two mating bores is of utmost importance, and there must be room for thermal expansion of the seal. Also avoid sharp edges so that seals are not damaged.

Table 7.2 Overview of market available pipe couplings and seal arrangements.

7.7.3 Piping Insulation

To insulate piping, styrofoam (maximum continuous service temperature is 60–65°C), foam glass or another rigid foam (e.g., foamed nitrile butadiene rubber) are preferred over fibrous materials. The problem with fiberglass batting is that this material has already proven to be an excellent harborage of dust, insects and rodents, and can be a clean-up and maintenance nightmare if not properly installed and maintained. Therefore, it is highly recommended to install vapor-tight aluminum or stainless steel cladding of appropriate thickness, which resists tear and abrasion (Fig. 7.36). The exterior of the insulation protection should be fully welded and smooth, with joints facing downwards. Proper sealing of the insulation cladding prevents the ingress of dust, insects, liquor, air, and moisture that could promote corrosion between the walls,

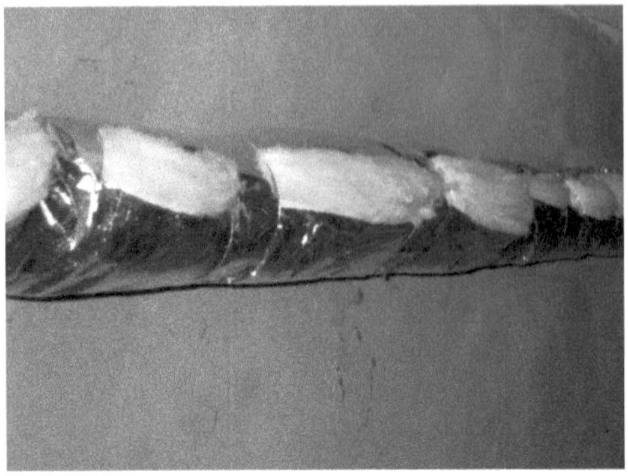

FIGURE 7.36 Fiberglass batting has been proven to be an excellent harborage of dust, insects and rodents (Moerman & Wouters, 2015).

FIGURE 7.37 The cladding which surrounded the insulation of the refrigerant supply and removal piping was probably lost during high-pressure cleaning operations. Now, worn off particles of insulation may be entrained by the airflow into the product (Moerman & Fikiin, 2016).

which could be assisted by possible microbial growth. Damaged or wet insulation should be repaired or immediately replaced (Fig. 7.37). Insulated lines should be kept away from food products, while pipes frequently soiled by food products or requiring periodic disassembly must be left uninsulated (Moerman and Kastelein, 2014).

7.7.4 Application of Hoses

The use of hoses is less recommended, because (1) failure of hoses can occur due to overstretching, kinking, rough handling, mechanical impact, aging, fatigue, abrasion, corrosive atmospheres, etc. and (2) the chance of leakage of

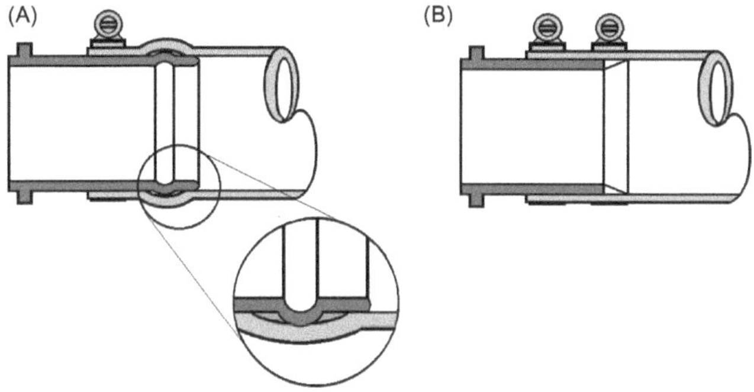

FIGURE 7.38 (A) Incorrect installation of hoses. (B) Correct installation of hoses on fixed pipes. Attached to stainless steel pipes, they should be clamped at the very end of the pipe. In this manner, the amount of dead space between the clamped portion and the end of the pipe is minimized. *Courtesy of Huub Lelieveld, personal communication.*

liquid occurring is much higher than when fixed piping is used. Attached to stainless steel pipes, they should be clamped at the very end of the pipe to minimize the amount of dead space between the clamped portion and the end of the pipe (Fig. 7.38). Braided (woven wire or fabric) covers on hoses should not be used.

Hoses need regular inspection for damage, deterioration and cleanliness, and should be cleaned and maintained in good condition. For ease of inspection, cleaning and maintenance, hoses should not exceed 3 m in length. Hoses out of service shall be pendant without touching the floor, and may never hang over open process equipment. When not in use, the ends of the hoses should be covered or capped to maintain proper hygienic conditions.

7.7.5 Hygienic Integration of Process and Utility Piping in Food Factories

Adequate pipework support must take into account provisions for thermal expansion during CIP operations. Welding of attachments on food processing support piping is not recommended, as they can cause stress on the pipe and the part of the supporting anchoring structure. All hangers and supports have to be designed in such a way that they either move together with the pipe (roll or slide) or that they can swing without exposing any stress either on the pipe or on the part of the supporting anchoring structure. Furthermore, process and utility piping should be grouped together in pipe trains whenever possible. All of this process and utility piping should preferably be positioned in such a way that all exterior surfaces are readily accessible, to allow cleaning from all sides (Fig. 7.39). The points of use should also be grouped, in an attempt to minimize individual ceiling drops. Vertical entrance of piping into the equipment is more hygienic than horizontal piping runs. Running of process and utility

FIGURE 7.39 For suitable cleaning, process equipment and piping must be installed at sufficient distance off the wall, ceiling, and floor. *Courtesy of CSI Central States Industrial, www.csi-designs.com.*

piping over open equipment in food preparation areas cannot be accepted, and nesting of ductwork should be avoided. Process equipment should be sited at a distance of ca. 4–10 cm from the nearest ceiling, wall and adjacent equipment so as to allow access for cleaning. Eventually greater gaps may be required. The distance depends on the diameter of the pipe. In addition, the equipment should be raised at least 20 cm off the ground to facilitate inspection, cleaning and maintenance.

7.8 HYGIENIC DESIGN OF PUMPS

7.8.1 Centrifugal Pumps Versus Positive Displacement Pumps

Many types of pump are used in the food industry for applications such as filling, emptying, transferring, and dosing. They are also often an integral part of a CIP system. The choice of pump for any given application depends primarily on the characteristics of the product and liquids to be pumped. Of particular importance are its rheological properties and its sensitivity to shear. Broadly speaking, there are two classes of pumps widely employed in the food industry:

- Centrifugal pumps used for transferring low viscosity liquids at relatively high flow rates but at comparatively low pressure heads. Liquid is directed into the eye of an impeller rotating at around 3000 rpm, which elevates both fluid pressure and velocity. On leaving the impeller, it is directed into a volute-shaped casing from which it is discharged tangentially.
- Positive displacement pumps (e.g., lobe pumps, gear pumps, vane impeller pumps, and progressive cavity pumps) capable of handling

high-viscosity and shear-sensitive liquids. Several positive displacement pumps can pump liquids containing suspended solids with minimal damage.

Design for cleanability is an important criterion in pump selection, but in many cases the functional requirements of the pump, such as delivery pressure and flowrate, determine the type of pump used, even if they are not the most hygienic option available. Positive-type pumps are usually considered to be less easily cleaned than the other due to the nature of their design.

7.8.2 Basic Hygienic Design Requirements

Pumps must meet the basic hygienic requirements:

- Materials of construction commonly used for product-contact surfaces should be stainless steels AISI 304L or 316L, as well as food grade plastics and elastomers. Materials of construction must comply with relevant legislation relating to materials in contact with food. Besides being compatible with the foods to be handled, they must withstand the corrosive effect of detergents and disinfectants. Materials of construction also must withstand the maximum envisaged process temperature, which may exceed 100°C if steam sterilization is employed.
- The internal surface finish of pumps does not have to meet specific standards, although a roughness similar to that recommended for pipes (less than 1 μm R_a) is commonly employed.
- Passage shapes must be smooth; sharp changes in cross-section should be avoided.
- Dead spots in which product can be retained or liquids held-up are not allowed. Even hygienically designed pumps can contain pockets if incorrectly reassembled after cleaning.
- Corners and edges must be rounded.
- Screw threads in contact with food are not allowed.
- Fastenings for pump bodies, clamp rings or bayonet couplings of smooth "clean" shape are preferred over bolts. No embossing and socket head screws should be used.
- Moving components should be fixed by flats rather than by keyways or splines.
- Bearings (sealed) must be located outside the product zone, although self-lubricating bearings are allowed in the product area for applications other than pumps.
- The pump casing must be drainable, either self-draining or capable of being operated so as to be self-emptying.
- Where pumps are cleaned by means of CIP, all parts must be intensively in contact with cleaning and disinfectant solutions.

- Where manual cleaning is used, pumps must be designed for quick dismantling with a minimum of tools and skills. There should be good access to all product contact surfaces to facilitate cleaning.
- Parts requiring periodic replacement should be easily replaceable and designed so that they cannot be wrongly assembled.
- Joints and seals should be so designed that a leakproof assembly is obtained. This is easiest to achieve with the correct components and without recourse to sealing compounds. Connections and seal areas must be crevice free.
- Wherever possible, shaft seals should be of the mechanical type and accessible for inspection, adjustment, and maintenance.
- Any leakage from the pump body must be easily visible.
- All external pump parts should be easily cleanable and capable of withstanding frequent hosing or similar cleaning-down procedures, as commonly used in the intended location.

7.8.3 Hygienic Design of Centrifugal Pumps

Typical types of impeller and shroud (Fig. 7.40) are: (1) unshrouded impellers (least efficient from a mechanical viewpoint), (2) partially shrouded impellers, (3) fully shrouded impellers (most efficient from a mechanical viewpoint), and (4) fully shrouded impellers with a detachable front shroud. The unshrouded impellers is easy to polish, whereas the shrouded impeller with its channel design cannot be polished economically. The shrouded impeller with detachable front shroud features metal-to-metal joints between the impeller body and front shroud, which are very difficult to clean in-place.

Mechanical seals are widely used to seal the shaft. A single seal is adequate for hygienic design purposes but a double seal is required for aseptic duties. In this case, the space between the seals is continuously flushed with either steam or antimicrobial fluid. Antimicrobial fluids should not be used where products are produced by means of fermentation processes. Shafts

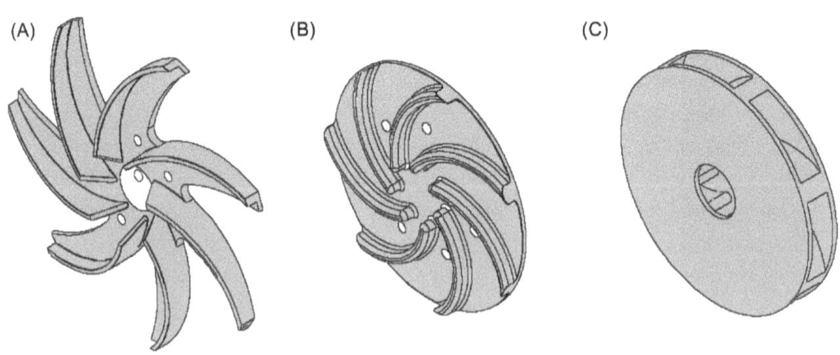

FIGURE 7.40 Types of impeller: (A) unshrouded, (B) partially shrouded, and (C) fully shrouded impeller (ASME BPE committee, © 2014).

must be so protected that food material cannot leak into the bearings or motor. Protection can be afforded by thrower rings and housing on the shaft between the motor and pump casing. However, if the shaft is enclosed at this point, inspection must be facilitated.

While it is often convenient for the arrangement of pipework to orientate the casing of a centrifugal pump so that the outlet port is pointing vertically up, this will result in the pump casing retaining liquid up to the level of the inlet port. The pump casing is drainable through the outlet port if the pump's outlet is arranged to point horizontally at the bottom (Fig. 7.41A), or the pump casing can be made drainable through its suction port if installed in vertical execution (Fig. 7.41B). Alternatively, a casing with a drain valve at the lowest point of the centrifugal chamber can be used (Fig. 7.42). The drain pipe section of the drain valve must be as short as possible, because the drain valve otherwise becomes a dead leg on its own.

Fig. 7.43A illustrates a number of design features that could create hygiene problems in a centrifugal pump, while suggested improvements are shown in Fig. 7.43B.

7.8.4 Hygienic Design of Rotary Lobe Pumps

Rotary lobe pumps having unhygienic design features (Fig. 7.44A) can only be cleaned effectively after dismantling. To avoid any introduction of contaminants into food products and to allow for CIP without dismantling, rotary lobe pumps should be hygienically designed (Fig. 7.44B and C). Metal-to-metal joints should be eliminated by hygienic O-ring assemblies. The O-ring groove design should be improved and O-rings should be positioned more appropriately. Alternatively, gaskets having controlled compression should be used. Sharp corners must be rounded to a minimum radius of 3 mm.

(A) (B)

FIGURE 7.41 (A) The pump casing is drainable through the outlet port if the pump's outlet is arranged to point horizontally at the bottom. However, one must keep in mind that with this manner of installation, air may become trapped in the upper part of the casing (courtesy of Unilever). (B) Now, the centrifugal pump is installed in a vertical position and is fully drainable through its suction port. *Courtesy of GEA Hilge.*

FIGURE 7.42 A casing with drain valve at the lowest point of the centrifugal chamber can be used. This solution where the pump's outlet is pointing vertically upwards while being self-draining prohibits the accumulation of air in the upper part of the casing. The drain pipe section of the drain valve must be as short as possible, because otherwise the drain valve becomes a dead leg on its own. The front cover can be removed to allow cleaning of the impeller area. *Courtesy of Packo.*

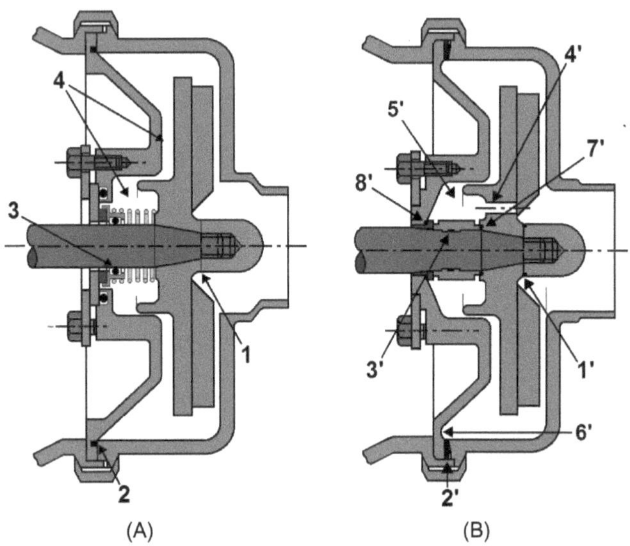

FIGURE 7.43 (A) Hygienically unsatisfactory centrifugal pump design with the following shortcomings: (1) metal-to-metal joints allowing the ingress and retention of product and micro-organisms, (2) retention of product residues in the crevices of the seal zone, even if CIP is employed, (3) retention of product residues in the annular space after cleaning, (4) dead volume unswept by fluid volume, resulting in a too-low turbulence to properly clean the surrounding surfaces in-place; (B) satisfactory hygienic centrifugal pump design: (1′) cap nut O-ring closing the metal-to-metal joint, (2′) translocation of the gasket away from the corner, (3′) slide ring (mechanical seal) encapsulates the spring, (4′) holes in the shroud increasing the flow of liquid behind the impeller, hence improving the cleanability of the surface, (5′) area behind the impeller containing fewer low-turbulent zones, (6′) radiusing of the corner, (7′) shaft O-ring and (8′) O-ring between the pump housing and the counter ring acting as slide ring holder. *Courtesy of Campden BRI.*

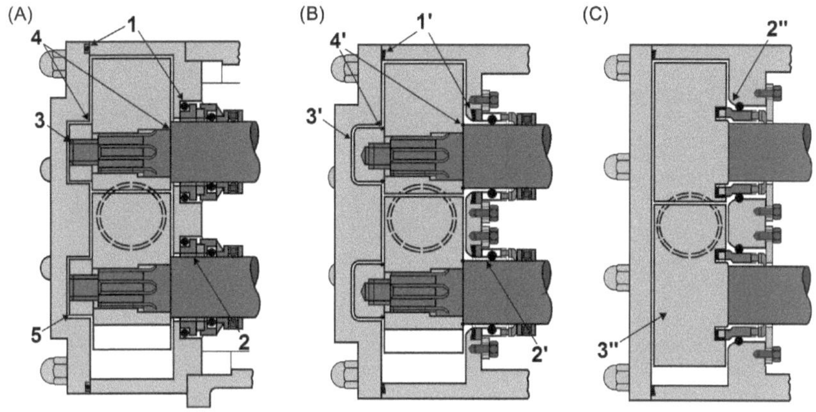

FIGURE 7.44 (A) Hygienically unsatisfactory lobe pump design with the following shortcomings: (1) metal-to-metal joints allowing ingress and retention of product residues and/or microorganisms, (2) long annular space within the mechanical seals which may retain food products and provide microorganisms a niche to grow, (3) retention of product residues in the threads of the rotor retaining nuts, (4) metal-to-metal joints and crevices accumulating dirt and microorganisms, (5) retention of product residues in the space around the rotor nuts within the end cover. (B) Improved hygienic lobe pump design: (1′) metal-to-metal joints in these areas have been eliminated by means of gaskets having controlled compression, (2′) the length of the annular space is reduced by changing the design of the mechanical seal, (3′) the exposed screw threads of the shaft are enclosed by the use of special nuts, (4′) metal-to-metal joints eliminated by the hygienic application of O-rings. (C) Further improvements to the hygienic design of lobe pumps: (2″) further reduction in the length of the annular space by reversing the elements of the mechanical seal and increasing the radial distance to give improved in-place cleanability; (3″) rotor retaining nuts and associated metal-to-metal joints are eliminated if the rotors and shafts are of integral construction (CFCRA, 1997). *Courtesy of Campden BRI.*

The length of the annular space within the mechanical seals should be reduced by changing the design of these mechanical seals (e.g., the elements of the mechanical seal should be reversed and the radial distance increased). Any exposed threads (e.g., threads of the rotor shafts, Fig. 7.45A) should be covered by crevice-free domed retainer nuts and have O-rings eliminating metal-to-metal joints. As a further improvement, the rotors and shafts should be designed as an integral construction so that rotor retaining nuts and associated metal-to-metal joints can be eliminated. As such, the inside of the front cover can be made completely flat and free of space holes for rotor retainers.

Some types of rotary lobe pumps are traditionally positioned in such a way that draining is impossible without dismantling. The inlet and outlet ports of these rotary lobe pumps are arranged in the horizontal position, as this has again been convenient for connecting the pipework. This results in the retention of liquid in the casing up to the level of the inlet and outlet ports. However, nowadays, there are hygienically well-designed rotary lobe pumps available with the ports arranged in the vertical plane (Figs. 7.45B and 7.46) so that it is possible to drain the casing.

FIGURE 7.45 (A) Ingress and retention of product residues and/or microorganisms in the threads of the rotor retaining nuts should be avoided by making use of crevice-free domed retainer nuts and by application of O-rings. (B) In a more improved version, the rotors and shaft should be designed as an integral construction. With the ports arranged in the vertical plane, it is possible to drain the lobe pump casing (Burggraaf and Partners B.V.).

FIGURE 7.46 Nowadays there are hygienically well-designed rotary lobe pumps available with the ports arranged in the vertical plane. Slots in the rotor retaining nuts create turbulences in the flow passing the narrow gap between end cover and lobes. They provide enhanced cleanabilty of the end cover and the front surfaces of the lobes. *Courtesy of Alfa Laval AB.*

7.9 HYGIENIC DESIGN OF VALVES

Valves are used to shut-off (isolate) processes or services, to prevent backflow (nonreturn), to change the direction of the flow of product or cleaning solutions (flow diversion in function of the product/cleaning routing selected), to regulate the flow and pressure, or to protect a process or service system against overpressure. Valves must be selected for the function of the chemical and physical characteristics of the product (does the product contain solids?), the envisaged application (aseptic or hygienic?) and the manner of operation (automated or remote?). Valves can be actuated manually, electromagnetically, pneumatically or with electric motors, and may have visual and/or electrical position indicators (limit switches). Any power-actuated valve is subject to the EU Machinery Directive 2006/42/EC & 98/37/EC and CE marking. However, some actuators (e.g., pneumatic actuators) are not compact, which could be a problem in confined spaces or where small valves are required. Compact nonpneumatic actuators are available which may possibly be employed in these confined spaces. Cost implications, the frequency of inspection and/or maintenance, as well as the ease of access are other decisive factors in the valve selection process.

The cleanabilty of a valve is largely determined by its internal geometry, the way in which the inlet and outlet connections are made, and the seal between the fluid and the external environment. The seals may be under a static or dynamic load with linear or rotary motion. Valves must meet the following hygienic requirements (Cocker, 2004; Moerman and Partington, 2014):

- Materials for valve bodies and seal materials must be selected as a function of the exact conditions of use, such as temperature and chemical exposure. Valve bodies are usually made of stainless steel AISI 316L (or AISI 304L where applicable), while other valve components could be designed in plastic (e.g., discs, balls, plugs, etc.) or elastomer (e.g., seals). Users must take into account that where high temperatures and/or halides are present, drying out can lead to very corrosive conditions even though normal operational conditions are much less severe. The cost of using higher-grade materials may have to be balanced against the cost of failure during production.
- Surface finish of valve components in direct contact with the product must be less than 1 μm R_a, not only to facilitate cleaning but also to impose minimal friction on the fluids passing the valve.
- Selection of a strong, stiff valve-body construction is a key defense against torsion, compression and pipe stress, for example from thermal expansion and contraction or from poor installation practices. Instances are known where the valves were distorted and there was an intermittent failure to close fully.
- Operating conditions must be taken into account, as excessive temperatures, pressures, and hydraulic shocks are common causes of hygienic failure.
- Marked changes in piping cross-section must be avoided, with the bore of the valve being similar to that of the adjacent pipes. With plug, stem, or diversion type valves, there must be sufficient turbulence to ensure adequate cleaning.

- Valves must have a minimum of dead volumes, and be free of pockets and crevices.
- Keep in mind that valves may give rise to dead ends (Fig. 7.47).
- Without dismantling, the valve body must be fully drainable in at least one installation position. Pooling liquid (Fig. 7.48) not only may give rise to microbial growth, the collection of condensate during steam sterilization will also result in cold spots, and hence an increased risk of microbial survival.
- Where valves are cleaned in-place, all parts must be intensively in contact with cleaning and disinfectant solutions.
- Where manual cleaning is used, valves must be designed for quick dismantling with a minimum of tools and skills. There should be good access to all product contact surfaces to facilitate cleaning.

FIGURE 7.47 Valves may give rise to dead legs (www.ourfood.com; Karl Heinz Wilm, © 2016).

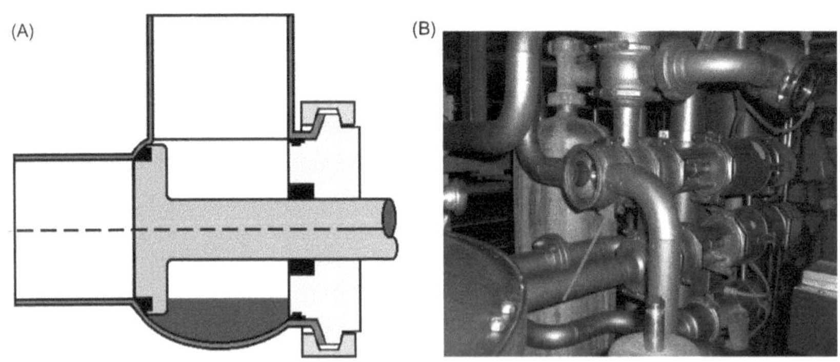

FIGURE 7.48 Nondrainable valve body (Matthias Schäfer, GEA group AG, © 2016).

- Valve seats should ensure an effective seal where full shut-off is required. Usually valve seats are sprung closed and opened by powered actuators, so that in the event of power failure the valve fails in the closed position.
- Don't use metal-to-metal seat seals for hygienic application, as they are not bacteria-tight. Although weak in hygienic design, they can be used where the product is resistant to spoilage.
- Valves also should be designed to resist wear. Failing valves may lead to loss of production (food out of specification), loss in production time, as well as maintenance costs. However, regular maintenance will always be needed since wear and distortion are inevitable with time, affecting the cleanability and hygiene. It is important in an installation with many valves that each valve can be overhauled reliably and quickly to avoid expensive downtime. Useful enhancements here are the provision of single-armature multiseat valve modules (you can replace several seat seals at once), single-fastener body clamps and fail-safe aids such as fixed compression stops on all seals.
- Minimize the number of seals, positively retained to avoid distortion and flush with adjacent surfaces.
- Seals must be regularly inspected as they may get distorted or even ruptured. Selected materials for seals must withstand temperatures and pressures likely to be met during processing and cleaning.
- In aseptic processing, dynamic seals on valve shafts must maintain aseptic conditions, requiring an absolute barrier between the product and the environment to prevent microbial recontamination. This can be achieved by using a bellows-type (Fig. 7.57C) or diaphragm-type seal (Fig. 7.57D) or a double seal arrangement (Fig. 7.57E).
- Allow rapid detection of internal leakage. Continuous leak detection for the seat seal is normally required for assurance of safe operation and should be an integral part of the design. As a minimum, this takes the form of an atmospheric leakage port, though some automatic detection and feedback systems are in evidence, especially for aseptic barrier modules, where the barrier zone has to be pressurized.
- The valve mechanism must be isolated from the product. There must be no possibility of external contamination of the product through the valve mechanism.
- Where unavoidable, springs in contact with product should have minimum surface contact area.

7.9.1 Diaphragm Valves

Diaphragm valves are suitable for aseptic operations. They use a flexible diaphragm clamped between the valve bonnet and valve body. The diaphragm may be of a variety of polymeric materials to suit the product and operating temperature. For closing the valve, the diaphragm is pressed by means of an external closure element (also called compressor) against the weir (Fig. 7.49A) located between the inlet and outlet port. Diaphragm valves without weir (Fig. 7.50) are also available from some manufacturers, but then the distortion of the diaphragm is much greater than for the weir type. Due to the short diaphragm movement, the lifetime of the diaphragm in a weir-type diaphragm valve is longer than in a weirless diaphragm valve. Because the diaphragm may rupture after a period of service in either type, diaphragm valves need visual detection of leakage (usually there are leakage holes in the valve bonnet). Damage to the diaphragm can result in product leaking through into the nonproduct side, which may give rise to contamination and also makes cleaning and disinfection nearly impossible. To avoid premature rupture, the diaphragm should be replaced at regular intervals depending upon the operating conditions. Maintenance, however, can be performed very quickly. Weir-type diaphragm valves with the actuating spindle vertical upwards in the horizontal position are not self-draining.

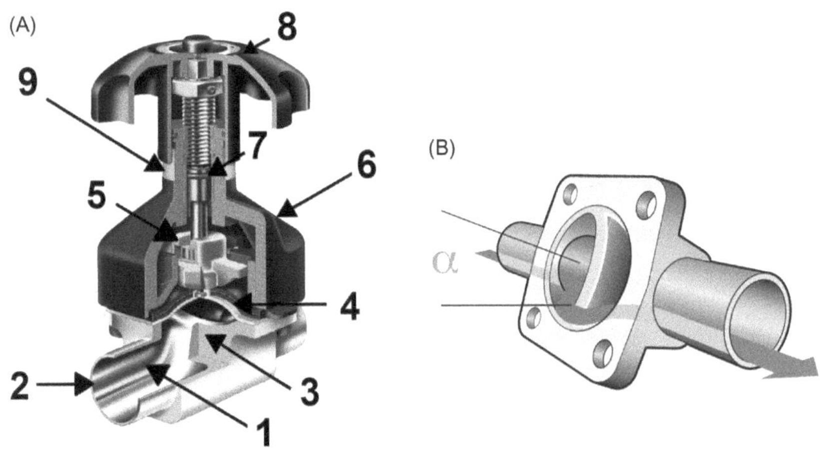

FIGURE 7.49 (A) Weir-type diaphragm valve with the following characteristics: (1) stainless steel 316 L valve body without cavities where soil could become entrapped, (2) butt well ends having sufficient turnback for orbital weld installation, (3) weir, (4) food grade rubber-backed PTFE diaphragm, (5) compressor supporting the diaphragm in both the open and closed positions for extended life, (6) bonnet made of temperature and chemical resistant plastic (e.g., polyethersulfone) covering the body fasteners to provide a clean exterior profile with minimum sensitivity to contamination, (7) O-ring seal preventing ingress of water and dirt from the outside environment into the bonnet, (8) contoured handwheel for optimal external washdown and cleanability, (9) valve position indicator. (B) In vertical piping, this type of valve is normally free draining, but when the pipe axis is approximately horizontal, the valve stem has to be inclined from the vertical to allow for drainage past the weir. An angle (α) less than 20 degrees above the horizontal is required, The angle may differ according to the valve size. *Courtesy of Crane Process Flow Technologies Ltd.*

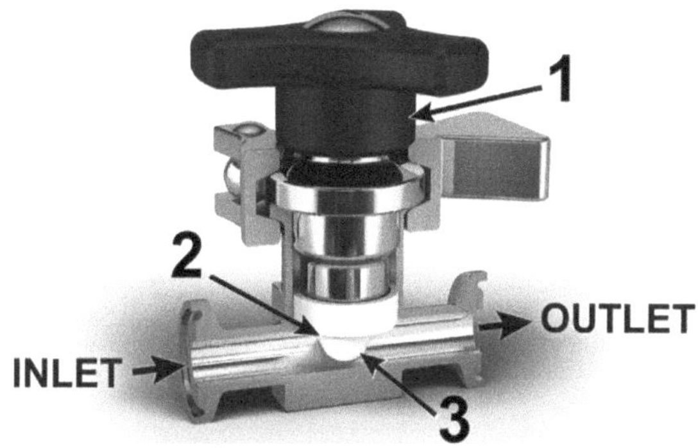

FIGURE 7.50 Weirless straight-through diaphragm valve: (1) manual actuator, (2) shoulder seal (static seal), and (3) shut-off seal (dynamic seal). *Courtesy of ASEPCO Corporation.*

For satisfactory drainability, the horizontally positioned diaphragm valve must be rotated about the port axes so that the actuating spindle is no more than 20 degrees above the horizontal (usually the drain angle is about 2–3 degrees) (Fig. 7.49B). Full self-drainability is always guaranteed with diaphragm valves installed in a vertical position. In addition to the diaphragm valve for on/off applications, there is also a flow-diversion variant. The design of this diversion diaphragm valve minimizes any dead spaces (advantageous if the valve is used for sampling) and ensures good drainability. However also for this diversion diaphragm valve, replacement of the diaphragm at regular intervals is required. Diaphragm valves are suitable for both hygienic and aseptic applications (CFCRA, 1997; Schonrock, 2005; Moerman & Kastelein, 2014).

7.9.2 Back-pressure Valves

In certain process operations, such as the continuous mixing of aerated products, it is necessary to maintain a constant back pressure. This is achieved by means of a back-pressure valve, which may be of the membrane or diaphragm type (CFCRA, 1997):

- A diaphragm type back-pressure valve (Fig. 7.51) comprises a diaphragm separating the product from the preset air pressure. The diaphragm actuates the valve controlling the product flow (Fig. 7.52). As the product pressure increases to a level greater than the preset air pressure, the diaphragm lifts the valve allowing for increased product flow. This reduces the product pressure, so as to maintain it at a constant level upstream of the inlet port to the valve. Conversely, the valve closes if there is a reduction in product flow to maintain a constant pressure upstream. The outer parts of the conical area are difficult to clean, particularly if viscous products are handled.

FIGURE 7.51 Diaphragm type back-pressure valve. *Courtesy of Alfa Laval AB.*

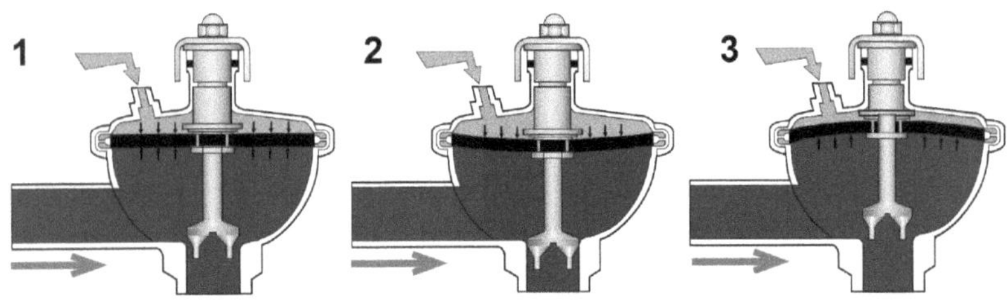

FIGURE 7.52 Diaphragm back-pressure valve in different situations: (1) product and air are in equilibrium; (2) the product pressure drops, the valve closes and the product pressure increases to the preset value; (3) the product pressure increases, the valve opens, and the product pressure drops to the preset value. *Courtesy of Tetra Pak Processing Systems AB; From Bylund, G., 2015. "Building-blocks of dairy processing", Ch. 6, section 6.8 – Piping, valves and fittings, Ch. 21. In: Teknotext, A.B. (Ed.), Dairy processing handbook. Tetra Pak Processing Systems A/B, Lund, Sweden.*

- Membrane type back-pressure valves (also called pinch valves) (Fig. 7.53) have a reinforced flexible elastomeric tube through which the product flows, further surrounded by an air chamber. Externally supplied compressed air may compress the tubular membrane to close it off. If the air pressure is constant, maintained by a preset regulator, the back pressure on the product upstream of the valve remains constant. Pinch valves are intrinsically hermetic and therefore suitable for both hygienic and aseptic applications. They can give full-bore flow and have the widest range of orientations for self-draining. They are not found in multiseat versions, and for hygienic applications currently are only available in small sizes.

7.9.3 Butterfly Valves

Butterfly valves (Fig. 7.54) comprise a plastic or stainless steel disc rotating through 90 degrees within the valve body, and further a rubber seal clamped

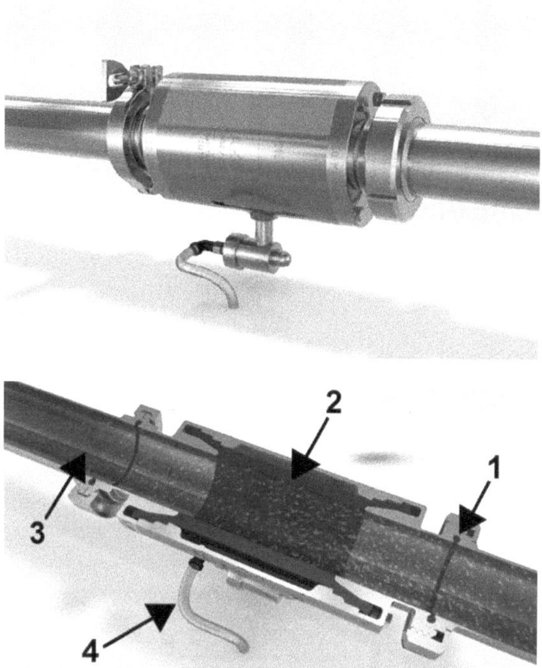

FIGURE 7.53 Membrane type back-pressure valves (also called pinch valves) can be installed in piping by means of flange, tri-clamp or weld-in connections, as well as couplings (1). The elastomeric tube (2) through which the product (3) flows is compressed by means of air at constant pressure (4) so as to keep the back pressure on the product upstream of the valve constant. *Courtesy of AKO Armaturen & Separationstechnik GmbH.*

between the halves of the body providing both a seat for the disc to close on and a seal for the disc spindles. Butterfly valves are suitable for on/off operation only, although they also may serve as throttle valves in vacuum systems. If properly designed, they are hygienic low-cost valves, with as main properties: low resistance to flow, appropriateness for automation and being cleaned in-place, as well as their very short length of pipe-run. Butterfly valves with a streamlined disk free of external ribs are hygienic. However, product containing fibrous material may build up on the leading edge of the disc and cause a cleaning problem. Moreover, the circular rubber seal can wear (Fig. 7.55) and break down after a period of time due to the frequent opening and closing of the butterfly valve, as well as cleaning and disinfection solutions. With time, uncleanable cavities will appear. Another weak point is the rotating shaft which passes in the valve body. Due to product pressures in the system, product can migrate along the shafts. Therefore, butterfly valves should preferably be disassembled for manual cleaning.

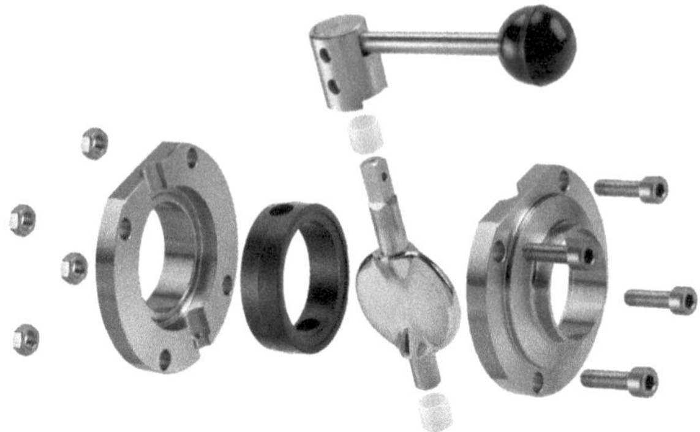

FIGURE 7.54 Butterfly valve having a highly polished stainless steel disc with a surface roughness $R_a \leq 0.8\ \mu m$. The bearing bushes are clipped onto the disc stems, avoiding any metal-to-metal abrasion and ensuring smoother disc movement. The rubber seal clamped between the halves of the body provides both a seat for the disc to close on and a seal for the disc spindles. However, as they are not hermetically sealed, they are not suitable for aseptic duties. *Courtesy of Alfa Laval AB.*

FIGURE 7.55 (A) Swelling of the seal, with tear, in conjunction with imprecise closing of the butterfly valve. (B) Seal abrasion at the disk (Wiedenmann, 2013). *Courtesy of Krones AG.*

If butterfly valves are in use, appropriate cleaning and maintenance schedules must be implemented (CFCRA, 1997; Schonrock, 2005; Moerman and Kastelein, 2014).

7.9.4 Ball Valves

Ball valves provide smooth, uninterrupted flow passage with minimal turbulences and pressure drop, which can be an advantage if the liquids contain large or delicate particulate material. However, they are not appropriate to control flow or pressure, and are suitable for on/off operation only. Traditional ball valves are considered to be unsuitable for process installations that are cleaned in-place. Due to the presence of inner-body cavities and crevices, the area between the ball, housing, and seal face is uncleanable. Food product is transferred into the annular dead space when the valve is operated from its open to its closed position. When the ball valve is then rotated back from its closed to its open position to allow CIP, the food product trapped in the annular space between the sphere and the housing will not be removed by CIP. As a result, the debris trapped in the inner-body cavity may start to ferment (putrefaction), cause damage to the seating surface and even block the valve operation. Finally, ordinary ball valves may also retain condensate in their internal cavities.

Often ball valves have special seat design incorporating cavity fillers (usually polytetrafluoroethylene) (Fig. 7.56) or encapsulating seals to prevent product flow around the exterior of the ball. However, product still may find its way under the seat surface and become an area for bacterial growth. Ball valves in existing installations must be disassembled completely for manual cleaning. But their design and construction do not always allow easily dismantling for cleaning. Certain ball valves with improved design allow for CIP, especially in a half-open position. However, this must be carefully checked. For some applications, steam-purged versions are available. Having connections to the housing, the annular space may be continuously purged with steam throughout production (CFCRA, 1997; Schonrock, 2005; Moerman and Kastelein, 2014).

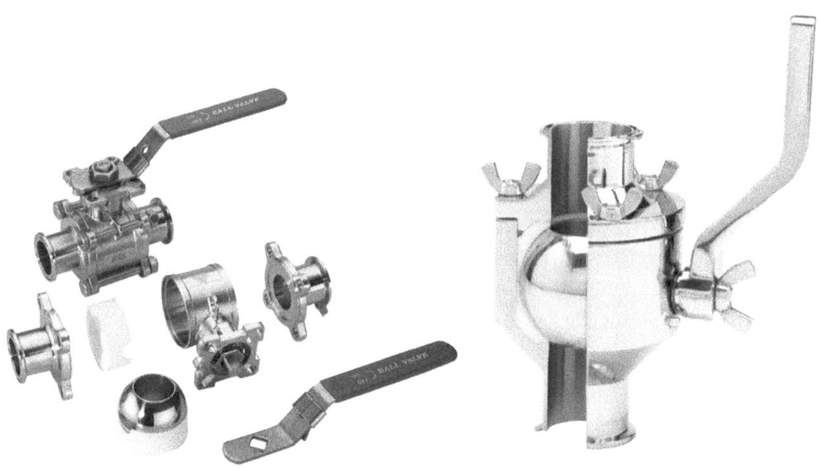

FIGURE 7.56 Cavity filled ball valves have special seat design that fills the gap (dead space) around the ball. Surfaces of the ball and valve body in contact with the product have a smooth finish with $R_a < 0.8$ μm. The ball valves are easy to dismantle allowing for inspection, cleaning, and maintenance. *Photo left, courtesy of Kaysuns Industry Ltd.; photo right, courtesy of Lee industries Inc.*

7.9.5 Linear Plug and Stem Valves

Linear plug and stem valves are actuated either manually or automatically, with typical applications being on/off and flow diversion operations. Linear plug and stem valves look quite similar to globe valves with a reciprocating shaft moving the valve head in open or closed position. This valve head is fitted with either a rubber or polytetrafluoroethylene (PTFE) seal. Various types of stem seals are available: lip, O-ring, diaphragm, bellows, and a double seal arrangement with steam barrier. Lip seal and O-ring arrangements are suitable for hygienic operation only. For aseptic processing applications where ingress of microorganisms must be prevented, the shaft must be sealed by means of a diaphragm and bellows (CFCRA, 1997; Schonrock, 2005; Moerman and Kastelein, 2014).

Characteristics of linear plug and stem valves:

- Linear plug and stem valves may incorporate a lip seal (Fig. 7.57A) to limit microbial contamination via the reciprocating shaft. However, although easily cleanable, the lip seal will not prevent the ingress of microorganisms because the stem passes from the atmosphere to the product side. Moreover, when the lip seal wear becomes excessive, product leakage will occur. A hole is required to detect product leakage.
- Arrangements incorporating an O-ring seal (Fig. 7.57B) are less hygienic because product can enter the clearance around the stem and become trapped in the O-ring groove. This debris cannot be removed by in-place cleaning.
- In the design of the bellows sealed linear plug and stem valve, stainless steel or PTFE bellows (Fig. 7.57C) are sealed to the collar of the valve body and the valve shaft. The bellows isolate the process interior from the outside mechanical part of the valve (mechanical activation elements located outside the process area can be lubricated without contamination of the process) and provide a bacteria-tight seal for aseptic processing applications. However, if the product contains large fibrous (e.g., rhubarb) or chunky products (e.g., nuts), these foodstuffs can become lodged in the creases. This particulate material trapped in the convolutions of the bellow is difficult to remove during cleaning operations, leading to long cleaning times. The waves of the bellows may not have an omega shape to avoid the formation of ring chambers, which are difficult to access for cleaning or sterilization. The circumferential grooves also may hamper the self-draining capability, especially when such a bellows is positioned vertically. Furthermore, the bellows are sensitive to pressure peaks, and dimples on the bellows edges due to inferior inflow from the side may lead to malfunction of the bellows. Finally, the bellows may rupture after a period of service (especially where cycle rates are high) (Fig. 7.58), requiring costly replacement at regular intervals (Ladenburger, 2015).

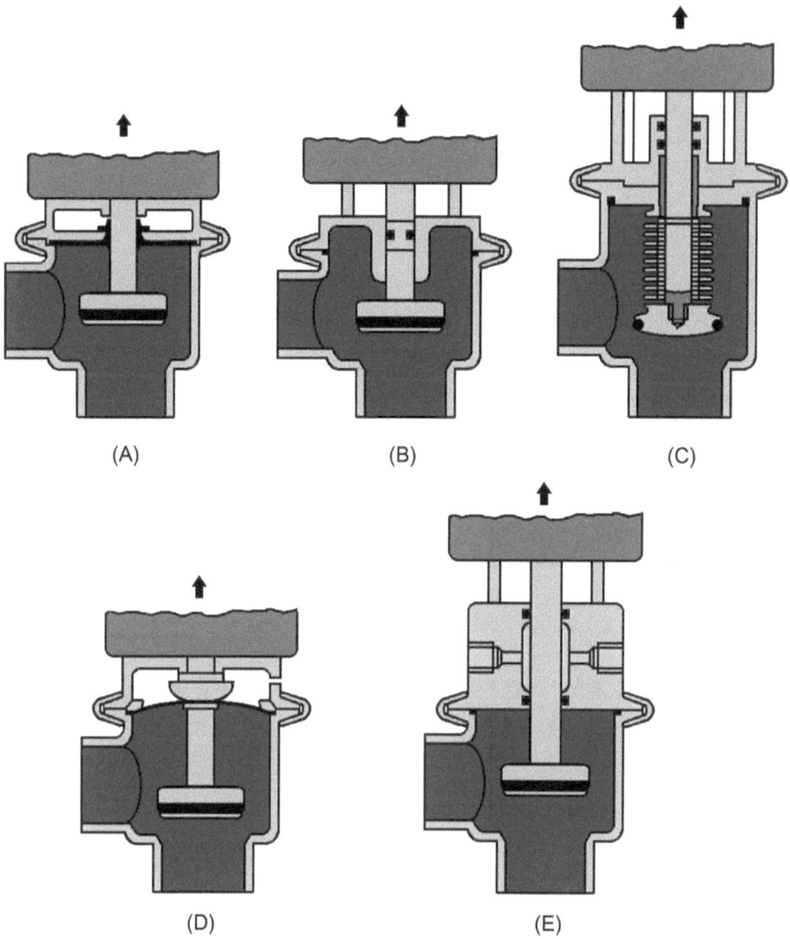

FIGURE 7.57 Linear plug and stem valves may incorporate: (A) lip seal, (B) O-ring seal, (C) stainless steel or PTFE bellows, (D) diaphragm, and (E) double seal arrangement, in between flushed with pressurized steam (CFCRA, 1997). *Courtesy of Campden BRI.*

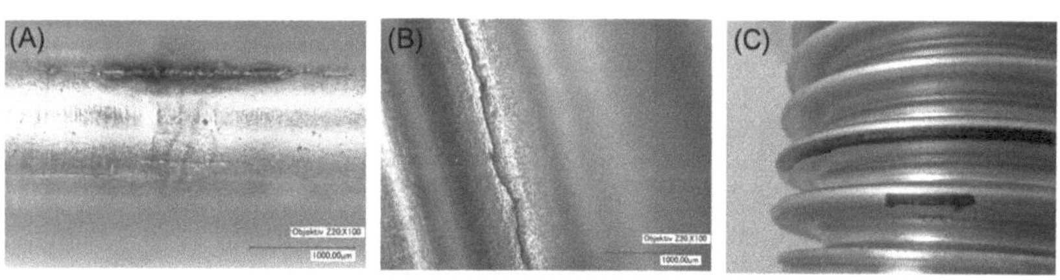

FIGURE 7.58 Metal bellows prone to (A) crack formation on the product-facing side (outside area) without complete breakage and recognition of a leak from the metal bellows. (B) and (C) Prematurely broken metal bellows (Ladenburger, 2015). *Courtesy of Pentair Südmo GmbH.*

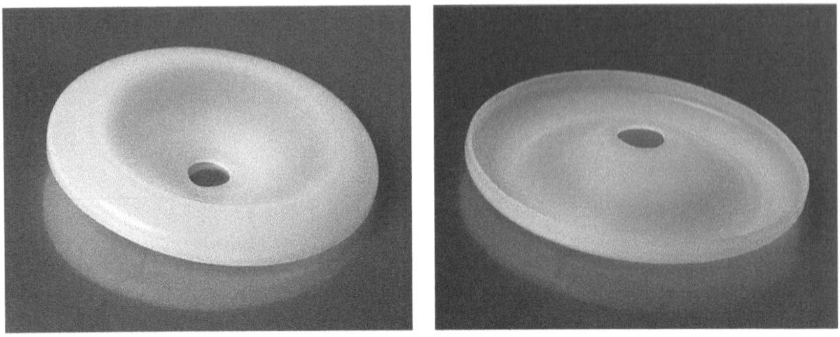

FIGURE 7.59 Diaphragm made from uniform and flexible PTFE with improved cold flow performance, and suitable for a high number of load changes. The risk of pocket or crack formation, typical for multicomponent systems, is absent (Ladenburger, 2015). *Courtesy of Pentair Südmo GmbH.*

- A diaphragm, usually of PTFE or PTFE-faced rubber (Fig. 7.57D), is clamped to the stem and in the housing to provide a bacteria-tight seal. A leakage hole must be provided to indicate failure of the diaphragm. The diaphragm must be replaced at regular intervals. For years efforts have been made to replace the bellows (described just above) with a diaphragm. These attempts, however, have failed due to the lack of an appropriate material. Nowadays, plastic materials with superior mechanical characteristics (Fig. 7.59) are available, allowing development of linear plug and stem valves with a diaphragm that provides hermetic protection of the spindle travel during aseptic applications (Fig. 7.60) (Ladenburger, 2015).
- Where the stroke of the shaft is too great to enable such a seal to be used, it is necessary to use a double seal arrangement (Fig. 7.57E). Such a seal has the following requirements: (1) the distance between the two seals must be greater than the distance moved by the shaft, in order to prevent microorganisms being brought into the aseptic zone during operation; (2) the space between the seals must be capable of being sterilized prior to production and to maintain asepsis during production. Hence, the space between the seals is usually flushed with pressurized steam, with the steam jacket having a length greater than the stroke of the valve. By using a steam barrier between the atmospheric and product sides of the valve stem, ingress of microorganisms is not allowed. However, the steam barrier is not widely used because of its high cost.

7.9.6 Mixproof Valve Systems

Mixproof valve systems (Fig. 7.61) are an essential part of automated processing. Double-seated mixproof valve systems have two independently operating shut-off valves, allowing separation of incompatible media such as liquid food product and cleaning/disinfectant solutions at flow path intersections

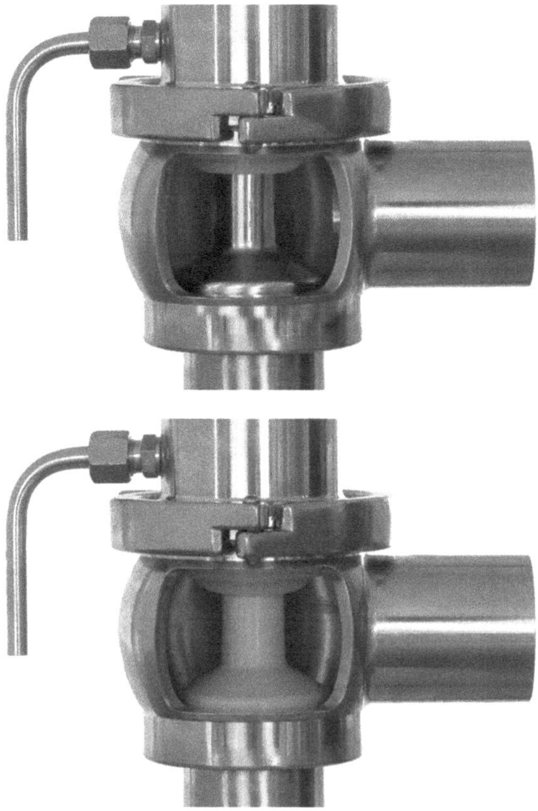

FIGURE 7.60 Linear plug and stem valves where a PTFE diaphragm with superior mechanical characteristics provides hermetic protection of the spindle travel during aseptic applications. Contrary to bellow sealed linear plug and stem valves, this valve is suitable for process operations handling liquid food with abrasive particles or products that crystallize in the atmosphere, such as lactose or instant coffee. *Courtesy of Pentair Südmo GmbH.*

within the pipe system. The operation principle of a mixproof valve is explained in Fig. 7.62. In the closed position of the mixproof valve system (nonactuated position), two seals are always located between the pipes, hence avoiding product contamination from cleaning fluids during CIP. Note that a single valve seat must never be relied upon to protect the product from the cleaning fluid, even if the product is maintained at a higher pressure (Moerman et al., 2014).

The valve heads of these two valves, held on their seat by spring pressure, are separated by a self-draining opening to the atmosphere. If one of these seals fails, leakage may drain via the thus-provided leakage outlet to the atmosphere (usually the drain pipe in the bottom shaft of the lower closure device) without intermixing with the product being in the second pipe. The vent space also must avoid a pressure buildup in case of a leak from a seal. The outlet from the vent line must be visible to detect any leakage.

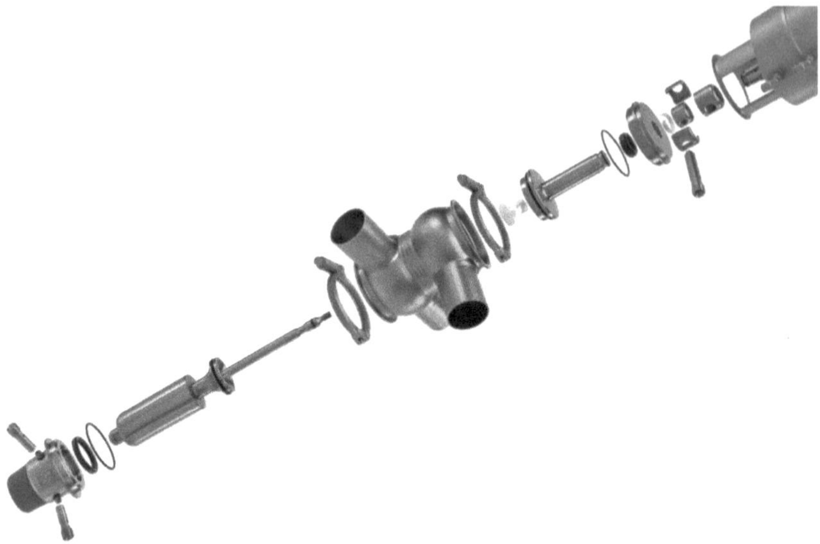

FIGURE 7.61 Mixproof valve system. *Courtesy of Alfa Laval AB.*

A steam or sterile barrier may also be applied in the atmospheric opening (vent) to prevent ingress of microorganisms (Moerman et al., 2014).

Also, double-seat mixproof valves need cleaning: both the upper and lower chamber of the valve housing soiled by the product being conducted through the pipeline, the seat area between the two chambers soiled when the valve is in the open position, and the cavity with the drain pipe in the bottom shaft due to operational leakage and leakage as the consequence of worn seat seals. The housing chambers can be cleaned in-place independently from each other, limited by the shaft seal on the one side and the seat seal on the other side. The seat seal and leakage chamber can be cleaned by seat lifting, that may occur periodically during each cleaning phase. The duration of the lifting pulses and intervals between them depend on the level of soiling, and are generally between 10 and 60 s in duration, with 3–5 min between the pulses. As an alternative to cleaning by means of seat lifting, cavity spray cleaning via an external CIP line connected to the leakage chamber can be done (Moerman et al., 2014).

7.9.7 Plug Cock Valves

A plug cock (Fig. 7.63) is a manual valve that has a conical plug rotated in the tapered bore of a valve body. The conical plug has either a straight-through port (on/off) or ports arranged in a tee configuration (three-way ports). Three-way plug cock valves allow 90-degree changes in flow direction of both food product and cleaning solutions. Although they have good internal flow design, they are relatively expensive and have the disadvantage that

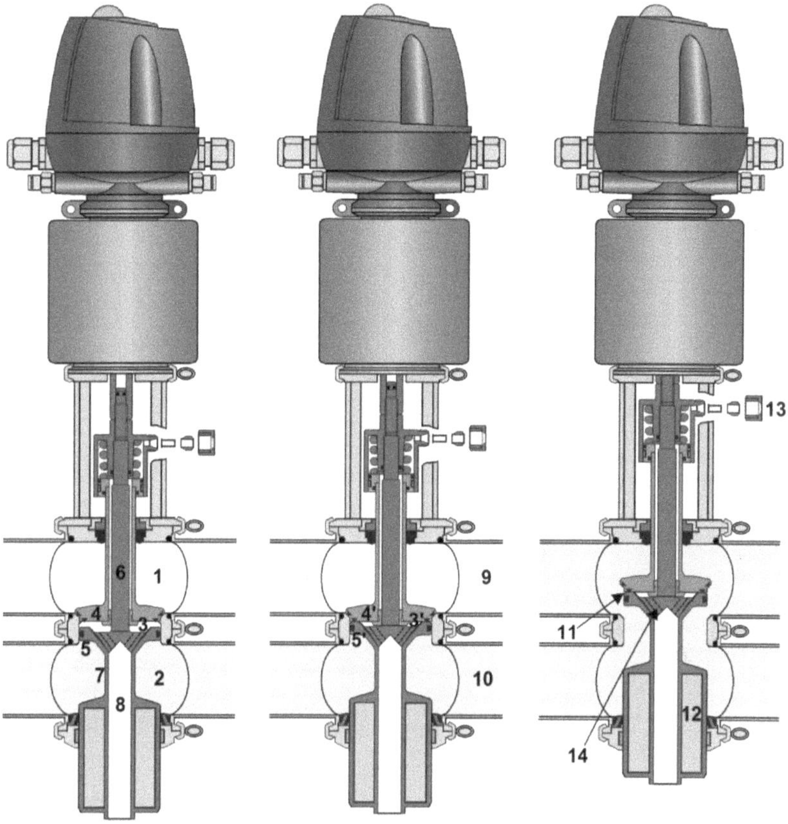

FIGURE 7.62 (A) A typical design of a double-seated mixproof valve consists of a valve housing with an upper valve chamber (1) and lower valve chamber (2). Between the two chambers the valve seat area is arranged with two seats, usually one on top of the other with a separation cavity (3) (vent space) in between. The seats consist of an upper closure device (4) and a lower closure device (5), typically a disc, which are connected independently to the upper shaft (6) and lower shaft (7) for opening, closing and individual seat lifting. The cavity acts as a leakage chamber (3) and is open to the atmosphere via a drain pipe in the bottom shaft (8) for leak detection (outlet of the vent pipe must be visible for leakage!!). In the closed position, the upper valve chamber (1) and the lower valve chamber (2) are each sealed by the two valve disks, held independently on their seat by spring pressure. (B) To open the connection between the upper pipeline (9) and lower pipeline (10), the actuated lower valve disk (5′) is raised off its seat first and then moves upwards a short distance before contacting the upper valve head (4′). As a consequence, the drainage chamber (3′) between the upper and lower body is gradually decreased. (C) Both valve disks then move further together into the open position. Meanwhile, in the more modern double-seated mixproof valves, the remaining cavity between the upper and lower valve disks remains sealed against the product area (11). It is important that the lower plug is hydraulically balanced (12, balancer) to prevent pressure shocks from opening the valve, hence allowing products to mix. When the valve closes, first the upper plug seals and then the lower plug seals. Both opening and closing of double-seated mixproof valves may show very small product losses getting into the cavity between the two valve discs during operation. However, this cavity can be flushed clean with cleaning fluid via a hose connection (13). The cleaning fluid will drain to the outside via the bores (14) and drain pipe (8) of the lower closure device. In aseptic applications, steam or a sterile barrier may be applied in the atmospheric opening (vent) to prevent ingress of microorganisms. *Courtesy of GEA group AG.*

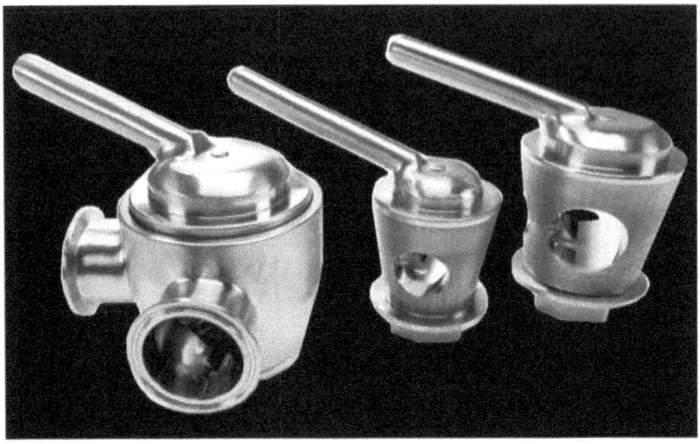

FIGURE 7.63 Plug cock valves are unsuitable to be cleaned in-place, because product can get caught around the clearance between the plug and the valve body during rotation of the plug in the turned-off position. But, as they are easy to dismantle, they can be cleaned manually. *Courtesy of Sanitary Solutions, Inc.*

FIGURE 7.64 With respect to plug cock valves, product can get caught as a thin stagnant film around the clearance between the plug and the valve body during rotation of the plug, and finally may leak to the outside. *Courtesy of Mondelēz International, © 2016.*

they neither can be automated nor cleaned in-place. They are unsuitable for CIP, because product can get caught as a thin stagnant film around the clearance between the plug and the valve body during rotation of the plug in the turned-off position (Fig. 7.64). Bacteria can also gain access via this route. However, due to their simple design, plug valves are easily to dismantle for manual cleaning. They are also self-draining (CFCRA, 1997; Schonrock, 2005; Moerman and Kastelein, 2014).

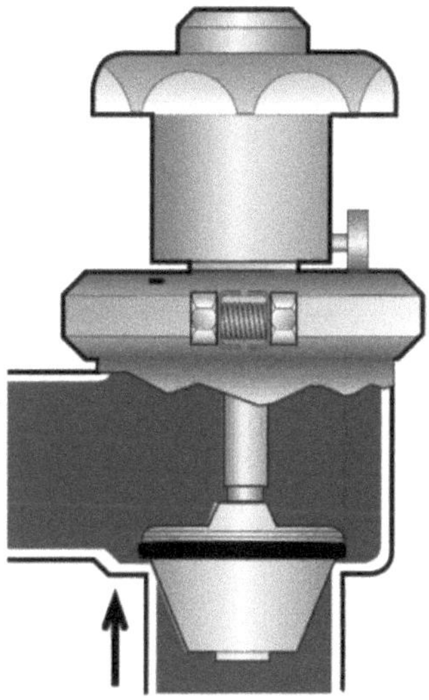

FIGURE 7.65 Manual control valve with variable-flow plug. *Courtesy of Tetra Pak Processing Systems AB; From Bylund, G., 2015. "Building-blocks of dairy processing", Ch. 6, section 6.8 – Piping, valves and fittings, Ch. 21. In: Teknotext, A.B. (Ed.), Dairy processing handbook. Tetra Pak Processing Systems A/B, Lund, Sweden.*

7.9.8 Flow Control Valves

Control valves with variable-flow plug (Fig. 7.65) are either operated manually or automatically and remotely with the aid of a pneumatic actuator. When the regulating handle is turned, the plug moves up or down, varying the passage and thereby the flow rate or the pressure. A scale on the valve indicates the setting. The plug-and-seat arrangement is similar in construction to an on/off linear plug and stem valve but has a tapered valve head. Its suitability for hygienic or aseptic applications is determined by the design of the valve stem seal (Bylund, 2015; CFCRA, 1997).

7.9.9 Nonreturn Valves

Nonreturn valves (also called check valves) are used to ensure that liquid flows in one direction only. When the flow is in the desired direction, the drag causes the valve head (disc), ball or shutter to move away from its seat. When the flow stops, the valve head or ball returns to the seat, thereby preventing flow in the reverse direction. Nonreturn valves must be installed in a position that allows full drainage (CFCRA, 1997; Schonrock, 2005; Moerman and Kastelein, 2014). Available types are:

- Nonreturn valves with a spring (Fig. 7.66) work on the principle that a light spring loading closes the valve head onto its seat once the flow stops. In this manner, flow in the reverse direction is prevented. As the

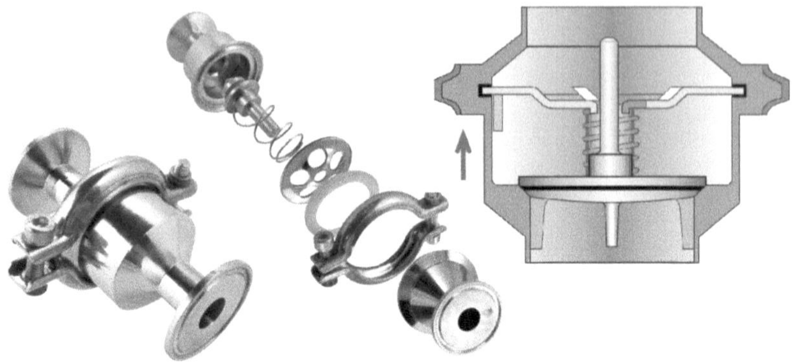

FIGURE 7.66 Stainless steel nonreturn valve with spring (tri-clamp type). *Figure right: Courtesy of Tetra Pak Processing Systems AB; From Bylund, G., 2015. "Building-blocks of dairy processing", Ch. 6, section 6.8 – Piping, valves and fittings, Ch. 21. In: Teknotext, A.B. (Ed.), Dairy processing handbook. Tetra Pak Processing Systems A/B, Lund, Sweden.*

disc is forced against the flow by the spring, resistance to the inflowing fluid and pressure drop is high. Such valves are unsuitable for use with viscous liquids and may be a cleaning nightmare. There is also no indication as to whether the nonreturn valve is working. When spring-loaded nonreturn valves are used, the coil spring(s) having product contact surfaces shall have at least 2 mm openings between coils, including the ends when the spring is in a free position. Spring-loaded nonreturn valves must be fully disassembled for manual cleaning (CFCRA, 1997).

- Nonreturn valves of the swing type (Fig. 7.67) use a hinged disc which swings open when the flow travels in the right direction. The disc closes toward the seat when the flow goes in the reverse and wrong direction. The spring-assisted closure tension holds the disc in-place. As the flappers, hinges and springs quickly become contaminated and could give rise to cleaning problems, nonreturn valves with hinged flapper should be avoided. In most cases, on a few exceptions, they are only applicable in horizontal pipe sections.
- Springless floating ball nonreturn valves are more preferable. When the fluid enters the inlet of the Y-type ball check valve (Fig. 7.68), an elastomeric ball is pushed upward into the "Y" branch of the valve. When the flow stops, the pressure within the valve equalizes, and the ball will return from the "Y" branch of the valve, and rest itself against the smaller diameter of the valve near its inlet. The opposing pressure of the reverse flow will seat the ball firmly against the inlet of the valve. Springless in-line floating ball check valves are mounted vertically, with simply the weight of the ball or poppet holding it against its seat. This ball-type check valve is hydraulically highly efficient, with flow passing straight

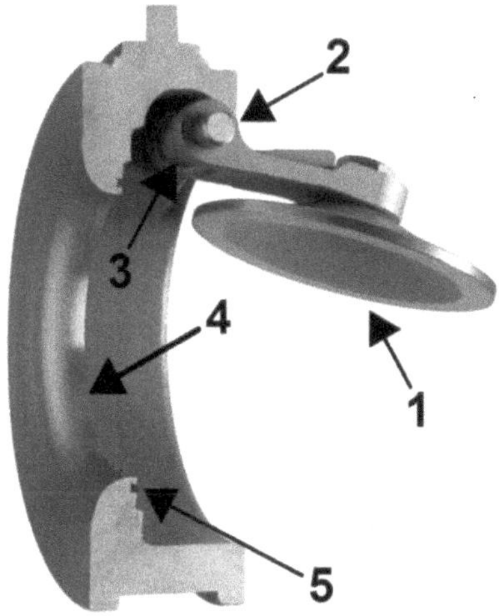

FIGURE 7.67 This nonreturn valve of the swing type consists of a disc (1) which opens in the right direction of flow by means of a hinge (2). When the flow goes in the reverse and wrong direction, the spring (3) forces the disc toward the seat (4). As an O-ring is contained in the groove of the body's seat, a uniform zero leakage seal is obtained. This design contains plenty of dead areas, and the hinge and spring are sensitive to contamination. Also the valve's drainability can be questioned.

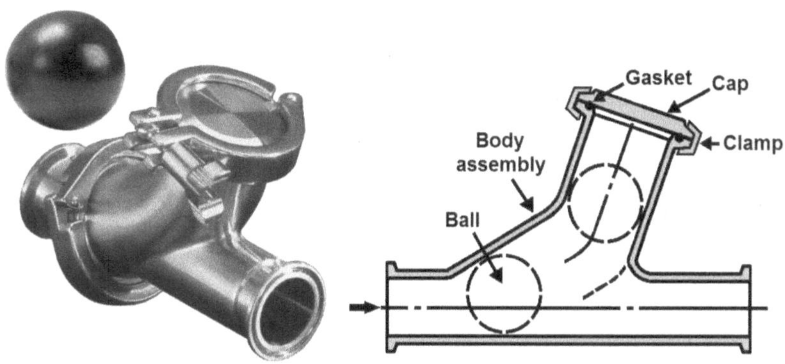

FIGURE 7.68 Y-type ball check valve. *Photo left, Dixon Valve & Coupling, © 2016.*

through vertically. Its streamlined internal design reduces the potential for material to clog or hang up.

- A magnetic nonreturn valves with floating ball (Fig. 7.69A) or shutter (Fig. 7.69B) are also available. Magnets built in the valve body keep the floating ball or shutter in a closed position. The nonreturn valve opens when the inflow pressure exceeds that of the combined pressure of the outflow and the magnetic field. As the ball or shutter moves away from the magnet, it is less attracted to the seat and therefore starts to provide lower resistance to flow. The valve closes when the difference in pressure

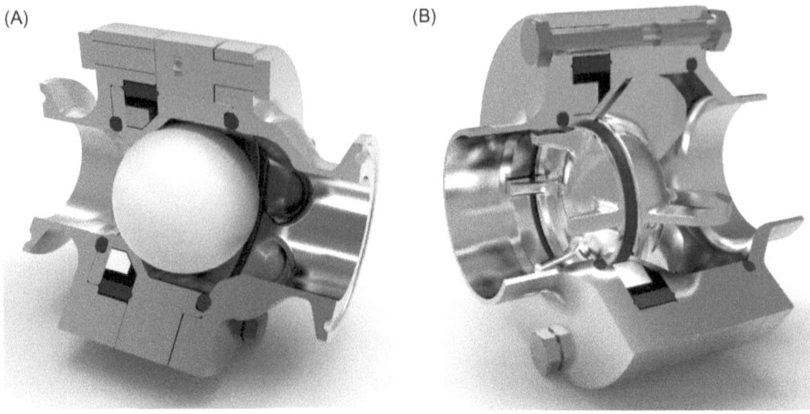

FIGURE 7.69 Springless magnetic nonreturn valves with butt-welded end or clamp connections, suitable for liquids, steam, and food gases. The risk of contamination is minimal due to the springless design and the lack of flow obstructing components. The check-valve can be installed horizontally and vertically (up and down), although full drainage of the valve in horizontal position may be compromised. The minimum opening pressure is ≤ 0.1 bar. (A) floating ball-type, (B) shutter type. *Courtesy of Carollo SRL, division Ygros valves.*

ceases or in cases of back pressure. The magnet will attract the ball or shutter back to its seat, to finally push it against the seat. Due to the design of the valve without flow-obstructing components, resistance to flow and pressure drop are low, as there are no springs, hinges, discs, or other components and there are no contamination or stagnation points. These check valves can be installed in horizontal as well as in vertical up-and-down positions.

7.9.10 Tank Outlet Valves

Tank outlet valves should be installed as close as possible to the product vessel to reduce the dead leg formed by the stub pipe that connects the bottom valve with the vessel. They may be manually or mechanically operated, and cleaned depending upon their design features. Two types of tank outlet valves are available (CFCRA, 1997; Schonrock, 2005; Moerman and Kastelein, 2014):

- Rising stem tank outlet valve (Fig. 7.70) is a modified version of the on/off linear plug and stem valve. It is usually fitted with a flange incorporating the valve seat, which is welded into the base of the tank. In the open position, the valve head projects into the vessel. This type of valve is widely employed for hygienic and, where necessary, aseptic applications.
- Falling stem tank outlet valve (Fig. 7.71) is similar to rising stem valves. However, in this design, the valve head drops down from its

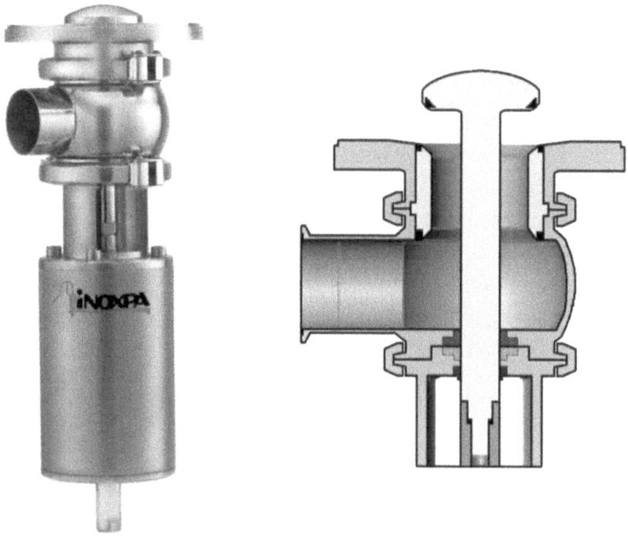

FIGURE 7.70 Rising stem tank outlet valve: the valve disk opens into the tank. This design avoids accidental openings in case of excessive pressure in the tank. *Courtesy of INOXPA S.A.*

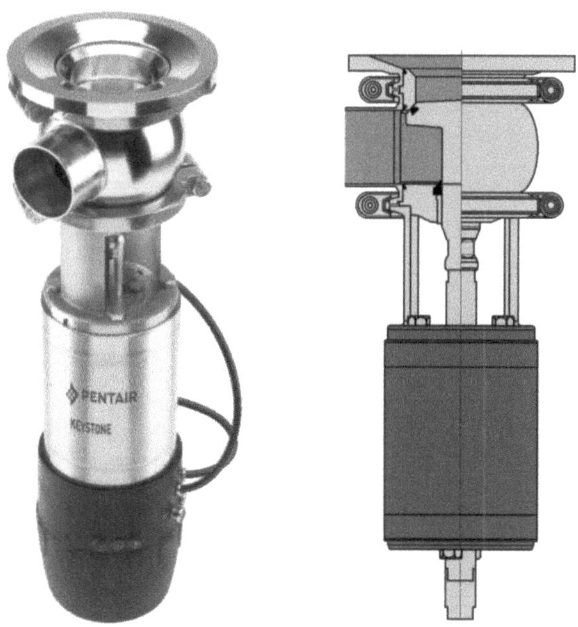

FIGURE 7.71 Falling stem tank outlet valve: the valve head drops down from its seat. *Courtesy of Pentair Südmo GmbH, division Keystone.*

seat. It may be used where the agitator blades are in close proximity to the bottom of the tank. Versions are available for both hygienic and aseptic applications.

At present, bottom outlet valves can be purchased with side ports to allow flushing of the body cavity. But according to the Pasteurized Milk Ordinance (FDA, 2007) the bottom outlet valve cavity is not allowed to be pressurized during cleaning when there is product in the tank. Experimental cleaning trials have shown that cleaning of the bottom outlet valve cavity, even with stationary spray devices, is very difficult. In a sophisticated and automated form, a retractable or permanently mounted cleaning device can be used. It allows cleaning of the inside cavity, stem, and plug of the bottom outlet valve without pressurizing the cavity. However, due to the shadowing effect caused by internals in the bottom valve cavity (e.g., valve stem), even that option is not completely successful. A patented solution with a spray device rotating around and fed by cleaning solution via a hollow stem (Fig. 7.72) can efficiently clean the bottom valve cavity without shadowing effects (Jensen et al., 2011).

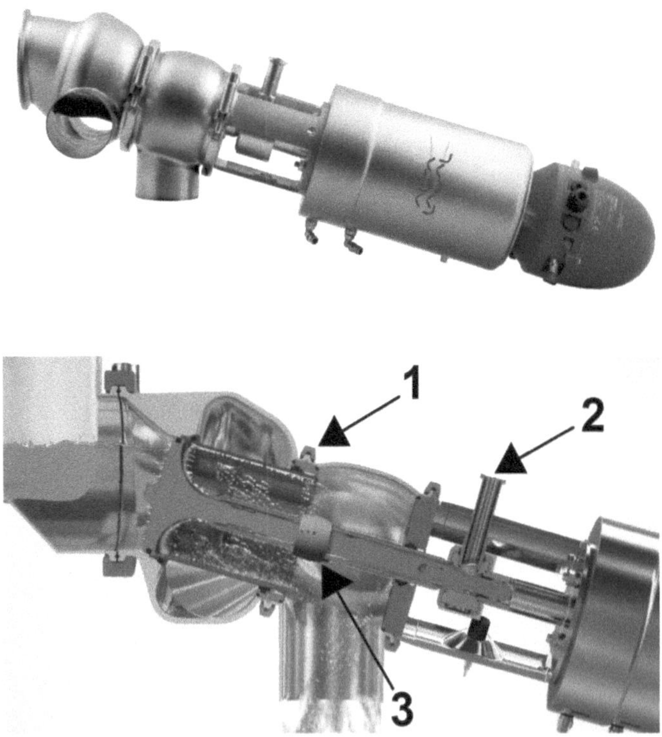

FIGURE 7.72 PMO mixproof valve for horizontal tank outlets: (1) sealing element, (2) liquid supply line for cleaning and disinfectant solutions, (3) cleaning nozzle (Jensen et al., 2011; Moerman and Kastelein, 2014). *Courtesy of Alfa Laval Tank Equipment A/S.*

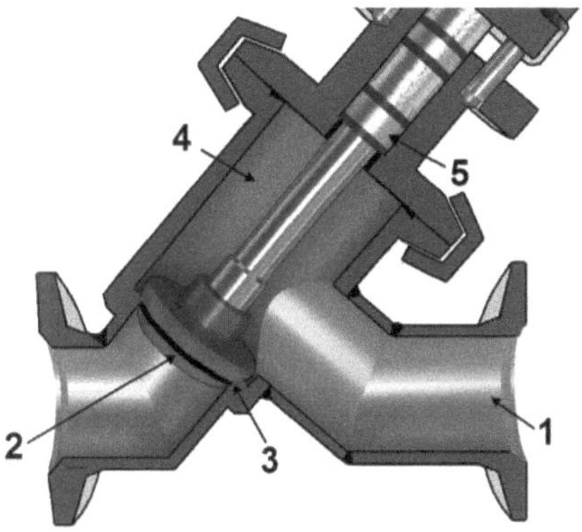

FIGURE 7.73 Y-type globe valve: (1) valve body, (2) profiled valve disc with plug seal, (3) seat designed for minimum pressure drop and dead zone, (4) valve disc retracts completely in the body, and (5) packed glands.

7.9.11 Globe Valves

Globe valves (Fig. 7.73) should not be used unless particular attention has been paid to hygienic design. They incorporate packed glands making cleaning extremely difficult and, furthermore, their flow characteristics are often unsatisfactory.

7.9.12 Membrane Sampling Valves

Sampling valves are used to take small samples of the fluid inside a tank or pipe. The sampling valve as shown in Fig. 7.74 is designed to be cleaned and sterilized during production, once the sample has been taken. The V-shaped valve has an outlet port for the sampled fluid and an inlet port for chemical cleaning (CIP) and/or steam cleaning (SIP). After flushing the fast draining valve chamber with cleaning solution, it can be rinsed and subsequently disinfected or steam sterilized if required. Via the outlet port, cleaning/disinfectant solutions and condensate are allowed to drain from the valve body.

In some occasions, the plunger is spring-loaded. On release of the plunger, the spring pressure then moves the plunger against the valve seat, hence closing the sample valve opening in front of the plunger.

7.9.13 Pressure Relief Valves

Pressure relief valves (Fig. 7.75) are fitted in food processing systems to ensure that unforeseen increases in pressure do not create a safety hazard or damage equipment. The valve head of the pressure relief valve is lifted off its seat when the product pressure exceeds that at which the valve has been set. Pressure relief valves are often variants of the on/off linear plug and stem valve fitted with a spring having a much lower rating, and which can be varied to suit the requirements. When the valve opens to relief

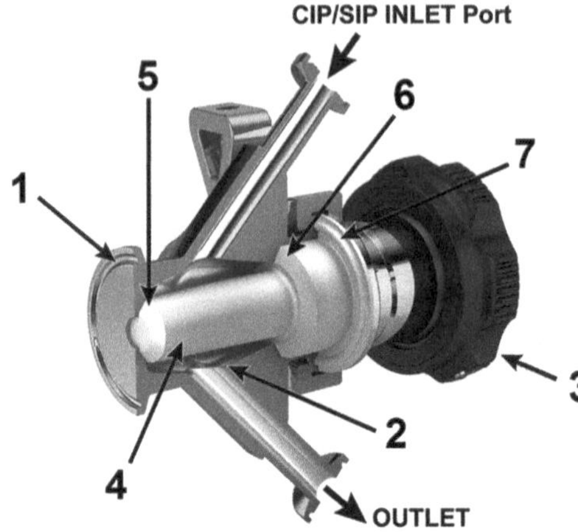

FIGURE 7.74 The sample valve is provided with (1) a ferrule for mounting to the outside of a tank or pipe by means of a hygienic clamp connection, although welding onto a short upstand is also possible. It is important that no dead leg is formed during mounting. The valve chamber (2) allows fast draining. Samples are taken by turning the handwheel (3) counterclockwise, allowing the plunger (4) enclosed by a rubber sleeve (diaphragm) (5) to be withdrawn from its seat. Once the sample is taken, the plunger can be moved forward on turning the handwheel clockwise, to finally make a seal with the valve body. In this closed position, the diaphragm sits flush against the inner tank wall or piping. *Courtesy of ASEPCO Corporation.*

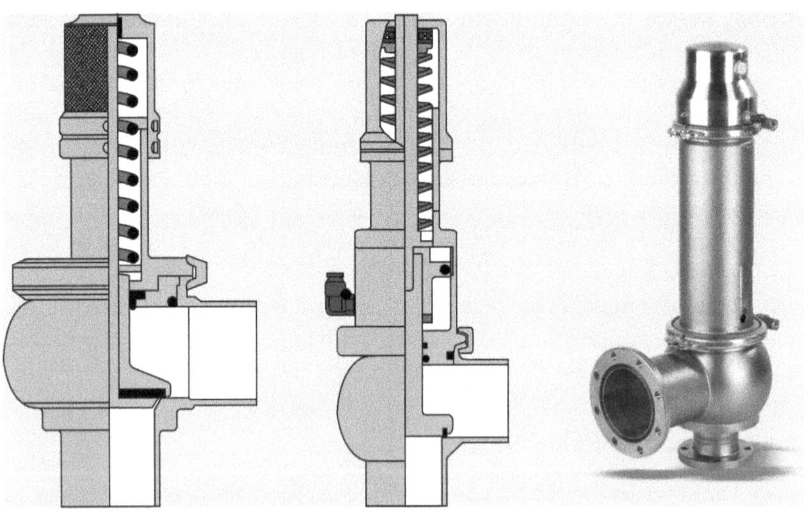

FIGURE 7.75 Hygienic pressure relief valve. *Photo right, courtesy of GEA group AG.*

pressure, product may be discharged to drain through the discharge port. To flush the inside of the valve body and the discharge port during CIP, it must be possible to lift the valve seat. Furthermore, the valve body must be installed in a position so that it is fully drainable to the outlet side, and should be mounted on a short tee to avoid a large dead leg in which product can be retained throughout the production. For tank applications, it is necessary to ensure that spray devices are so installed that cleaning fluid can access the valve (CFCRA, 1997; Schonrock, 2005; Moerman and Kastelein, 2014).

7.10 PRESSURE MEASUREMENT DEVICES

7.10.1 Selection of the Appropriate Pressure Gauge for Accurate Pressure Measurement

Pressure measurement devices serve to measure the pressure in the food factory's process and utility infrastructure, allowing detection of system malfunctioning, unsafe conditions and leaks.

7.10.1.1 Liquid-Filled Pressure Gauges

Open manometers measure pressure relative to the local barometric pressure, and therefore are not very accurate. On the other hand, the closed manometer is a simple and exact, low-cost gauge for measuring pressure independent of the atmospheric pressure. Unfortunately, its use in food plants is limited because of its size and being prone to breakage. If the gauge fluid is mercury (usually), it may contaminate the process, but the use of food grade oils instead of mercury may reduce that contamination risk. These oils also may increase the sensitivity of a closed manometer. When liquid-filled vacuum pressure gauges are used for pressure measurement in vacuum systems, low vapor pressure oils must be used. High vapor pressure oils may contaminate vacuum processes because the working liquid may evaporate if its vapor pressure exceeds the pressure within the vacuum system (Moerman, 2013a).

7.10.1.2 Mechanical Gauges

Mechanical pressure gauges measure the pressure directly by recording the force that liquids, air, or steam exert on the gauge surface in contact with these media (Moerman, 2013a).

- The Bourdon gauge is the oldest type of mechanical gauge. Its pressure reading depends on the external pressure and is only moderately accurate ($\pm 2\%$ of span). There are also Bourdon gauges that give their readings in absolute pressure. When the whole gauge is subject to mechanical vibration, the entire case (containing the pointer and indicator card) can be filled with an oil or glycerin to dampen the vibration of the pointer. Furthermore, it leaves no room for humid ambient air to enter. As a result, water cannot condense and accumulate within the gauge.

Bourdon gauges are less likely to contaminate the system than liquid-filled gauges.

- Capsule diaphragm pressure gauges use a hermetically sealed beryllium-copper capsule as basic component. This capsule, manufactured by fusion-welding of two diaphragms together at their peripheries, is evacuated to a pressure several orders of magnitude below the lower limit of the gauge pressure range. A capsule diaphragm pressure gauge indicates the pressure of a gas on a linear scale, independent of the external atmospheric pressure. Because capsule gauges are sensitive, they should not be used for pressure measurements in liquids.
- Differential diaphragm gauges use a flexible disc as diaphragm, which is made from a sheet of special steel or metal, either flat or with concentric corrugations. The diaphragm separates the interior of the differential diaphragm gauge - in which the lever system of the gauge head is located - from the process. The diaphragm deflection is again transmitted to a pointer. Due to its capacity to measure over a larger range of pressures, a differential diaphragm gauge is more suitable than a capsule diaphragm gauge, although the differential diaphragm gauge is more sensitive to vibration. When food residues accumulate in the mechanical diaphragm gauges—both the capsule and differential type—their accuracy may be compromised, while backflow of the food debris may cause the food to be at risk.

7.10.1.3 Electronic Gauges

In electronic gauges, the pressure signal is converted into an electrical signal that can be transmitted, recorded, or displayed. The most commonly used electronic gauges to monitor pressures in process and utility systems in the food industry are the (oil-filled) capacitance manometers (Moerman, 2013a).

- This is a pressure gauge in which the diaphragm makes up a part of a capacitor. A change in pressure leads to the flexure of the diaphragm, which results in a change in capacitance. Capacitance diaphragm gauges are rugged measuring devices, immune to contamination, because the gauge electronics never come in contact with the process. The only part in direct contact with the process is the diaphragm, which separates the reference cavity from the process or service system. The sensor body is usually fabricated from stainless steel or Inconel. The diaphragm that is exposed to the process pressure on one side and to the reference pressure on the other side is made from stainless steel, high-nickel steel alloys (e.g., Inconel and Hastelloy, for corrosive service), tantalum (for highly corrosive and high-temperature applications) or ceramic material with a vacuum-metalized coating. The use of ceramic diaphragm material can minimize the influence of temperature. Variable-capacitance sensing increases the accuracy, repeatability, and sensitivity of diaphragm gauges by several orders of magnitude, but changes in ambient temperature or process temperatures require either electronic compensation of the known temperature drift or sensor heaters to maintain the sensor at elevated temperature (typically 40–80°C).

- Liquid-filled capacitive transmitters make use of a diaphragm seal. That diaphragm seal protects the sensing element of the capacitance manometer by placing an isolating diaphragm between the gauge sensor and the process media that it is measuring. The cavity between the gauge-sensing diaphragm and isolating diaphragm is filled with a liquid. One side of the sensing diaphragm is evacuated and sealed to provide a reference for measuring absolute pressure. As the measured pressure changes, the outside isolating diaphragm (also called process diaphragm) deflects slightly, shifting the position of the fill fluid. The fill fluid transmits the deflection of the process diaphragm to the sensing diaphragm which causes a change in capacitance, which is translated to a stable dc voltage or current signal.

7.10.2 Hygienic Design of Pressure Gauges

The casing of pressure gauges (Fig. 7.76) should preferably be made of stainless steel AISI 304, 316(L) or food grade plastic, and should be watertight to protect the gauge against cleaning and disinfectant solutions. The outside, nonproduct contact area of the casing may have a roughness $R_a \geq 0.8\ \mu m$ if test results have shown that the required cleanability is achieved. Weep holes (Fig. 7.76) can be provided in the bottom of the case to ensure proper drainage of condensate, cleaning solutions, etc. The type of glass used as gauge window or gauge tube should be carefully evaluated with regard to crack risk and sensitivity to corrosion, as hydrolysis, especially at higher temperatures and pH values, may occur. Polycarbonate (PC), poly(methyl

FIGURE 7.76 Weep holes can be provided in the bottom of the vacuum gauge case to ensure proper drainage of condensate, cleaning solutions, etc.

methacrylate) (PMMA) and polysulfone (PES) are the most commonly used materials for gauge windows (Höhler et al., 2007; Cole-Parmer, 2009).

To avoid buildup of food pathogens and spoilage microorganisms and/or to prohibit the formation of biofilms, product contact surfaces (e.g., metallic diaphragm) should be free of microscopic faults and electropolished to a surface finish of 0.8 μm R_a or better. When in contact with process media, they need to be passive and free of pits and crevices, to reduce the likelihood of unwanted particles adhering to these surfaces. When a ceramic diaphragm separated from the metallic sensor body via an elastomer seal is used, there is an increased risk of pores. Therefore, this design must be considered carefully. Finally, all wetted surfaces should be fully drainable (Hauser et al., 2004a).

The housings of electronic pressure instrumentation (Fig. 7.77) (transducers, transmitters, switches, etc.) are made from stainless steel or plastic (either fully or partially). This plastic should be stress-crack resistant, resistant to cleaning agents and disinfectants, and possess high hydrolytic stability in the presence of hot water and steam. External housing and wiring may not collect dust or soil, and must be easily cleanable and self-drainable. The recommended ingress protection rating for electronic pressure instrumentation in the food industry should be IP65 or better. However, when exposed to CIP or SIP an ingress protection rating of IP67 or better is required (Höhler et al., 2007; Cole-Parmer, 2009).

7.10.3 Proper Installation of a Gauge

7.10.3.1 Installation for Visibility

When possible, pressure instrumentation should be installed in visible, readily accessible locations. Readouts should be located at eye elevation. Headroom should be provided for instrument removal, as well as any space for tools and test equipment that might be needed (Moerman, 2013b).

7.10.3.2 Measures to Eliminate Temperature Effects

Ambient temperatures above or under the specific acceptable temperature operating range of the gauge (due respectively to the vicinity of a source of heat or cold) must be avoided. The external housing must be checked for sufficient temperature decoupling from the process or utility system, as gauges may be affected by the temperature of the fluids flowing within the process or utility system. If the process/utility system or lines become too warm, the effects of changes in process fluid temperatures can be minimized by means of a cooling element (Fig. 7.78), a diaphragm seal with capillary tubing for remote mounting, a siphon, a loop seal or by purging. Loop seals and siphons are less hygienic solutions. Alternatively, the gauge housing can be

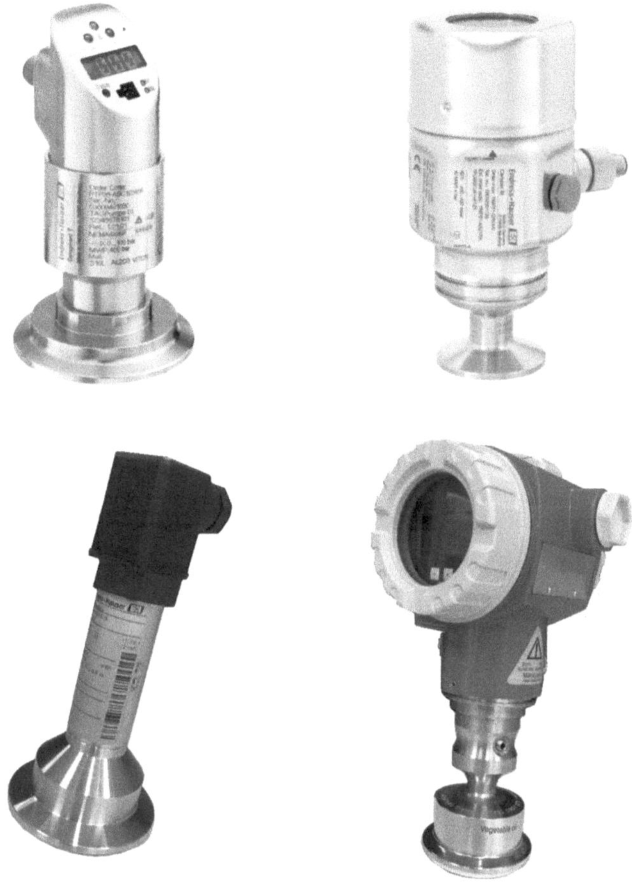

FIGURE 7.77 Hygienically designed electronic pressure measurement devices with stainless steel or plastic housing. The external housing and wiring may not collect dust or soil, requiring a design with smooth transitions and free from pockets. To be self-draining the instrument housing can be made with a sloped top surface, although it is possible to install pressure measurement devices with angles of more than 30 degrees from the vertical. *Courtesy of Endress + Hauser AG.*

cooled electrically (Peltier effect) or by means of water. If the process/utility system or lines have become too cold, sensor heaters may keep the sensor at elevated temperature (typically 40–80°C). If there is downward temperature drift due to the process/utility system being frozen, freeze protection may occur by means of steam tracing or resistance heating (electronic compensation) in combination with thermal insulation. If a diaphragm seal is used, the oil filling in the diaphragm seal housing not only may prevent humid ambient air to enter the gauge but it also may prevent a cold process/utility fluid from freezing any condensate within the gauge (Moerman, 2013b).

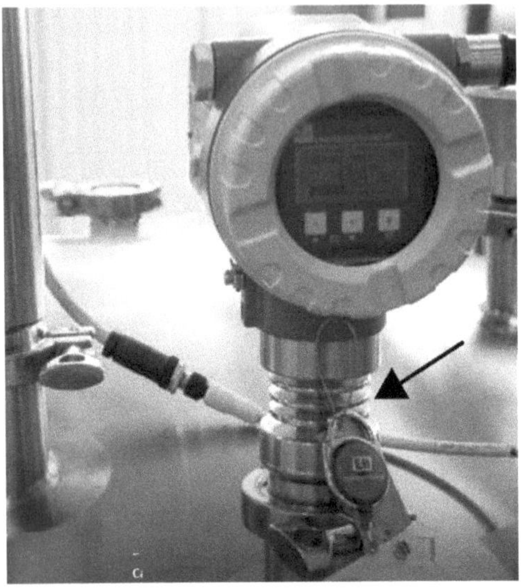

FIGURE 7.78 Changes in ambient temperature around the measurement system due to high process/utility temperatures can be minimized by means of a cooling element mounted between the gauge and the hot process/utility system. *Courtesy of Endress + Hauser AG.*

7.10.3.3 Measures to Eliminate Pressure Misreadings Due to Dirt

Pressure measurements may become falsified, especially for all types of gauges in which high-sensitivity and high-accuracy measurement systems are particularly susceptible to soiling. Besides incorrect pressure readings, unhygienic conditions may be created in the food contact area. It is nearly impossible to prevent the measurement system in a gauge from becoming soiled. Also, considerable quantities of continuously or intermittently liberated process gases or vapors may pass into the pressure measurement system. In this manner they may damage and destroy the pressure sensing mechanisms due to condensation, corrosion or deposition of contaminants. Especially in mechanical gauges, delicate links, pivots, and pinions are sensitive to condensation of water vapor. That water condensate, in colder climates, may even freeze and damage the gauge housing. Therefore, measures should be taken to ensure that the influence of contamination on pressure measurement remains as small as possible. Where possible, the gauge should be installed in such a way that product cannot come in contact with the gauge, especially because positioning a gauge at the same height as the process equipment may result in its contamination (Moerman, 2013b).

Sensors should only be installed inclined upwards (Fig. 7.79) with the flange at its bottom to keep condensate, debris, suspended solid particles, flakes, etc. from clogging the sensor port or from falling into the sensor and the measurement system. The user of the gauge can attempt to protect the measurement systems against contamination by providing suitable shielding but this solution—although clean—often leads to pressure readings deviating considerably from the pressure actually prevailing in the system. Cleaning the

FIGURE 7.79 An instrument branch facing downward instead of inclined upwards creates a difficult-to-clean dead area where condensate, debris, suspended solid particles, flakes, etc. may accumulate. Fluid velocities in the dead leg are much lower, and thus the area of entrapment may not be reached by cleaning or sterilizing procedures, hence leading to contamination of the product. *Courtesy of Joe Stout, Commercial Food Sanitation LLC - Intralox, © 2016.*

measurement system is another solution but certain gauges (e.g., liquid-filled gauges, mechanical gauges, etc.) are too fragile to be cleaned (Moerman, 2013b).

7.10.3.4 Measures to Eliminate Vibratory Effects

Install gauges at those points in the process/utility system that will remain free of vibration during operation, because liquid-filled and mechanical gauges are prone to failure by vibration. Glass tubes of the liquid-filled gauges may break, and the delicate links, pivots, and pinions of the mechanical gauges may become damaged. Pulsation dampeners can help to absorb pressure shocks and average out pressure fluctuations. To dampen pointer vibration and gauge lifetime, the pointer gauge housing is usually filled with a viscous oil (Section 7.10.1.2 on mechanical gauges) (Moerman, 2013b).

7.10.3.5 Measures to Eliminate the Effect of Magnetic Fields and Electrical Potentials

Measurement cables that are too long (connector cables between the sensor and the pressure gauge control unit) should be avoided because strong interfering magnetic leakage fields or electrical potentials can falsify pressure measurements. Wireless transfer of data between pressure measurement instrumentation and control equipment is highly recommended (Moerman, 2013b).

7.10.3.6 Installation for Hygiene

Most pressure instruments (gauge, switch, transducer, or transmitter) are designed without hygienic process connections. However, the connector threads of standard sensors provide a space where bacteria could grow. Installation of pressure gauges on piping and equipment processing sensitive food should occur in agreement with EHEDG guideline documents Nos. 10, 16 and 37. Typical hygienic connections free of crevices, threads, metal-to-metal contact, etc. are DIN 11864A & B, ISO 2852, NEUMO BioConnect, and VARIVENT. Flush-welded versions are also available (Cole-Parmer, 2009).

Incorrect mounting of sensors in process lines will result in large dead areas in the instrument branches, which are unacceptable (Figs. 7.80 and 7.81). So, instrument branches must be installed inclined upwards, especially to avoid air pockets. Moreover, the length of the dead area must be as short as possible and its cleanability must be demonstrated. For all pipe diameters the length of the up-stand should be smaller than its diameter ($l \leq d$). Furthermore, it is possible to avoid such dead areas by mounting, e.g., the pressure transmitter on a swept tee (Fig. 7.82), but the dimension l must be as short as possible relative to dimension d: maximum $l = d$. However, as swept tees in a horizontal pipeline could hamper the drainability, the swept tees preferentially should be mounted in a vertical pipeline. Alternatively, pressure transmitters with tubular membranes, with the same inner diameter as the adjacent pipelines, can be installed by means of clamp fittings in spherical valve bodies welded into the piping. The stainless steel diaphragms are sealed by O-rings fitted into grooves such that there is no metal-to-metal joint on the product side (Fig. 7.83). This way of mounting of pressure transmitters provides a dead space–free, flush transition from the process line to the pressure transmitters (Moerman, 2013b). In a similar manner, pressure transmitters with tubular membranes can be directly installed in the wall of a tank, again providing a dead space–free, flush transition, more specifically from the tank wall to the pressure transmitters (Fig. 7.84).

7.10.4 Retractable Measurement Instruments

Automatic retractable assemblies (Figs. 7.85–7.87) permit removal and installation of the sensor while protecting both the process and operating

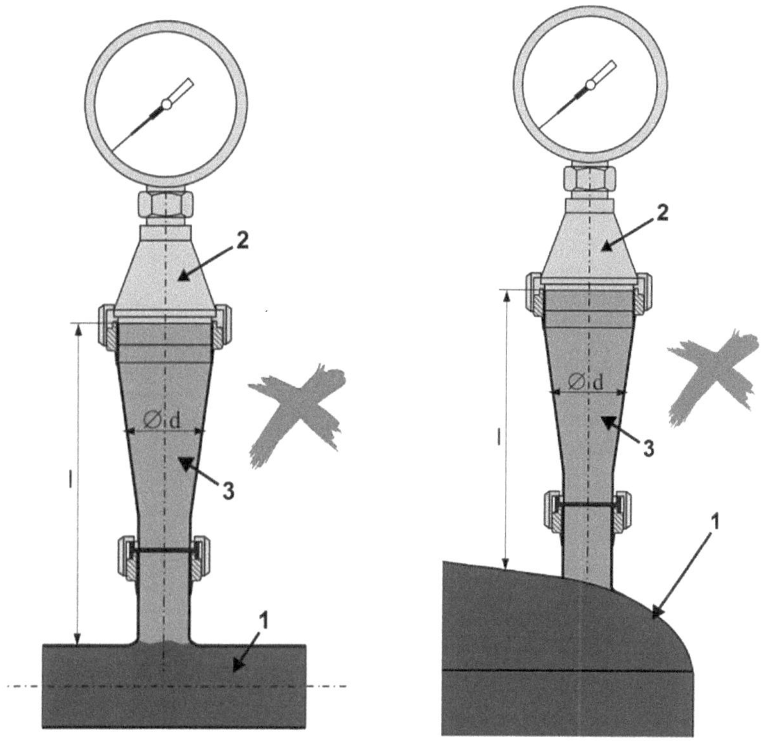

FIGURE 7.80 Product (1) is present in a pipe (left) or tank (right) with a pressure gauge (2) mounted on a too-long tee branch. An unacceptably large dead area (3) is created. The length of the up-stand should be smaller than its diameter ($l \leq d$) (Hauser et al., 2007).

(A)

FIGURE 7.81 Pressure measurement devices installed on a too-long instrument branch.

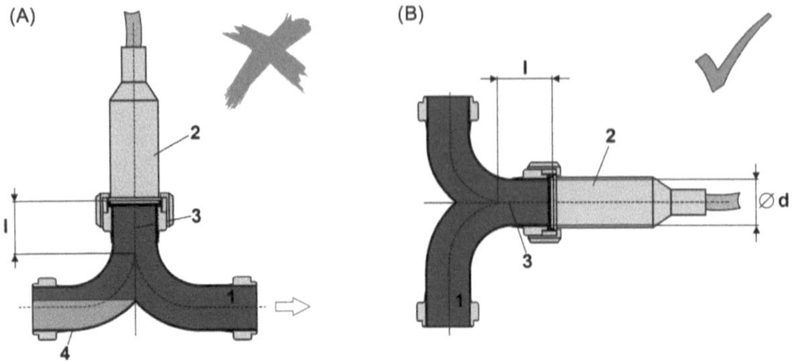

FIGURE 7.82 Incorrect mounting of sensors (2) in process lines (1) may give rise to tees with closed ends (3), which may create large dead areas if too long. (A) But a swept tee if mounted in a horizontal pipeline may impede adequate drainage (4). (B) Swept tees should be mounted in a vertical pipeline. Dimension "l" must be as short as possible relative to dimension "d," maximum l = d (Lelieveld et al., 2003; Hauser et al., 2007).

FIGURE 7.83 A pressure transmitter with tubular membranes, having the same inner diameter as the adjacent pipeline, can be easily integrated into the process line. Installation usually occurs by means of a clamp fitting, in a standard spherical valve body welded into the piping. The stainless steel diaphragm is sealed by O-rings fitted into grooves so that there is no metal-to-metal joint on the product side. This way of mounting of pressure transmitters provides a dead space–free, flush transition from the process line to the pressure transmitter. At the bottom side a temperature measurement device is installed, again by means of a clamp fitting. *Courtesy of WIKA Alexander Wiegand SE & Co. KG.*

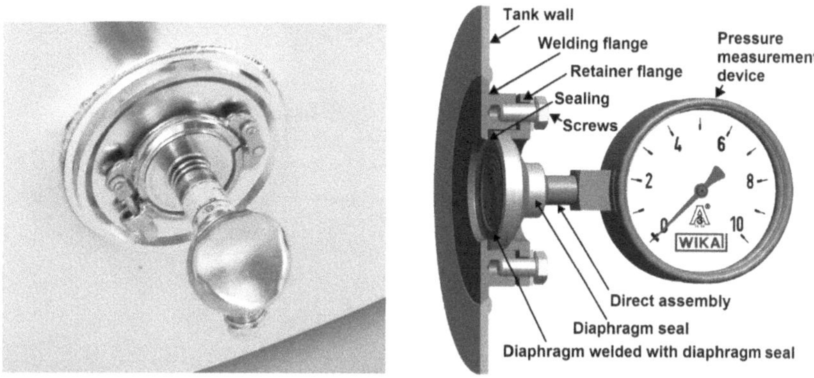

FIGURE 7.84 Pressure measuring instrument directly installed in the wall of a tank by means of a clamp fitting allows a flush transition from tank wall to diaphragm of the pressure measurement instrument. *Courtesy of WIKA Alexander Wiegand SE & Co. KG.*

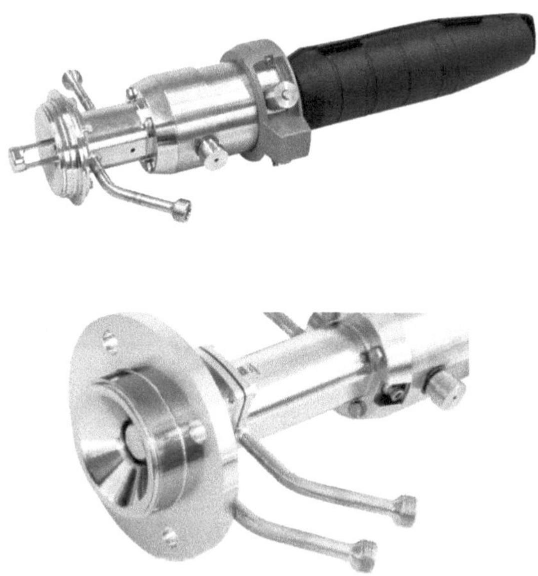

FIGURE 7.85 A retractable assembly permits removal and installation of a sensor while protecting both the process and operating personnel. *Courtesy of Endress + Hauser AG.*

personnel. Moreover, when moving the sensor from the measuring position to a service position, retractable assemblies allow the sensor to be cleaned, calibrated, or replaced without interrupting the process. Fig. 7.86 shows a retractable assembly with only one chamber (service chamber) separated from the pressure cylinder by means of two fixed seals. Fig. 7.87 shows the other option with two chambers (front and service chamber) where the

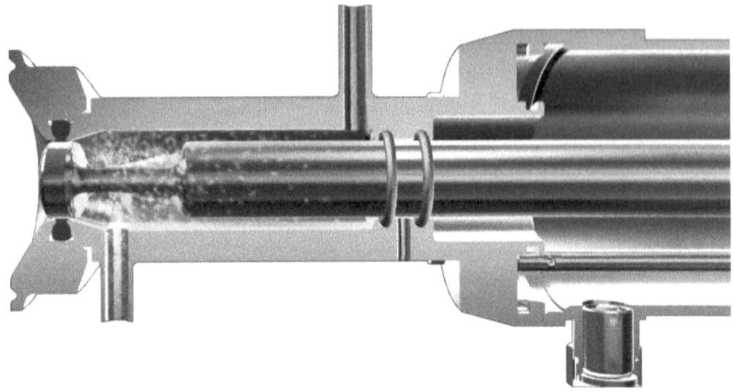

FIGURE 7.86 Retractable sensor assembly with two fixed seals separating the service chamber from the pressure cylinder. By moving the sensor from the measuring position to a service position, the retractable assembly allows the sensor to be cleaned, calibrated, or replaced without interrupting the process. It is also possible to clean and sterilize the seals' contact surfaces, which ensures that the sensor remains free from contamination when it is reinserted into the process. *Courtesy of Endress + Hauser AG.*

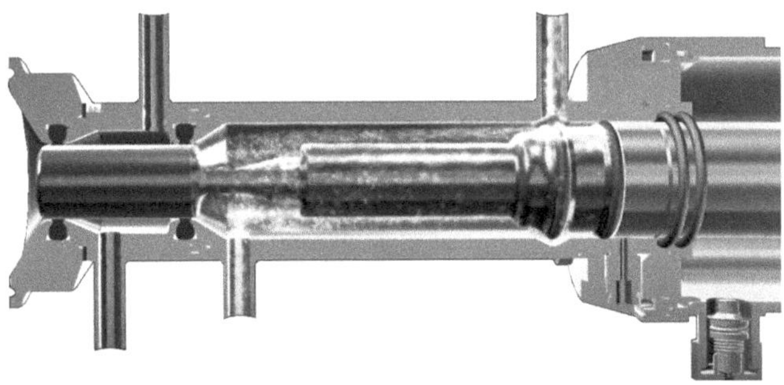

FIGURE 7.87 This assembly consists of a dual chamber system, with a front chamber being used as an extra protective barrier to the process. The front chamber is located between two fixed seals and can be cleaned and sterilized. The service chamber is now separated from the pressure cylinder by means of two dynamic seals. The front dynamic seal is cleaned and sterilized together with the service chamber. During the cleaning of the service chamber, it is possible to continuously flush the front chamber (near the surface) with sterile water, and therefore excluding any potential entry of cleaning solution from the service chamber into the process, even when the seal between the front and service chamber is damaged. The front chamber can also be used to isolate the service chamber temperature from the process. After cleaning and sterilization, the sensor guide as well as the two dynamic seals move toward the front chamber. During this action, the sterile front space and the medium remain untouched by unsterilized parts, ruling out the possibility of cross-contamination. *Courtesy of Endress + Hauser AG.*

service chamber is separated from the pressure cylinder with two dynamic seals. The later, more advanced retractable sensor assembly can better maintain aseptic conditions, which is optimal for situations where very sensitive food is manufactured.

In both versions, the retractable immersion tube, process connection, and service chamber (as well as front chamber in the advanced version) could be made of stainless steel or other alloys, while the seals can be ethylene propylene diene terpolymer (EPDM), fluoro-elastomers (FKM) or perfluoro-elastomers (FFKM). The materials selected depend on the application. The service chamber (as well as front chamber in the advanced version), process connections and couplings must be designed for minimal dirt buildup and maximum cleanability, as well as sterilizability where required. The alignment of the inlet and outlet of the service chamber (and front chamber in the advanced version), as well as the flow configuration, facilitates the loosening and removal of built-up solid deposits in the chamber(s) and from the sealing surfaces during the cleaning process. The chamber(s) also must allow liquid media to drain freely and completely.

7.10.5 Diaphragm Seals

7.10.5.1 Single Diaphragm Seal

Without a diaphragm seal, process fluids accumulated in the port or dead-ended sensor cavity of pressure gauges may compromise the physical and microbiological integrity of the process fluid. As an example, milk getting into the pressure port of a pressure gauge may spoil and contaminate the milk in a vacuum pan during its *in vacuo* evaporation and concentration. Furthermore, certain metals in electronic pressure sensors may contaminate the fluid with lead, zinc, copper, cadmium, etc. Therefore, to protect the process, the use of a diaphragm seal is recommended. In addition, the isolating diaphragm between the gauge sensor and process media prevents the pressure gauge port from plugging up with debris and liquid condensate, as well as protects the sensing element of the gauge by prohibiting corrosive, abrasive and noxious process materials from reaching the dead-ended sensor cavity (Fig. 7.88) (Moerman, 2013b).

A single diaphragm seal is composed of three main parts (Moerman, 2013b):

- The housing containing the fill fluid and process and sensor connections for the diaphragm seal.
- The isolating diaphragm, a flexible membrane that separates the fill fluid and mechanical or electrical sensing element from the process fluid or process aid. Pressure effects are allowed to cross the isolating diaphragm from the process/utility system and are hydraulically transmitted by the fill fluid to the sensor's measuring element. Isolating diaphragms have a thickness of approximately 3 mm, and are made of food process compatible materials such as plastics, rubbers, or metal plate. Metal diaphragms of stainless steel (several grades), Hastelloy, Monel, Inconel, tantalum, titanium, and other metals should be welded flush to the housing.

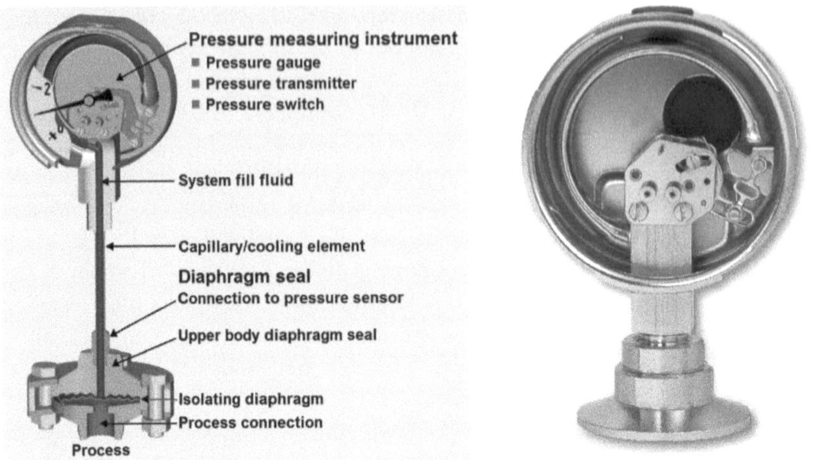

FIGURE 7.88 Mechanical gauges in particular, such as the bourdon gauge, require a diaphragm seal. Diaphragm seal gauges can cope with a greater range of process temperatures and aggressive products. *Courtesy of WIKA Alexander Wiegand SE & Co. KG.*

Capacitive sensors consist of a ceramic diaphragm of Al_2O_3, separated from the metallic sensor body via an elastomer seal. This design must be considered carefully as there is a risk of pores.

- A fill fluid (in the cavity between the gauge sensing diaphragm and the isolating diaphragm) is application specific, which means that it varies for food, beverage, or industrial applications. In food applications, a stable, food grade, noncorrosive, low thermal expansion, and viscosity fluid should be used (Table 7.3). For high-temperature applications, a sodium-potassium eutectic is commonly used, while a mixture of glycerine and water is recommended for ambient temperatures. Ethyl alcohol or silicon oil are applied for low temperatures. Water-based glycerine is not an appropriate fill fluid in vacuum and high-temperature applications due to its risk of vaporization, which can destroy the diaphragm seal, while at low temperatures it becomes too thick to produce accurate readings due to the extremely slow response time.

With regard to diaphragm seals, besides the common pressure gauge mounting to existing fittings such as T-pieces or welding sockets, there are also pressure measurement device installations using flow-through diaphragm seals (Fig. 7.89). These diaphragm seals consist of a body with an internal cylindrical thin diaphragm, which can be made of a variety of plastics. The diaphragm in-line seals can be installed directly in the pipeline between two flanges. A variety of nominal pipe diameters enables adaptation to the cross-section of the particular pipeline. These diaphragm in-line seals are ideal for use with flowing process media. With the seal being completely

TABLE 7.3 Fill Fluid in Pressure Gauges (Cole-Parmer, 2009)

Fill Fluid	Suitable Temperature Range		Kinematic Viscosity at Temperature	
	P_{abs} <1 bar (°C)	P_{abs} >1 bar (°C)	(10^{-6} m^2/s)	(°C)
Glycerine	N/A	+15 to +240	1110	+20
Glycerine/water	N/A	−10 to +120	88	+20
NEOBEE M-20	−28 to +160	−25 to +205	9.8	+25
Food grade silicone oil	N/A	−15 to +300	350	+25
Mineral oil	−20 to +170	−20 to +250	56	+20
Vegetable oil	−10 to +120	−10 to +200		
Propylene glycol/water		−18 to +93		

N/A = not applicable.

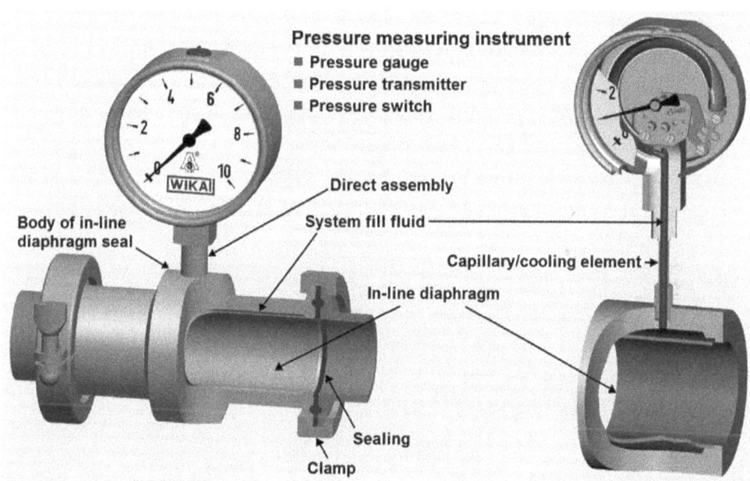

FIGURE 7.89 Diaphragm in-line seals. *Courtesy of WIKA Alexander Wiegand SE & Co. KG.*

integrated in the process line, measurements are not affected by any turbulence, corners, sharp edges, dead-ended cavities (where solids could accumulate), or other obstructions in the flow direction. Moreover, whereas other designs with grooves or noncircular geometry are more critical to clean, diaphragm in-line seals with their perfectly circular form are less likely to plug, are self-cleaning and are easy to drain. All product residues or films can be easily cleaned, even by pigging in certain applications. However, if maintenance is required, the process has to be shut down.

7.10.5.2 Double Diaphragm Seals With Diaphragm Monitoring System

Unforeseen process disturbances can damage or even destroy the isolating diaphragm. If a seal with only one diaphragm is used, then the fill fluid will find its way into the process. As the process operators are not warned of diaphragm failure, large amounts of product can be contaminated. Therefore the process operators are required to remove all pressure-measuring instruments from the process after every batch to check the diaphragm for possible damage. If no failure of the diaphragm has occurred, only then can the product batch be released for further processing. On the other hand, in the case of diaphragm failure, the contaminated batches have to be quarantined or discarded. Because of the time delays, unplanned shutdowns and product contamination due to failure of the diaphragm in single diaphragm seals, a more advanced design was required to avoid all these problems. Therefore, a double-diaphragm seal annex diaphragm break-monitoring system has been developed.

In a double-diaphragm seal, the space between the two diaphragms is evacuated, and in this manner a vacuum is created between the two diaphragms (Fig. 7.90). The pressure in this space is continuously monitored with a pressure gauge, pressure switch, or pressure transmitter (Fig. 7.91). If for any reason the primary diaphragm fails, the vacuum is compromised (so, pressure increases), and the monitoring system alerts the process operators by giving a visual, acoustic, or electrical warning. Although the wetted

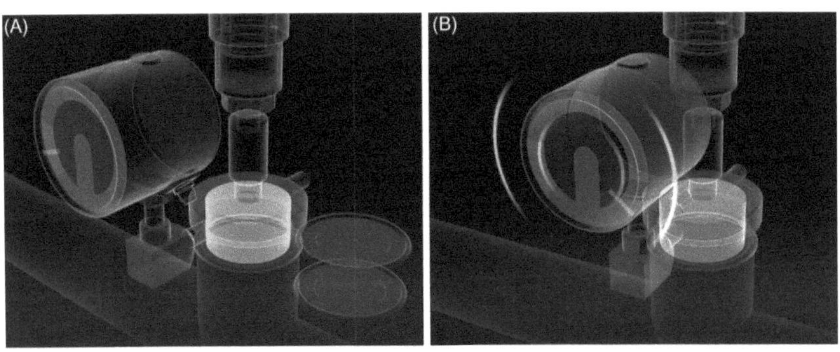

FIGURE 7.90 (A) Double-diaphragm seal: vacuum is created between two intact diaphragms. The pressure in this space is continuously monitored with a pressure gauge, pressure switch, or pressure transmitter. (B) Wetted diaphragm breaks but the second still forms a reliable seal for the process. The space between the two diaphragms is no longer under vacuum. The monitoring system alerts the process operators by giving a visual, acoustic, or electrical warning. *Courtesy of WIKA Alexander Wiegand SE & Co. KG.*

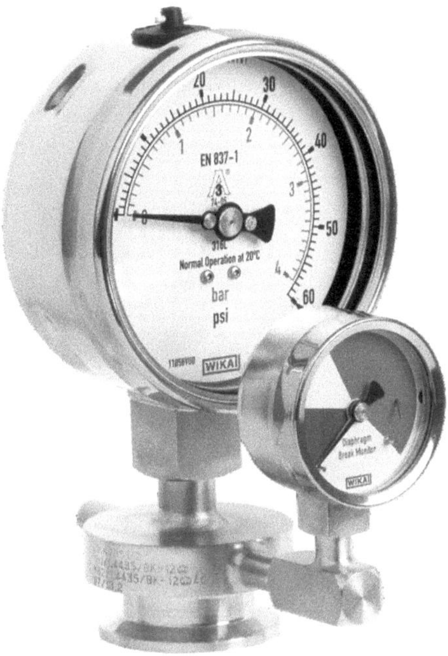

FIGURE 7.91 Diaphragm monitoring system. With regular on-site inspection, a pressure gauge with green-red display will be sufficient to detect failure of the diaphragm in contact with the process. When the pointer goes in the red (right corner of the diaphragm monitoring device), the operator knows that the diaphragm in contact with the process is broken. *Courtesy of WIKA Alexander Wiegand SE & Co. KG.*

diaphragm is damaged, the second still forms a reliable seal to the process and maintains the pressure monitoring until the damage has been rectified. So, there is no need to stop the process immediately to make repairs. As the process integrity is not affected when the first layer of protection (wetted diaphragm) fails, the food manufacturer is not required to stop immediately. Using a double-diaphragm system can thus prevent unplanned shutdowns.

7.10.5.3 Calibration of Pressure Gauges and Pressure Sensors

Pressure gauges and pressure sensors require scheduled, periodic maintenance, and/or recalibration. Pressure transducers can be recalibrated in a calibration laboratory or online.

7.11 TEMPERATURE MEASUREMENT DEVICES

7.11.1 Hygienic Design of Temperature Measurement Devices

To measure temperatures in a hygienic application, bimetal, resistance, or thermocouple technology can be used. Bimetal thermometers can only supply local readings, while resistance thermometers and thermocouples are primarily used to obtain an electrical output for remote readings. However, devices exist that combine resistance and bimetal elements and provide both local and remote capabilities in one package. This allows the user to tap into the process only once, reducing the potential for contamination. Temperature measurement based on electronic detection of a change in resistance is the most common method. The actual temperature sensor elements used integrate either platinum thin-film resistors (Pt100, etc.), or employ other sensing elements with a varying electrical resistance against temperature (NTC or PTC resistors). Also, semiconductor devices are common (Cole-Parmer, 2009; Moerman and Kastelein, 2014)

The temperature-sensing element itself is inserted in a thermowell (Fig. 7.92), which is a closed-end reentrant tube provided with means for a pressure-tight attachment to a particular process equipment component. The dimensions of the tapered and straight thermowells must be chosen as a function of the vibration or stress caused by the process medium flowing through the pipeline. The protective sleeve (thermowell) is typically made of stainless steel and highly polished to a surface finish of $R_a \leq 0.8\,\mu m$. However, special metallic overlays or polymer coatings can be applied to the surface of the thermowell for use in processes involving high-velocity particulates and acidic solutions that may cause, respectively, erosion and corrosion. A paste with high thermal conductivity is used inside the thermowell to bring the temperature sensor in close thermal and mechanical contact with the liquid (in a pipe, recipient, etc.) from which the temperature must be measured.

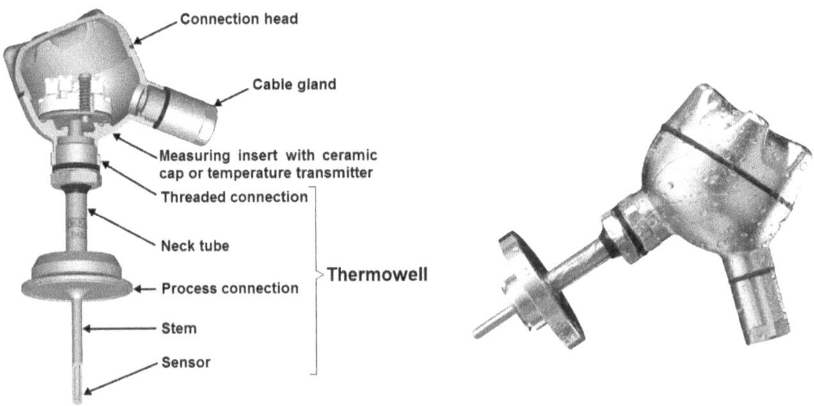

FIGURE 7.92 Design of an electrical thermometer with thermowell. *Courtesy of WIKA Alexander Wiegand SE & Co. KG.*

7.12 INSTALLATION OF TEMPERATURE MEASUREMENT DEVICES

Temperature measurement devices may not be mounted on a tee branch that is too long (Fig. 7.93) because an unacceptably large dead area is then created. Thermowells with flanged process connection (Fig. 7.94) can be integrated in process lines. By means of clamp fittings, they can be installed in standard spherical valve bodies welded into the piping. The sheath of the probe is welded into one of the two blanks which are sealed to the spherical valve body by O-rings fitted into grooves. Thus there is no metal-to-metal joint on the product side. This way of mounting a temperature measurement device provides a dead-space free, flush transition from the process line to the blank containing the thermowell. Thermowells also can be directly orbital welded into the pipeline (Fig. 7.95).

In addition, the thermowell can be directly fitted via an orbital welded pocket (Fig. 7.96). Attention should be given to the quality of the weld, which must be smooth and continuous. Furthermore, to avoid shadow areas, the direction of the flow must be as indicated.

For temperature measurement in tanks and larger vessels, the thermowells can be continuously welded to the tanks with welding balls or welding collars, after which the inner welding seam is polished and passivated. Temperature measurement devices can also be installed via a hygienic process connection sandwiched (detachable seal joints such as O-rings) into

FIGURE 7.93 Temperature measurement device mounted on a upwards sloped tee branch. The length of the up-stand is not larger than its diameter (l ≤ d). Sloped mounting of the instrument branch decreases the risk of reduced turbulent flow conditions in the instrument branch during cleaning operations. As turbulence in the sloped instrument branch is higher, stagnation of air in the upwards sloped tee is also less likely to occur. *Courtesy of Endress + Hauser AG.*

(A) (B)

FIGURE 7.94 (A) This thermowell fitting, having the same inner diameter as the adjacent pipeline, is integrated into the process line. The sheath of the probe is welded into one of the two blanks, which are sealed to the standard spherical valve body welded into the piping. Installation in the spherical valve body is done by means of a clamp fitting and O-rings fitted into grooves. In such a way, a metal-to-metal joint on the product side can be avoided. (B) In this example, the temperature measurement device is fitted in a tank by means of an O-ring and clamp fitting. Also here, a dead-space free, flush transition from the temperature measurement device fitting to the tank wall is obtained. *Courtesy of WIKA Alexander Wiegand SE & Co. KG.*

FIGURE 7.95 This flow-through thermowell is directly orbitally welded into the pipeline. The measuring insert can be withdrawn from the thermowell to calibrate the thermometer on-site. *Courtesy of WIKA Alexander Wiegand SE & Co. KG.*

the pipeline (Fig. 7.97). The dimensions of the O-ring and the design of the groove to be used for mounting sensors are critical in achieving controlled compression of the seal. The O-ring needs periodic maintenance with an inspection of the O-ring upon dismantling. Used O-rings should not be reinstalled.

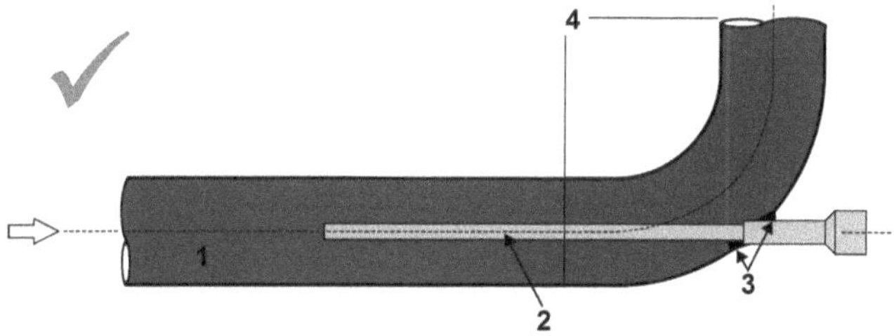

FIGURE 7.96 To avoid dead areas, the temperature probe (2) can be welded flush in the piping, in which the product flows (1) in the opposite direction. This may occur via an orbitally welded pocket. Attention should be given to the quality of the weld (3), which must be smooth and continuous. Welding of the temperature probe into the bend may be done off-line, after which the bend can be built permanently (by welding) or with dismountable joints into the piping system. In the latter case, the bend section is detachable (4) (Lelieveld et al., 2003; Hauser et al., 2007).

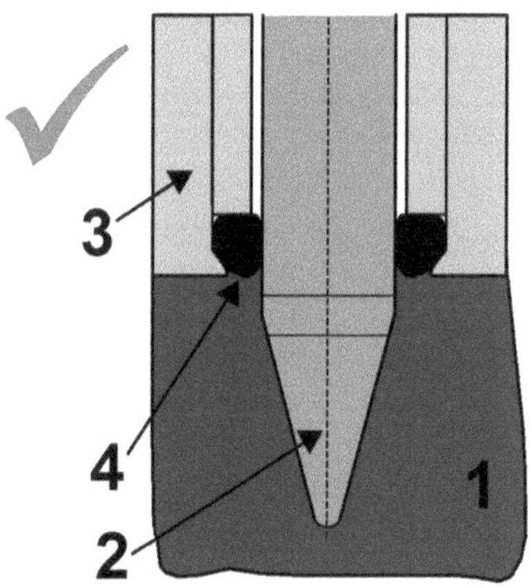

FIGURE 7.97 In the product area (1), a sensor (2) can be installed via a weld-in adapter (3) into and flush with the tank wall. The detachable seal joint (e.g., O-ring, 4) is almost completely enclosed with the surrounding metal protecting the non-product side from the product contact area (Lelieveld et al., 2003; Hauser et al., 2007) (photo, right, courtesy of Weed Instrument Co., Inc. d/b/a Ultra Electronics, Nuclear Sensors & Process Instrumentation).

7.13 CONCLUSIONS

Problems caused by microbial contamination of foods tend to be expensive, particularly if these result in consumer recalls. As a result of the development and application of increasingly mild preservation technologies, processed foods become more sensitive to microbial (re)contamination, requiring greater control of the manufacturing process. One way to achieve this added control is to "build in" hygiene into the equipment used in the food manufacturing facility from the start. The hygienic design of equipment plays an important role in controlling the safety and quality of the products made. Good hygienic design practice has proven to be a powerful tool in reducing or excluding microbial (e.g., pathogens, spoilage microorganisms, toxins of microbial origin), chemical (e.g., lubricating fluids, chemicals used for cleaning and disinfection, coolants, antimicrobial barrier fluids) or physical (e.g., glass, wood, packaging materials, and insects or other vermin) (re) contamination of foods.

Many of these problems can be alleviated by good basic hygienic design practices, including: (1) the selection of appropriate materials of construction, which must resist the harsh conditions encountered in the food industry (low or high temperatures, corrosive cleaning chemicals, steam, etc.), (2) the correct finishing of their internal surfaces (smooth surfaces with low surface roughness or dirt-repelling coatings can reduce biofilm formation), (3) correct joining of equipment parts (preference for welds instead of fixings), (4) use of hygienic fasteners where applied, (5) seals preventing leakage of liquid food and cleaning/disinfectant solutions to outside as well as ingress of microorganisms from the outside, (6) no niches such as pits, cracks, crevices, open seams, gaps, lap seams, inside threads, holes that may accumulate dirt and hamper the cleanability adequate drainage), (7) no dead areas where product may accumulate or removal of residues is prohibited, (8) excellent drainage of all liquids (liquid food, cleaning, and disinfection solutions), (9) maximum access for manual cleaning or a design that allows removal of all soil by means of CIP, (10) measures to avoid contamination of food by non-food grade lubricants. Furthermore, specific hygienic design requirements are given for specific components applied in closed food processing equipment. At first sight, designing closed equipment for producing contaminant-free liquid foods being safe to consume seems a complex task. However, several international standard-setting organizations, for example 3-A, EHEDG, NSF, USDA, as well as national associations, have developed specific hygienic design standards and guidelines for food-equipment manufacturers and the food industry. Today, many if not most manufacturers of equipment for the food industry can supply process equipment that meets the requirements for hygienic and/or aseptic processes.

ACKNOWLEDGMENTS

I am grateful to H.L.M. Lelieveld, formerly Unilever R&D, The Netherlands, for commenting on the manuscript. Furthermore, I thank all the companies that have supplied photographic material of their products, as well as several food manufacturers who delivered photos either directly or indirectly: Kraft Heinz Company, General Mills, Inc., Kellogg Company, MOM Brands, McCain Foods Limited, Ocean Spray Cranberries, Inc., PepsiCo Quaker Oats Company, Nestlé, Unilever, Mondelēz International, Inc.

REFERENCES

ASME BPE committee, 2014. Bioprocessing Equipment, ASME BPE-2014. American Society of Mechanical Engineers (ASME), New York, 324 p.

Baking Industry Sanitation Standards Committee (BISSC) of the American Society of Baking, 2003. American National Standard for Baking Equipment – Sanitation Standard, ANSI/BISSC/Z50.2-2003, Chicago, 35 p.

Bylund, G., 2015. Building-blocks of dairy processing, Ch. 6, section 6.8 – Piping, valves and fittings, Ch. 21. In: Teknotext, A.B. (Ed.), Dairy processing handbook. Tetra Pak Processing Systems A/B, Lund, Sweden.

Campden Food & Chorleywood Research Association (CFCRA), 1997. Hygienic Design of Liquid Handling Equipment for the Food Industry. In: Timperley, A.W. (Ed.), Technical Manual N° 17, pp. 1–204.

Cocker, R., 2004. Hygienic design and assessment of food processing valves. New Food 7 (1), 8–15.

Cole-Parmer, 2009. The Sanitary Instrumentation Primer: keeping your process application contamination-free, Illinois, 12 p.

FDA, 2009. Grade "A" Pasteurized Milk Ordinance, Silver Spring, Maryland, 382 p.

Haga, R., Murakami, S., Ostrove, S., Weiss, S., 1997. Cleaning mechanism study for biopharmaceutical plant design. Pharm. Eng. 17 (5), 8–21.

Hauser, G., Curiel, G.J., Bellin, H.-W., Cnossen, H.J., Hofmann, J., Kastelein, J., et al., 2004a. Hygienic Equipment Design Criteria, EHEDG Guideline N°8, second ed. EHEDG working group Design Principles, EHEDG, Frankfurt, Germany, pp. 1–16.

Hauser, G., Curiel, G.J., Bellin, H.-W., Cnossen, H.J., Hofmann, J., Kastelein, J., et al., 2004b. Hygienic Design of Open Equipment for Processing of Food, EHEDG Guideline N° 13, 2nd ed. EHEDG working group Design Principles, EHEDG, Frankfurt, Germany, pp. 1–24.

Hauser, G., Curiel, G.J., Bellin, H.-W., Cnossen, H.J., Hofmann, J., Kastelein, J., et al., 2007. Hygienic Design of Closed Equipment for Processing of Food, EHEDG Guideline N°10, 2nd ed. EHEDG working group Design Principles, EHEDG, Frankfurt, Germany, pp. 1–22.

Hauser, G., Eastwood, C.A., Woodall, D.L., Timperley, D.A., Curiel, G.J., Peschel, P., 1993. Welding stainless steel to meet hygienic requirements, EHEDG Guideline N° 9, first ed. EHEDG working group 'Design Principles', EHEDG, Frankfurt, Germany, pp. 1–21.

Höhler, A., Schumacher, B., Schrodt, T., Bühler, H., Nalbach, U., Gasparetti, M., et al., 2007. Hygienic Design and Application of Sensors. EHEDG Guideline N° 37. EHEDG working group "sensors", EHEDG, Frankfurt, Germany, pp. 1–35.

Holah, J., 2000. Food processing equipment design and cleanability. In: Gormley, R. (Ed.), Flair-Flow Technical Manual 377A/00. National Food Centre, Dublin, 39 p.

Jensen, B.B.B., Nielsen, J.B., Falster-Hansen, H., Lindholm, K.-A., 2011. Tank cleaning technology: innovative application to improve cleaning-in-place. EHEDG Yearbook 2011/2012. EHEDG, Frankfurt, Germany, pp. 26–30.

Kopitzke, T., Barnickel, M., Gasparetti, M., Merhof, P., Wahlers, J., 2006. Hygienic Welding of Stainless Steel Tubing in the Food Processing Industry, EHEDG Guideline N° 35, first ed. EHEDG working group 'Welding', EHEDG, Frankfurt, Germany, pp. 1–29.

Ladenburger, D., 2015. The hygienic advantages of the P3-diaphragm in aseptic processing. EHEDG Yearbook 2015/2016, pp. 108–113.

Lelieveld, H.L.M., Mostert, M.A., Curiel, G.J., 2003. Hygienic Equipment Design. In: Lelieveld, H. L.M., Mostert, M.A., Holah, J., White, B. (Eds.), Hygiene in Food Processing, 1st ed. Woodhead Publishing, Cambridge, United Kingdom, pp. 122–166. Book N°88 (Chapter 8).

Moerman, F., 2013a. Vacuum pressure measurement (part1): direct vacuum pressure gauges. Maint. Mag. 21 (114), 28–31.

Moerman, F., 2013b. Vacuum pressure measurement (part 3): vacuum gauges suitable for the food and pharmaceutical industry. Maint. Mag. 22, 61–64.

Moerman, F., Fikiin, K., 2016. Hygienic design of Air Blast Freezers. In: Lelieveld, H.L.M., Holah, J., Gabrić, D. (Eds.), Handbook of Hygiene Control in the food industry, second ed. Elsevier, Cambridge, United Kingdom, pp. 271–316. (Chapter 22).

Moerman, F., Kastelein, J., 2014. Hygiene design and maintenance of equipment. In: Motarjemi, Y., Lelieveld, H.L.M. (Eds.), Food Safety Management: A Practical Guide for the Industry. Academic Press, San Diego, pp. 673–739. (Chapter 26).

Moerman, F., Partington, E., 2014. Materials of construction for food processing equipment and services: requirements, strengths and weaknesses. J. Hyg. Eng. Design 6, 1–37.

Moerman, F., Wouters, P., 2015. Hygiene concepts in food factory design. In: Leadley, C., Sykes, R. (Eds.), Innovation and future trends in food manufacturing and supply chain technologies, 1st ed. Woodhead Publishing, Cambridge, United Kingdom, pp. 81–133. (Chapter 4) N ° 293.

Moerman, F., Rizoulières, Ph, Majoor, F.A., 2014. Cleaning-in-place. In: Lelieveld, H.L.M., Holah, J., Napper, D. (Eds.), Hygiene in Food Processing: principles and practice, 2nd ed. Woodhead Publishing, Cambridge, United Kingdom, pp. 303–383. N° 258. (Chapter 10).

Schonrock, F.T., 2005. Improving the hygienic design of valves. In: Lelieveld, H.L.M., Mostert, M.A., Holah, J. (Eds.), Handbook of Hygiene Control in the Food Industry, 1st ed. Woodhead Publishing, Cambridge, United Kingdom, pp. 263–272. Book N° 116. (Chapter 16).

Wiedenmann, W., 2013. Damage scenarios for valves: identifying the potential for optimisation. EHEDG Yearbook 2013/2014, European Hygienic Engineering & Design Group, Frankfurt, Germany, pp. 87–91.

Chapter 8

Personal Hygiene and Good Maintenance Practices for the Servicing of Food Processing Equipment

F. Moerman
Catholic University of Leuven – KU Leuven, Leuven, Belgium

8.1 INTRODUCTION

Food processing equipment, like all industrial plant, is susceptible to failure through breakdown, deterioration in performance owing to wear and tear with time, and to obsolescence due to improvements in technologies. In the past, food manufacturers resorted to inefficient "breakdown" maintenance, which occurred shortly, or a considerable time, after detection of the failure. Breakdowns usually result in the contamination of foodstuffs with foreign bodies from broken parts, potential microorganisms growing in harborage sites such as cracks, crevices and pockets, and lubricating fluids from, e.g., broken bearings. As the failure may be detected too late in this type of maintenance, contamination may already have taken place, which may result in food safety problems, inferior product quality and, finally, costly product recalls. Therefore, food manufacturers now use predictive and preventive maintenance as tools to detect and prevent premature failure. As part of preventive maintenance, the equipment's overall condition and integrity are assessed, frequently requiring the dismantling of equipment. Subsequent servicing often requires further break-in to the system, with the result that preventive maintenance may in itself become a food contamination hazard.

Poor hygiene and bad maintenance practices during maintenance, repair and reassembly of the equipment may bring about food quality and food safety problems. Hence, during "preventive" and "breakdown" maintenance, the maintenance operatives must pay attention to their personal hygiene practices and must respect Good Maintenance Practices according to the principles of proper hygienic design. Training of maintenance operatives in all aspects of

Food Protection and Security. DOI: http://dx.doi.org/10.1016/B978-1-78242-251-8.00008-4

their job requirements is thus essential to ensure that equipment after reassembly will not compromise the product integrity when returned to service and for a predicted future interval.

This chapter aims to provide guidance to food manufacturers and maintenance operators in the application of appropriate hygiene procedures during the maintenance of food processing equipment and utilities. In Section 8.2 we want to emphasize the importance of maintenance of food processing facilities, which is also required by national and international legislation, as well as by many food safety certification schedules and programs. Because reducing the risk of food contamination during maintenance operations starts during the design process, installation and start-up of the processing equipment, these subjects are discussed in Section 8.3. As the hygienic performance of equipment components may be compromised long before it fails, we handle this subject in Section 8.4. Subjects of Section 8.5 are: purchase and acceptance of parts, tools, lubricants, etc., brought onto the site; hygienic design principles to respect during repair; lubrication according to the principles of hygienic design and hygienic recalibration of measurement devices. Section 8.6 deals with personal hygiene practices as this is of paramount importance in maintaining hygienic conditions in the food factory during service operations. Section 8.7 deals with proper hygiene measures that could be taken before, during and after maintenance operations. Evaluation of the quality of the performed maintenance work and record keeping is part of every maintenance program in the process industry, including the food processing industry, and therefore is the subject of Section 8.8. Maintenance practices must be regularly reviewed and adapted when necessary, requiring a discussion in Section 8.9.

8.2 MAINTENANCE AND REPAIR, A NECESSARY EVIL

Failure of equipment and its components is inherent to all machinery, including food processing equipment. Typical indications of failure of equipment and its components are increase in noise, higher lubricant consumption, temperature rises or increasing leakage. However, food manufacturers can handle this proactively by inspection (regular check of food processing equipment and/or its components with respect to their performance) and preventive maintenance. Equipment maintenance checks should include an assessment of the equipment's overall condition and integrity (e.g., whether it is working properly), the sources of physical contaminants (e.g., damaged, lost or worn parts; rust or loose/flaking paint; broken parts such as needles and blades; loose parts on equipment prone to vibration; polymeric deposits; friction, fatigue, chemical reaction) and the potential microorganism harborage sites (e.g., worn or frayed hoses, gaskets or belts, porous welds, roughened product contact surfaces). Fig. 8.1 shows a small piece of a damaged scraper that was found after a pump. Fig. 8.2 demonstrates a conveyor belt

FIGURE 8.1 Small piece of a damaged scraper found in a failed pump. *Courtesy of Joe Stout, Commercial Food Sanitation LLC - Intralox, © 2016.*

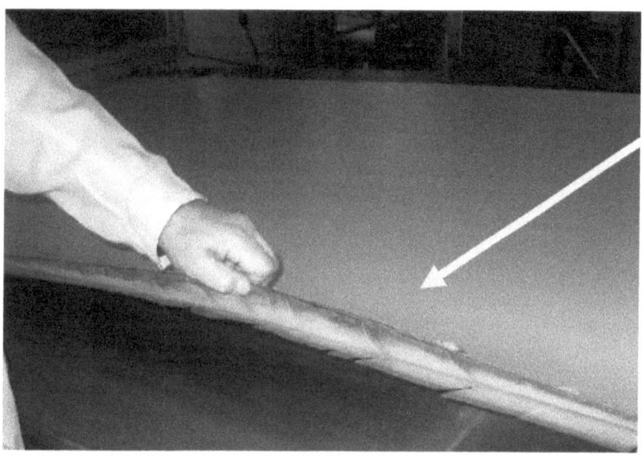

FIGURE 8.2 Damage at the edge of a conveyor belt. *www.ourfood.com, courtesy of Karl Heinz Wilm, © 2016.*

damaged at its edge. Fig. 8.3 is an example of an air blast-freezer in which ripped-off pieces of a plastic strip curtain are lost as foreign bodies in the product stream. Fig. 8.4 is an example of damage to the gasket and gasket grooves of a plate in a plate heat exchanger commonly caused by overtightening. Fig. 8.5 shows a crack in a plate of a plate heat exchanger due to overtightening or because the plate heat exchanger was opened or closed in the wrong way. Also the surface of heat exchanger plates may show surface damage, usually caused by a hard object getting in the plate heat exchanger (Fig. 8.6). In Fig. 8.7, inspection of the inside of a leaking hose revealed the presence of many cracks, which are quite common after a long period of service in contact with liquid food and aggressive cleaning/disinfecting solutions. Fig. 8.8 shows a seal moved out of position in the food area. Worn parts should be replaced as soon as practical, not only to ensure that production is maintained but also to prevent debris or worn or broken parts from entering the food product or contaminating the production line.

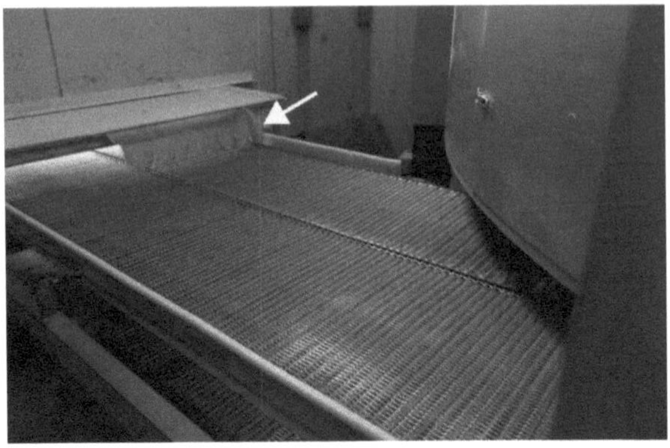

FIGURE 8.3 Ripped-off pieces of a plastic strip curtain have been lost as foreign body contaminants in the product stream, while the remaining plastic strips are extremely dirty. The plastic strip curtain is also no longer suitable to reduce the infiltration of warm humid air, and has become a serious source of microbial cross-contamination, especially for exposed products (Moerman and Fikiin, 2016). *(Frank Moerman, © 2016).*

FIGURE 8.4 Damage to the gasket and gasket grooves is caused by overtightening or overpressuring. Even a new gasket will not sit properly and is likely to leak. *Courtesy of PHEX LLC.*

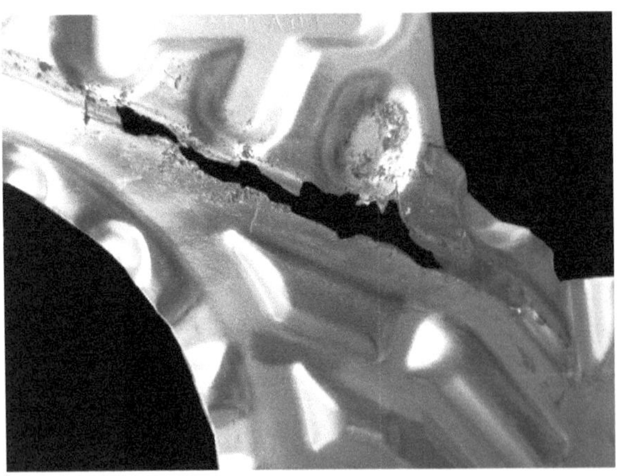

FIGURE 8.5 Plates in a plate heat exchanger may crack by overtightening or when the plate heat exchanger is opened or closed in the wrong way. *Courtesy of Thermo Logistics Ltd.*

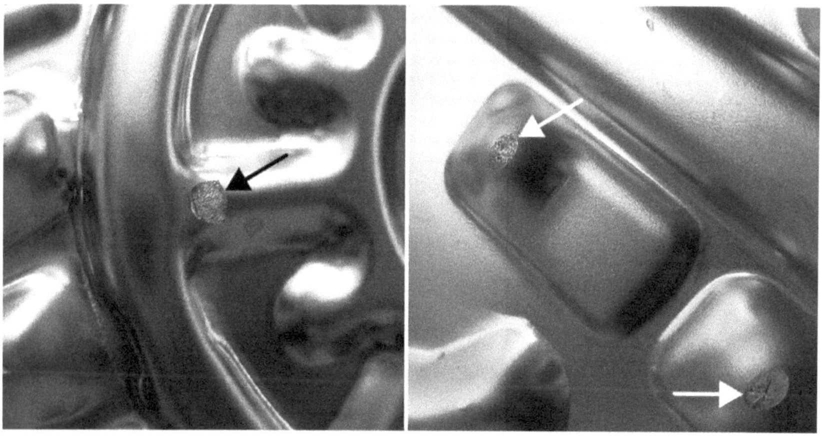

FIGURE 8.6 Plates in plate heat exchangers may suffer from surface damage, usually caused by a hard foreign body getting into the heat exchanger. The damaged spots show increased surface roughness which may promote the adhesion of food constituents (e.g., proteins in milk) and biofilm formation. *Courtesy of Thermo Logistics Ltd.*

FIGURE 8.7 Inside of a leaking hose, revealing the presence of many cracks which are quite common after a long period of service in contact with liquid food and aggressive cleaning/disinfecting solutions. *(Frank Moerman, © 2016).*

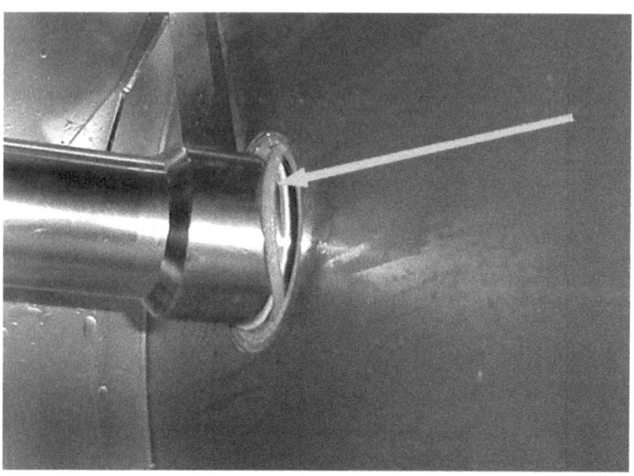

FIGURE 8.8 Worn seal may contaminate the product. *www.ourfood.com, courtesy of Karl Heinz Wilm, © 2016.*

8.3 SCHEDULED PREVENTIVE MAINTENANCE

Scheduled preventive maintenance should be preferred over inefficient "breakdown" maintenance and repetitive repair, more specifically for the following reasons (Jha, 2006):

- to prevent unscheduled downtime, and thus to provide higher throughput;
- to maximize the performance of all process and service equipment;
- to extend the useful life of the process equipment;
- to maintain and/or enhance the energy efficiency of the process and service equipment (consumption of less electricity, fuel, air power, etc.);
- to maintain and improve the quality and appearance of food products (reducing the risk for product recalls);
- to avoid the need to break into the system, hence reducing the introduction of food pathogens and spoilage microorganisms into the system;
- to save on spare parts, lubricants, maintenance chemicals, maintenance tools, etc.;
- to reduce or eliminate property hazards, such as fire;
- to enhance the safety of all food factory personnel;
- maintenance can occur at times during which no food is prepared.

No longer does the maintenance department have the luxury of extended periods of available equipment downtime in order to carry out maintenance. Nowadays the maintenance function is moving toward a more predictive approach. If the failure characteristics of the equipment are known, predictive maintenance can detect the failure well in advance and appropriate actions can be taken in a planned and organized manner. Predictive maintenance makes use of a group of emerging scientific technologies that can be employed to detect potential failures: vibration analysis, thermal imaging, ultrasonic measurement, and oil analysis. The maintenance technicians should be skilled in using these diagnostic tools, and they must have detailed knowledge of the operating characteristics of the equipment to make the correct failure diagnosis. By means of a risk analysis, the manufacturer may define which parts of the system are critical, ensuring that the necessary treatment (to which interval, to which time point, and with which measures) is undertaken. That maintenance schedule should be frequently reviewed during the initial operating period of an installation to establish the optimum maintenance frequency (Liggan and Lyons, 2011).

Unscheduled downtime and "breakdown maintenance" (maintenance when it breaks) can be reduced by:

- Proper selection of the materials of construction. Chemical, physical, and thermal resistance of the materials of construction have an enormous impact on the reliability of equipment (components) and the frequency of maintenance and repair. As an example, selecting wires and cables designed in materials which are not compatible with a specific aggressive environment may directly affect the performance and reliability of the food processing equipment. Wires and cables of the highest quality may extend the mean time between failures.

- Selecting and purchasing the right type of equipment for any specific job. Equipment must have sufficient capacity. If a machine that is of low capacity is consistently being forced to run at high capacity, no amount of preventive maintenance will cure it! From this point of view, training operators in the correct operation of the equipment and in the application of proper and efficient procedures is thus essential. Although time-consuming, it pays off in the long run.
- Equipment shall be appropriately designed within predefined tolerances of use and conform to the required specifications. Define for each piece of equipment or its components the working parameters and their minimum and maximum values. If the maximum operating limits of the equipment (components) are lower than those of the system in which it is being fitted, or if malfunction of the equipment (components) could result in a serious contamination, ensure that the system can handle these over-limit situations.
- Selecting electrical and mechanical equipment with high ingress protection for water and with high thermal resistance, both positive and negative temperatures. Acid or abrasive foods, water, steam, detergent and disinfectant solutions, and extreme heat or cold all create a hostile environment for electrical and mechanical equipment.
- Ensuring proper installation and assembly (e.g., bearings frequently fail due to misapplication, overloading, and misalignment).
- Conducting an inspection program of equipment after delivery.
- A short period of in-plant testing of the equipment to screen the entire population of equipment for leakage, to verify if all components function appropriately so as to ensure that they will fulfill their duty once brought into operation to produce food.
- Correct start-up.
- Avoiding mishandling during maintenance.
- Following the manufacturer's recommendations for scheduling preventive maintenance.
- Implementing regular and correct calibration practices.
- Periodic updating of computerized components. Up-to-date software may improve the equipment's efficiency and prevent possible problems that could cause disruption during production. Also updating of a machine's firmware can help prevent unplanned downtime.
- Replacing original equipment components with higher quality, more reliable counterparts designed to better withstand the harsh conditions encountered in the food industry (e.g., higher quality cables with longer expected life or higher stress resistance).
- Inspecting machines for unexpected signs of wear.

- Implementing a policy of cleanliness. Spills and dried-on gunk attract insects and vermin that may choose to snack on the equipment (e.g., wires and cables) or to nest within equipment (e.g., pipe insulation).
- Detailed recordkeeping of maintenance activities. Items that fail most must be more frequently inspected and replaced. For that purpose, there must be a sufficient stock of spare parts as backup (e.g., wires, cables, and shear pins). However, it is not always possible to keep a backup for every single unit in a plant due to a lack of space for storage or the high cost of the component.

8.4 HYGIENIC VERSUS OPERATIONAL PERFORMANCE

Preventive maintenance is primarily undertaken to make the equipment "failsafe," which means that the equipment is unlikely to break down during production periods. When the equipment is failsafe, downtime and production losses are reduced. Components and replacement parts, however, may become a hygiene risk before they physically fail. For example, as seals in pipe couplings become worn or lose elasticity resulting from extensive heating and cooling cycles, they can become microbial harborage sites before they physically fail and cause pipe leaks. If a component has a predicted failsafe life of 12 months, it may require changing after 9 months because it becomes a microbial harborage risk.

The hygienic performance of components and replacement parts may only be applicable to certain parts of the factory. For example, for ready-to-eat products or other products where microorganisms can lead to food safety or spoilage incidents in particular, hygienic performance may be relevant in equipment past any heat treatment processes. However, the hygienic performance is not always easy to predict, as it is likely to be different for all food manufacturing processes where product constituents, process conditions and cleaning and disinfection programs may all influence the changing physical condition of the components and replacement parts.

Following plant installation and commissioning, a risk assessment on each piece of equipment should be conducted in order to establish preventive maintenance schedules. For most pieces of equipment, there are several different maintenance schedules and levels of maintenance. The main question is: how important is this activity and this piece of equipment to product quality, product safety or legal compliance? So, besides the failure performance, the hygienic performance of components and replacement parts must be taken into account. Based on the risk assessment, the maintenance engineer can then make an estimation of the required frequency of specific maintenance activities. To perform the risk assessment, components in the new equipment must be examined at intervals, e.g., every 2–3 months and up until the predicted failsafe replacement time. During this examination, any signs of deterioration which could lead to a microbial hazard must be monitored. When the examination process is finished, the food manufacturer subsequently may establish which are the high-, medium-, and low-risk components and activities. Gasket inspection or replacement is deemed a high-risk activity and must be scrupulously adhered to. A maintenance schedule may call for replacing gaskets every month. With respect to low-risk components, a complete check can be done once a year, e.g., during a shutdown of the equipment.

8.5 MAINTENANCE ACCORDING TO THE PRINCIPLES OF HYGIENIC DESIGN

8.5.1 Purchase and Acceptance of Parts, Tools, Lubricants, etc., Brought Onto Site

8.5.1.1 Equipment

No equipment, spare parts, tools, etc., should be brought directly into a food production area. They should ideally be held in an external workshop or storage area, or if this is not possible, within the goods-in area. They should be inspected for any damage and to ensure that they meet the specifications as defined by the food manufacturer. This could include a visual inspection, as well as control for the presence of objectionable odors and physical contaminants. If appropriate, an assessment of the equipment's surface roughness using an appropriate stylus instrument may be performed.

Before entrance into the factory, additional information can be gained as to how the equipment must/will be cleaned or maintained in practice. The sanitation manager can examine the equipment and begin to devise cleaning schedules, which might involve the cooperation of the maintenance team to help in planned dismantling and reassembly. The design of any parts trolley (e.g., parts rack) that may be required to store dismantled parts during cleaning is also a task of the maintenance department.

Equipment should be physically cleaned and decontaminated, though ideally this should have been undertaken prior to the equipment arriving at the factory. For new equipment and components, this may be to remove any materials of construction deposits or lubricants used to machine or drive equipment components. For secondhand equipment this even could include food deposits, including potential allergens. Cases are known where the secondhand equipment has introduced strains of *Listeria monocytogenes* and *Salmonella* spp. (prevalent in their original factories) into their new home. Special care should be taken, therefore, before introducing equipment into areas where ready-to-eat foods are prepared. Once equipment has been brought into the food production area and installed in its operational positions, the equipment should be cleaned and decontaminated again.

Equipment finally can be assessed for its operational performance (if this is possible out of its intended point of installation) and an inspection made to ensure all parts are correctly installed and tightly fitted. The maintenance team can also begin to plan potential maintenance schedules and secure the provision of any specialist tools and spare or replacement parts.

8.5.1.2 Replacement Parts

All materials used for maintenance and repair (replacement parts, maintenance tools, reparation aids, etc.) shall be fit for the intended use and, if they are in direct contact with food, must be constructed of materials that will not contribute a food safety risk. Indeed many food manufacturers, particularly those subject to Global Food Safety Institute (GFSI, http://www.mygfsi.com/) audits, request certificates of conformity, or other evidence from the replacement parts manufacturer to confirm its suitability for use.

Further to the requirement for replacement parts intended for food contact to be of approved materials for such purpose, EC Regulation. No. 1935/2004 requires that replacement parts must be traceable. This is to ensure that the food manufacturer is in a position to recall any food products manufactured on lines that have been fitted with replacement parts that are subsequently found to be a health risk. Replacement parts must therefore:

- be traceable to a supplier;
- be traceable when in storage at the food manufacturer's site;
- be traceable as to which piece of equipment or line into which they were installed;
- suitably identified to facilitate their traceability.

Prior to acceptance, all replacement parts should be examined for damage and/or contamination and to ensure that they meet the appropriate specification. All nonconforming, damaged, or contaminated parts should be rejected.

8.5.1.3 Lubricants

Many items of food processing equipment require lubrication to prevent contact between moving and static surfaces, so as to avoid excessive wear and potential overheating. Within Europe, food processing machinery should be designed such that no ancillary substances (e.g., lubricants) can come into contact with foodstuffs (2006/16/EC). However, on occasion, some lubricants may inadvertently come into contact with foods or food contact surfaces and such lubricants must be food safe on incidental food contact. *Food safe* means that these lubricants will not harm the consumer's health if they accidentally come into contact with foodstuffs and are consumed. Historically in the United States, the US Department of Agriculture (USDA) approved lubricants if their ingredients were listed on the FDA Code of Federal Regulations (CFR), Title 21, section 178.3570 lubricants and other sections referenced therein. Lubricants with ingredients meeting this standard, which lists both approved ingredients and the quantity of that substance that is permissible in foodstuffs, were labeled H1 lubricants. The USDA's authorization and inspection program was suspended, however, in 1998. The National Sanitation Foundation (NSF) continued to register food-grade lubricants on a commercial basis as H1 lubricants, and a list of approved products can be found on their website (http://www.nsf.org/). More recently, INS SERVICES (UK) Ltd. have also offered this service (www.Insservices.eu). Alternatively, lubricants can be approved to ISO 21469:2006, which, in addition to requiring lubricants to be formulated from nontoxic materials as listed by the FDA or the European Food Safety Authority (EFSA), also requires them not to affect the organoleptic qualities of the food or to pose additional health risks, such as supporting microbiological growth.

H1 registered lubricants are available in the following categories:

- bearing greases for low temperatures, ambient temperatures, and high temperatures;
- chain lubes for low temperatures, ambient temperatures, and high temperatures;
- gearbox fluids (open and enclosed);
- assembly and antiseize compounds;
- hydraulic fluids;
- compressor oils;
- airline lubricants;
- penetrating fluids;
- can seamer lubricants;
- sugar-dissolving solutions;
- release agents;
- general-purpose sprays and lubricants.

As with all raw material, ingredient and service supplies into the food factory, lubricant manufacturers should be part of a Supplier Approval Scheme. The European Hygienic Engineering & Design Group (EHEDG) guideline N° 23 (Steenaard et al., 2009) includes a number of critical points, together with suggestions for their management. Lubricant hygiene is described in the guideline as: all measures necessary to ensure the safety and wholesomeness of food-grade lubricants. These measures shall cover all stages during preparation, processing, manufacturing, packaging, storing, transportation, distribution, handling, and offering for sale or supply to the customer:

- for H1 approved lubricants, only food safe materials can be used;
- manufactured hygienically;
- quality approval scheme;
- a batch registration system and a raw materials identification system are recommended;

FIGURE 8.9 A clearly damaged grease container on arrival in the factory site. This grease should not be used because it will no longer meet the requirements of a food-safe product (Steenaard et al., 2009).

- analyses can be used to check the purity of the raw materials. Stored raw materials must be systematically checked to ensure that they are not outdated.

All shipments should be delivered in clean containers suitable for the transportation and protection of their contents with respect to integrity and quality and in keeping with good commercial practices and be labeled properly. All packs should be sealed with tamper-evident seals fitted at the point of filling. Deliveries arriving without their seals intact must be rejected (Fig. 8.9).

8.5.1.4 Materials of Construction

Product contact surfaces are all the surfaces exposed to direct contact with the product as well as indirectly impacted surfaces from which splashed product, liquid or solid particles may drip, run off, drop off, or fall into the product. Materials of construction for food processing equipment, process piping and utilities must be compatible with the food product and must be homogeneous, hygienic (smooth, nonporous, nonabsorbent, nontoxic, easily cleanable, impervious, and nonmold-supporting surfaces), and inert (nonreactive to oil, fat, salt, etc. and may not adulterate the food by imparting deleterious substances to it nor affect its organoleptic characteristics), chemically resistant (corrosion-proof, nondegrading and maintaining its original surface finish after sustained contact with product, process chemicals, cleaning agents, and disinfectants), physically durable and mechanically stable (resistant to steam, moisture, cold, heat, cycling temperatures; resistant to impact, stress and fatigue; resistant to wear, abrasion, erosion, and chipping; not prone to cracks, crevices, scratches, and pits; unbreakable), and easy to maintain, in agreement with the guidance described in EHEDG guidelines N° 8 (Hauser et al., 2004a) and N° 32

(Partington et al., 2005). Additional requirements could be availability, welding ability and machinability in different shapes. Notice that materials of construction which are worked (for instance: bent, cut, sheared, extruded, or drawn) during manufacture may require additional treatment (such as surface finishing) following fabrication in order to render them corrosion resistant. Hence, materials of construction should be selected that are suitable for surface treatment (Hauser et al., 2004a). Materials used in the construction of components located in the nonfood contact area may be of a lower grade but must be corrosion resistant and able to withstand all the cleaning solutions normally used.

8.5.2 Hygienic Design Principles to Respect During Repair

Maintenance and repairs should occur according to the principles of proper hygienic design to ensure that safe food is produced once production is resumed. The following recommendations should be followed.

8.5.2.1 Design for Maximum Access

Equipment should be of such a design that cleaning or maintenance of it does not introduce food safety hazards, e.g., consideration should be given to eliminate or minimize the need for physical entry into the system. All equipment parts and components shall be readily and easily accessible for inspection, maintenance and troubleshooting. For that purpose, enough space and clearance should be provided around equipment, process and utility piping, equipment utility connections, etc.

8.5.2.2 Compatible Materials of Construction

Materials of construction used during maintenance and repair must be adequate to cope with the food product produced or process aids they are in contact with, as well as with the harsh conditions encountered in the food processing environment (detergent and disinfectant solutions, lubricants, etc.) (Moerman et al., 2014).

- Corrosion of metals, steels and alloys may result in leaks and impair the smoothness of the surface finish. This increase in surface roughness makes equipment materials of construction more prone to adhesion of food residues and bacteria. The latter finally can give rise to biofilms which could be very difficult to remove.

  Immersion tests with metal coupons (Fig. 8.10) or specific equipment components (Fig. 8.11) allow evaluation of the effect of food products, detergents and disinfectants on the materials of construction used in the manufacturing of food processing equipment and utilities. Static immersion tests of the candidate materials of construction are rapid screening tests. The large numbers of welds and the numerous transitions from one metal to another make process equipment also very sensitive to

FIGURE 8.10 The compatibility of several materials of construction with detergent and disinfectant solutions can be tested in the laboratory by means of immersion tests conducted with coupons. *Courtesy of Evapco Inc. Moerman & Fikiin, 2015.*

FIGURE 8.11 Bearings made from different materials of construction were subjected to immersion tests in salt brine. Bearings No. 1, 2 and 8 are thin dense chrome plated; bearings No. 3, 5, and 7 are 400 series stainless steel; bearing No. 4 is coated; and bearing No. 6 is black oxide coated. *Courtesy of John Butts, Land O'Frost, © 2016.*

aggressive cleaning and disinfectant chemicals. Hence, if the plant item is to be welded, it is prudent to subject welded coupons to similar tests, as the weld metal and heat-affected zones may have different corrosion resistances in comparison with the unwelded material. To assess the risk of crevice corrosion, a testing procedure that involves the use of castellated washers is often used (Moerman and Partington, 2014; Moerman and Fikiin, 2016).

Several experimental parameters can be changed, such as temperature, detergent/disinfectant concentration, water quality (pH, hardness, etc.), application frequency, etc. It is recommended to perform "challenge tests" under forced conditions, which means that coupons are immersed for several days or even weeks in highly concentrated cleaning and disinfectant solutions, and, if necessary, the food product. At the end of this immersion period, coupons should be rinsed and dried to evaluate the effect of these cleaning and disinfection solutions. Parameters that can be measured are visual appearance, weight loss, thickness, hardness, etc.

The removal of the galvanizing and the release of zinc make galvanized steel unsuitable for application in the product contact and splash area, not least because zinc often contains residual traces of cadmium and lead as impurities. Painted steel never should be used in the splash zone or any other area where food is exposed, because paint may peel off and can splash/fall onto the food products (Fig. 8.12). Paints especially may create a health risk because they often contain toxic substances such as zinc, lead, cadmium, and phenolics. Paint surfaces used in nonproduct contact areas also may crack or flake, and must be repainted immediately.

Care must be taken when selecting a replacement part, because experience has shown that many items that were supposed to be stainless steel

FIGURE 8.12 This motor is constructed from mild steel, and is used to impart vibration to a perforated stainless-steel bed laden with food products. During hose-down operations, the cleaning agents and high pressure typical for traditional manual cleaning procedures have ruptured the physical integrity of the paint and allowed peeled-off paint to splash onto the cleaned conveyor bed. As a result, food products subsequently produced on this process line may become contaminated. *Courtesy of Joe Stout, Commercial Food Sanitation LLC - Intralox, © 2016.*

316L turned out to be 304. Packages containing these items were often marked and labeled as stainless steel 316L. Even 316L stainless steel tanks with inlet ferrules and manway collar in 304 stainless steel have been found. All the raw materials of construction that enter the factory, as well as purchased items, can be tested on their elemental composition using the nondestructive X-ray fluorescence method as a Positive Material Identification technique. By bombarding the surface of the test material with high-energy X-rays or gamma rays, secondary (fluorescent) X-rays emitted from the material can be detected. Each element in a material emits its own unique fluorescent X-ray spectrum, allowing it to be identified. However, commercially available portable handle-held XRF guns (Fig. 8.13) are quite costly and are limited in their ability to precisely and accurately measure the abundance of elements with atomic number $Z < 11$ (e.g., atomic number Z of carbon is 6). For this reason, XRF can't be used to differentiate between stainless steel 316 and 316L. For small food manufacturers, the procurement of an XRF-analyzer can't be justified due to its high cost. They still have to rely on certificates delivered by their vendors, although they still can qualify their vendors by hiring a contractor that is in possession of an XRF-analyzer and experienced in elemental analysis by this technological means.

- Plastics must have good dimensional stability on exposure to high loads, corrosive chemicals, as well as high or low temperatures. Changes in dimension or shape, cracking, and breaking during operation may not occur, as they allow food residues access to areas where they will be

FIGURE 8.13 Quality technician doing Positive Material Identification on a stainless-steel vessel using an X-ray fluorescence gun. *Courtesy of Holland Applied Technologies.*

difficult to clean and pose a contamination risk. They must have high mechanical strength to withstand mechanical shocks, and resistance to aging, creep, brittleness, fatigue, erosion, etc. Excellent resistance to wear and abrasion is required in certain applications such as the transfer of solids, slurries or pastes (e.g., tomato concentrate). These food products may damage the plastic surface, promoting the accumulation of soils and the formation of biofilms, finally negatively affecting cleanability. The plastic material also must be chemically resistant to hydrolysis by steam, acids and alkalis, reducing and oxidizing agents, as well as cleaning agents and disinfectants. The equipment manufacturer should/can test the chemical resistance of the plastic material in the same way as described for metal, steels and alloys. They also must be tested on their thermal resistance (Partington et al., 2005; Moerman and Partington, 2014).

For use in the food contact area, it is important that plastics be odorless, nonporous, smooth and free from cracks, crevices, scratches, and pits, as they may harbor and retain product constituents and/or microorganisms after cleaning. Within the pores, microorganisms are also better protected against the bactericidal activity of disinfectants. And, spherical void expansion may cause changes in the chemical and physical properties of a certain plastic material, hence affecting its cleanability. In food contact applications, it is recommended to avoid additions into plastics, as additives incorporated in plastics may migrate into the product. But in addition to these additives, volatile remnants of monomers (e.g., styrene), oligomers, low molecular weight polymer fragments and certain organic solvents may leach from the polymer material into the food, inducing changes in the organoleptic qualities. It is recommended to delay any exposure of food to recently produced or processed plastics, so as to allow these plastics to release most of their chemical substances before application and during storage. Also, the first food batch should be sent to disposal.

While certificates of conformity are proof that the material is, in principle, safe for food contact and the component is made to specification, they are not evidence that the part specified is technically suitable for a particular application. It is good practice to require the supplier of plastic raw materials or components to provide documentation that the grade of material selected is certified for contact *with the product in question* (requiring studies to evaluate migration of specific substances).

- Rubber compounds are individually developed by the supplier, causing elastomers to differ greatly from one supplier to another. Rubber for food contact must comply with many different normative references (e.g., EN 1935/2004, BfR, FDA and 3-A), as well as the European Commission's REACH Regulation (Regulation on Registration, Evaluation, Authorisation and Restriction of Chemicals). It is a major recommendation to avoid ingredients which are not chemically bonded, as they may be released in the food product). The rubber parts supplier should assist food manufacturers in selecting the most suitable elastomer for their specific operation. This selection must occur in function of the physical and chemical resistance characteristics of the different types of elastomers and the in-use process conditions. Degradation of elastomers by product, detergents, disinfectants, and thermal and mechanical stress proceeds much faster. Moreover, elastomers are more prone to microbiological degradation because they facilitate extensive biofilm formation more than plastic materials do (Moerman and Partington, 2014).

A typical symptom of this elastomer deterioration—mostly due to a combination of aggressive chemicals and elevated temperatures during cleaning processes—is hardening of the material, leading to loss of elasticity and eventually loss of sealing function. The result could be: physical contamination of the product with elastomer particles (consequence of abrading and break-up of the rubber material); leakage of lubricants or refrigerants; loss of bacteria tightness; permanent product and process contamination due to increased adherence and retention of dirt and bacteria in crevices; and insufficient cleaning and problematic disinfection. Moreover, ingress of liquids containing chlorides may occur under partially destroyed gaskets and seals, so that a high chloride concentration may subsist between damaged seals and adjacent metal, favoring crevice corrosion even in stainless steel. To ensure that the rubber remains in good condition, regular inspection is required. Routine replacement of elastomers must be done in function of the physical and chemical stresses imposed on the material.

8.5.2.3 *Right Combination of Metals, Steels and Alloys to Avoid Bimetallic Corrosion (Galvanic Corrosion)*

In the assembly of food process equipment and services, the right combination of steels, alloys or metals must be used to avoid bimetallic corrosion. Bimetallic corrosion occurs between metals, steels, or alloys with considerably different standard reduction potential (Fig. 8.14). A current will flow from the less noble metal to the more noble metal, resulting in the oxidation (corrosion) of the less noble metal. To determine the compatibility of two metals, steels, or alloys, a galvanic corrosion metal compatibility chart (Fig. 8.15) or diagram (Fig. 8.16) can be used. Also galvanic corrosion of welds may occur if the weld metal is less noble than the surrounding materials joined. Higher alloyed filler metal in comparison to the welded materials may reduce the risk for galvanic corrosion. Note that replacement parts could have another chemical composition than the materials of construction used in the existing plant equipment.

FIGURE 8.14 Stainless-steel shell and knives of the roller are combined with roller ends in mild steel. As mild steel is less noble than stainless steel, it will corrode. Work in black steel and stainless steel must always be kept separate. *Courtesy of organization Sanitary Design Workshop, © 2016.*

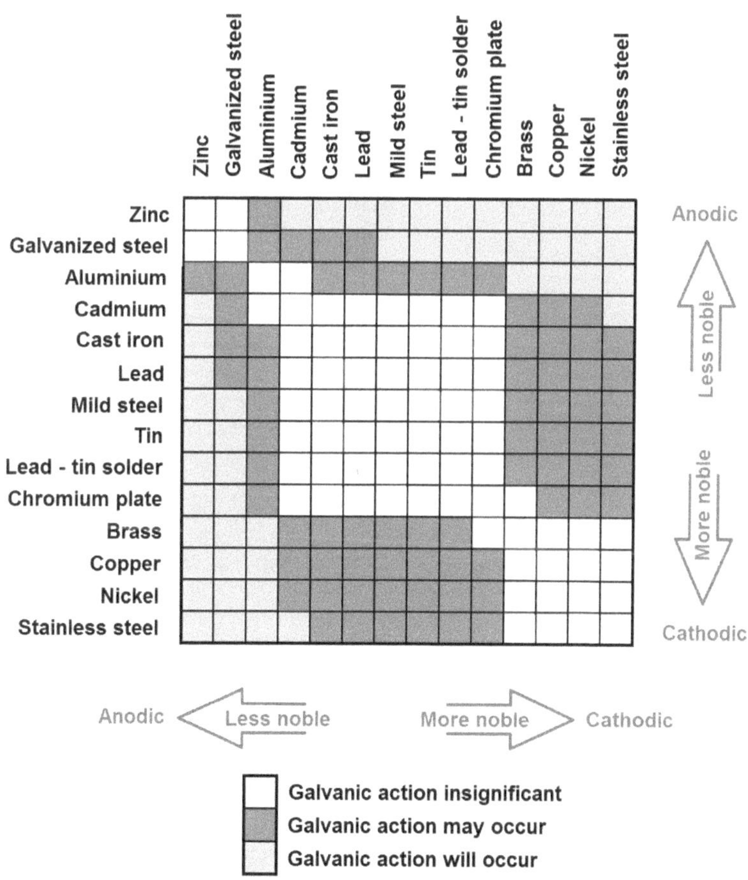

FIGURE 8.15 Galvanic corrosion chart for dissimilar metals, steels, and Dalloys. *Courtesy of Insertsdirect.com Ltd.*

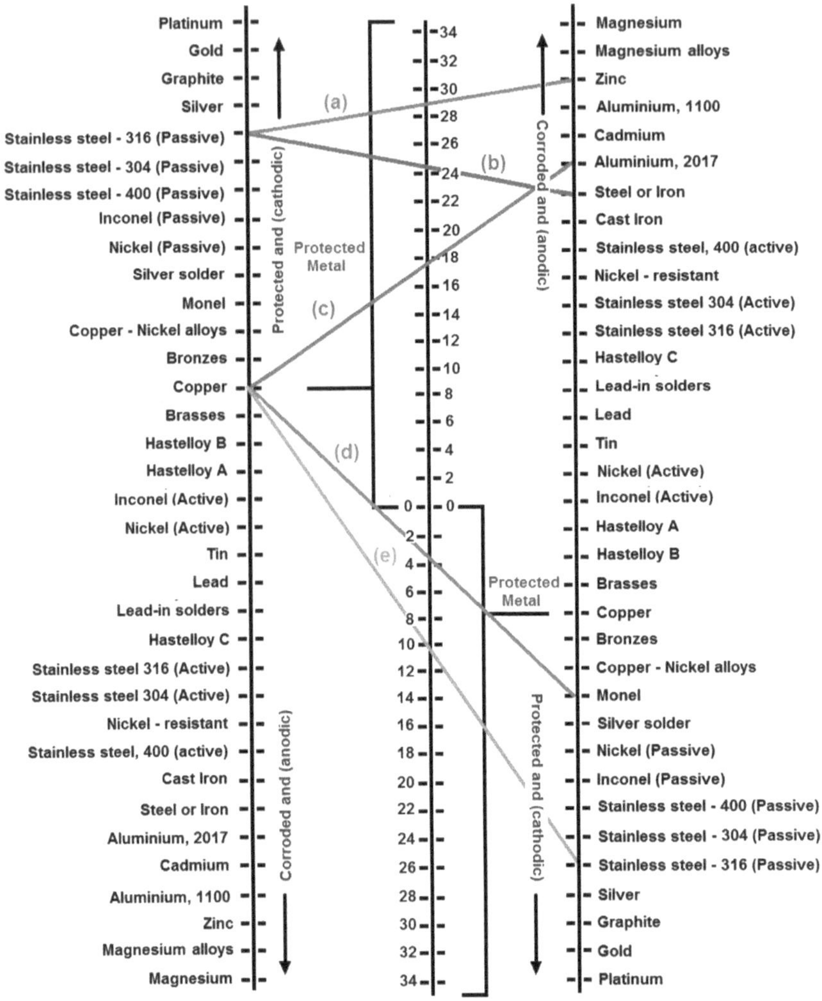

FIGURE 8.16 Galvanic corrosion metal compatibility diagram. Use a rule to line up the two metals, steels or alloys which are being combined. The closer to zero, the lower the risk of galvanic corrosion. (a) Combining stainless steel 316 with zinc (e.g., zinc coating of galvanized steel) will result in severe corrosion of the zinc coating. (b) Combining stainless steel 316 with mild steel (also called carbon steel or black steel) will result in the fast corrosion of mild steel. (c) Aluminum will quickly corrode when combined with copper. (d) Copper slowly corrodes in contact with Monel (contains copper). (e) Copper corrodes when combined with stainless steel 316.

8.5.2.4 Selection, Delivery and Storage of Replacement Parts

Minimize difficult-to-clean materials of construction, components, and spare parts, and avoid the use of piping, valves, joints, and fittings that may allow product build-up or hamper complete draining. Only parts and materials that are approved by the equipment manufacturer should be used to perform maintenance and/or to modify/repair equipment (temporary or permanent) in product-contact areas.

For optimum protection and cleanliness, the equipment supplier should deliver stainless steel 316(L) or 304(L) spare parts that are prepacked in plastic in a clean environment, and in accordance with proper Good Maintenance and Good Manufacturing Practices they should be stored segregated from other nonstainless steel products (e.g., black steel). For example, in Fig. 8.17, the stainless-steel equipment components are wrapped with plastic film, and their inlet and outlet connections are fitted with protective caps to prevent ingress of impurities, insects, and small animals.

Pipes, fittings, valves, components, etc. must be stored in dry, dust-free conditions, at a temperature corresponding to that of the mounting site. If this is not possible, the materials must be brought to the mounting site no later than 24 hours prior to the mounting so that they may achieve the temperature of the mounting room. This is to prevent condensation inside the pipes, which may cause welding defects and lead to rejection of the welds. Furthermore, precautions must be taken to prevent deformation of the stored materials through collision or insufficient support. The body and internal parts must be handled carefully to ensure that the machined surfaces are not damaged (Moerman et al., 2014).

8.5.2.5 Correct Installation Practices

Open profiles are preferred over hollow sections, but consideration should be given regarding their orientation. Open profiles installed in the horizontal plane preferably should have their folding turned downward, although outward turned constructions are allowed as long as they are cleanable. Installed in the vertical plane, open profiles must also have their folding turned outward. In the horizontal plane, open profiles should never have their folding turned upwards. If closed profiles are used, frequent inspection for

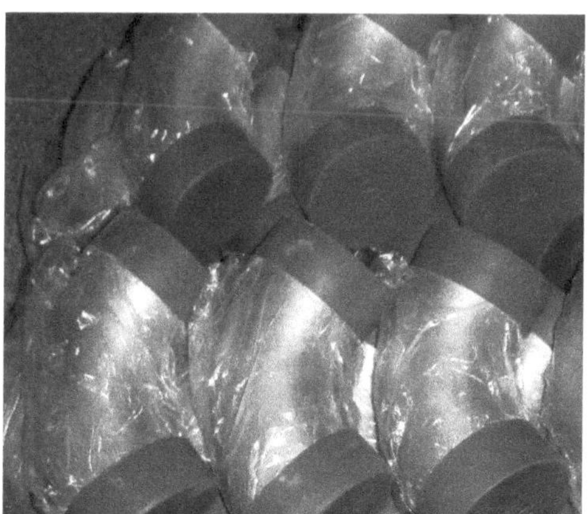

FIGURE 8.17 The stainless-steel equipment components are wrapped with plastic film, and their inlet and outlet connections are fitted with protective caps to protect them against corrosion, dirt, pests, etc. during transport and storage. *Courtesy of Zhejiang Jugang Pipe Co, Ltd.*

cracking should be carried out to prevent risk from contamination. Round or square section members turned through 45 degrees that provide sloping surfaces are recommended. Profiles may not create ledges, projections, and pockets where debris can accumulate.

Equipment must be self-draining (e.g., lines must be sloped). Always determine the correct installation situation and direction of fluid flow, so as to ensure appropriate cleanability and drainability. Especially, flow with respect to dead legs is of paramount importance.

8.5.2.6 Making Sheet Joints

Equipment is only as hygienic as the manner in which joints between equipment components are made. Use permanent joints (e.g., weld joints) rather than dismountable joints (O-ring or gasket joints), because the latter type of joints may give rise to projections, protrusions, edges, recesses, metal-to-metal contact, etc.

- Welded joints, ground and polished, are thus preferred over mechanical fixings, such as bolted or screwed joints. However, note that several types of common defects may arise in welded joints (e.g., misalignment, cracking, porosity, inclusions) which can act as a source of microbiological problems. All welds in the product contact area are recommended to be continuously welded and with sufficient weld seam protection (inert shield-gas protection at both sides). They must be smooth and free of pits, crevices, cracks, and pockets (Figs. 8.18 and 8.19), and—where required—must be polished to have the same surface finish ($R_a \leq 0.8$ μm) and appearance, etc., as the surrounding materials. Once the job is completed, any

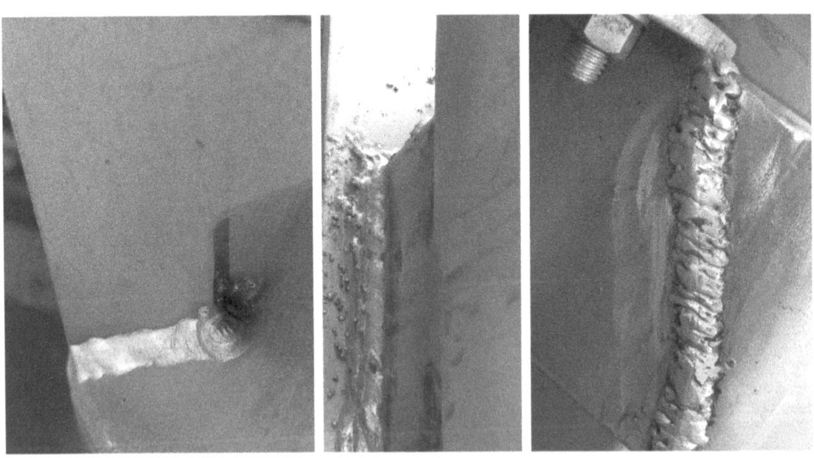

FIGURE 8.18 Poor weld repair. *Courtesy of Rudi Groppe, Heinzen Manufacturing International, © 2016.*

remaining debris should be brushed away and disposed of. Spot and unfinished welds (Fig. 8.20) are not allowed because they never will give hygienic constructions. They leave open spaces that may accumulate dirt, could provide a habitat for insects and microorganisms, and are difficult to clean.

- Dismountable metal sheet joints make use of fasteners (e.g., screws or bolts) to fix plates, appendages, etc. together, but only should be used if dismantling is unavoidable. Metal sheet joints made by means of fasteners (Fig. 8.21) may lead to metal-to-metal contact corrosion. Joining metal sheets with screws via the product zone is also not allowed, because they create gaps, dead areas, and/or exposed threads where microorganisms may accumulate and grow in the presence of food residues and nutrients. Also hexagon nut-and-bolt pairs (Fig. 8.22A) which protrude in the product zone should be avoided, because they give rise to

FIGURE 8.19 Poor weld repair. *Courtesy of Joe Stout, Commercial Food Sanitation LLC - Intralox, © 2016.*

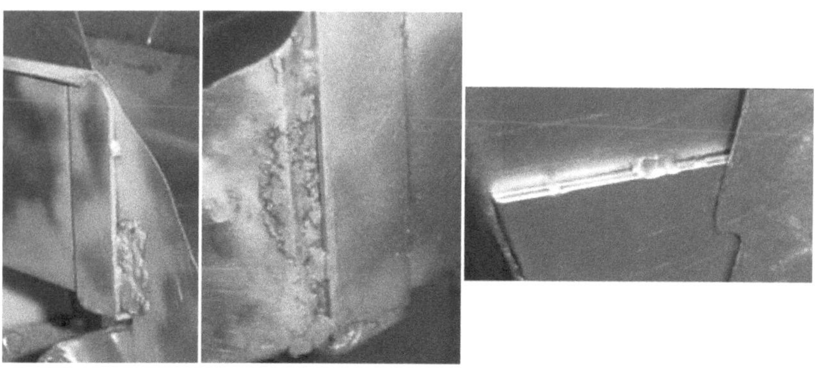

FIGURE 8.20 Noncontinuous welding of overlapping sheets of metal allow product debris and microorganisms to accumulate between the two sheets of metal. The overlapping sheets of metal also create a step. *Courtesy of John Butts, Land O'Frost, © 2016.*

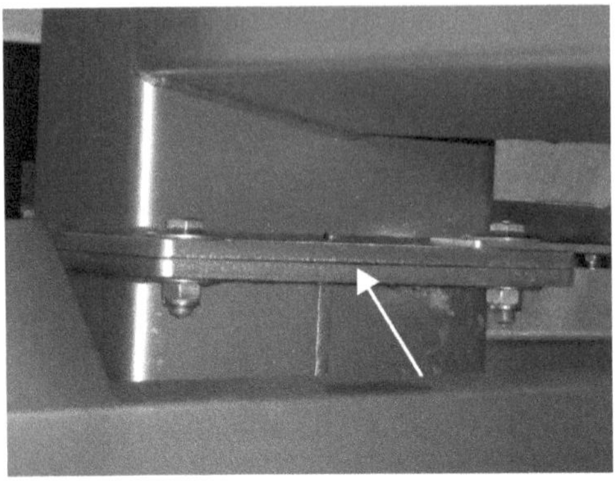

FIGURE 8.21 Metal-to-metal sandwiched area giving rise to a crevice where microorganisms may establish a niche to grow and the framework may rust. *Courtesy of Joe Stout, Commercial Food Sanitation LLC - Intralox, © 2016.*

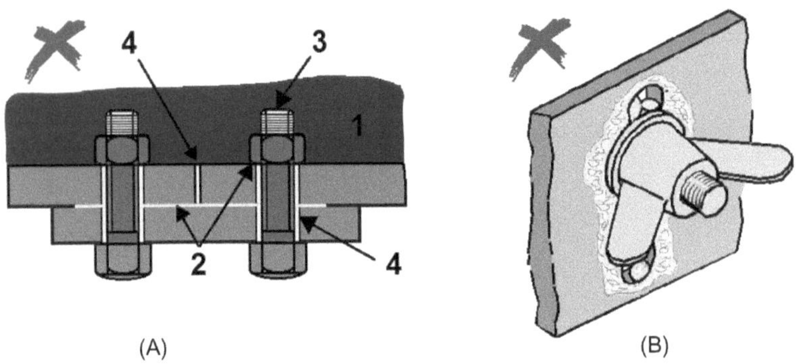

FIGURE 8.22 (A) Exposed bolt ends and nuts in the product zone (1) are not allowed because they give rise to metal-to-metal contact corrosion (2), exposed threads (3), and crevices (4). Debris also tends to adhere to and around fixings and provides nutrients for microbial slime growth. Exposed threads should be cut to the correct length or preferably domed nuts should be used. (B) Wing nuts are often used where adjustment is required but debris collects around and in the exposed portion of the slot behind the nut (CFPRA, 1983; Lelieveld et al., 2003; Hauser et al., 2004b).

metal-to-metal contact corrosion and create gaps, dead areas and/or exposed threads.

However, exposed threads sited on the product side can be covered with a domed nut and metal-backed elastomer gasket (Fig. 8.23). Hexagon-headed bolts with plain bolt head are allowed, although hexagon-headed bolts with domed bolt head are more preferable. Correct design of bolt heads and nuts and their effective sealing is thus essential to render the used bolts hygienic. Metal-backed elastomer gaskets not only may seal the crevice between the bolt-head or domed nut and the food-contact surface, they also will protect the annular clearance between the shaft of the bolt and the hole through which it passes (Figs. 8.23 and 8.24). Wing bolts and nuts, often used where adjustment is required, also should not be used on the product side to avoid metal-to-metal contact

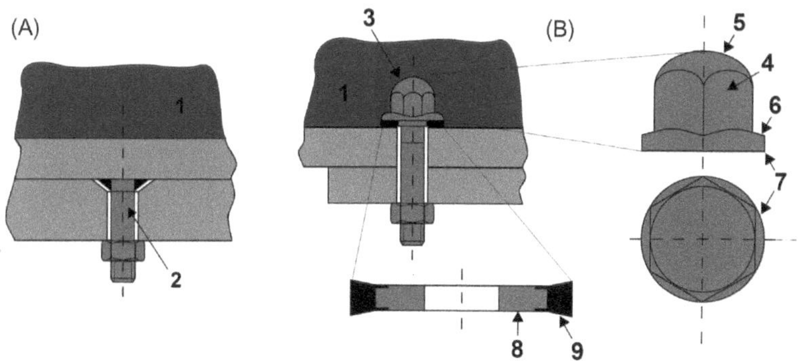

FIGURE 8.23 (A) To prevent crevices at the product side (1), screws, pins, or a stud welded on the nonproduct side (2) should be used. (B) A bolt head (3) that is hexagonal (4), domed (5) and provided with a sloped circular collar (6, 7) is easily cleanable, and the metal-backed (8) elastomer gasket (9) is used to seal the thread (Lelieveld et al., 2003; Hauser et al., 2004b).

FIGURE 8.24 Hexagon-headed bolts with plain bolt head are accepted, although hygienic bolts with domed bolt head are more preferred. The thread of the bolts can be covered with a hexagonal domed nut. Metal-backed elastomer gaskets may seal the crevice between the bolt-head or domed nut and the food-contact surface. *Courtesy of NovoNox KG.*

corrosion, as well as gaps, dead areas and/or exposed threads. Debris may collect in the exposed portion of the slot behind the nut (Fig. 8.22B), as well as around the wing head/nut and in the thread. Nowadays, more hygienic wing heads and nuts with metal-backed elastomer gaskets exist (Fig. 8.25). The hygienic wing nuts may completely cover the thread.

- Rivets only should be used where construction necessitates this type of fabrication. Pop rivets (Fig. 8.26) should not be used, certainly not in the product zone. Use solid rivets instead of pop rivets, even at the nonproduct side (CFPRA, 1983; Lelieveld et al., 2003; Hauser et al., 2004b).

8.5.2.7 Making Pipe Joints

It is strongly recommended that the number of joints, whether welded or detachable, is minimized. Cold bending of pipes is preferred over

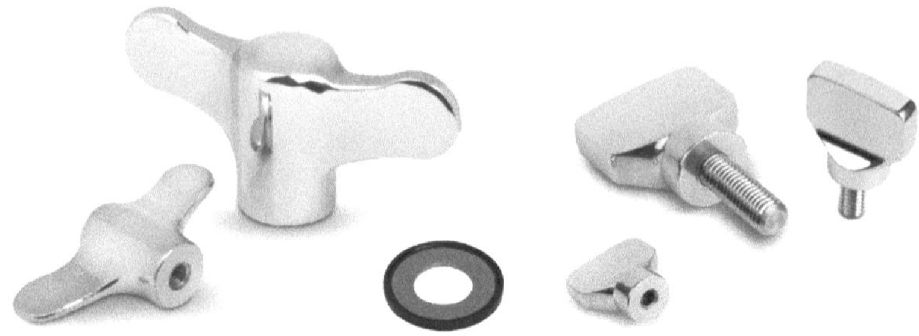

FIGURE 8.25 Cleanable wing nuts free of crevices in which food debris can accumulate. *Courtesy of NovoNox KG.*

FIGURE 8.26 The name plate is fastened with pop rivets. The small holes may accumulate dirt and become a niche for microorganisms to grow. Food debris, dirt, liquids, and microorganisms also may find their way behind the name plate. *Courtesy of Mondelēz International, © 2016.*

prefabricated bends, which have to be installed using joints. Although more hygienic, this is still true for welded joints as they also remain the weaker places in a process system.

- Stainless-steel hygienic tubing joints should be made by automatic orbital welding where possible and hand welding in those places that are difficult to access. However, those welds that are difficult to access should, wherever possible, be completed in the workshop prior to installation on the plant. Piping with the correct interior diameters should be applied because any mismatch in diameters or thickness may result in misalignment (Fig. 8.27) introducing a step in the wall or bore. If the diameters of the pipes to be joined are not the same, then the smaller pipe should be expanded to match the larger. Misalignment also can be due to incorrect fitting up (missed coincidence between the axes of the two coupled components) prior to welding. Alignment and clamping tools are available to ensure accurate alignment. Misalignment tolerance must be limited to less than 20% of the wall thickness. The two components also

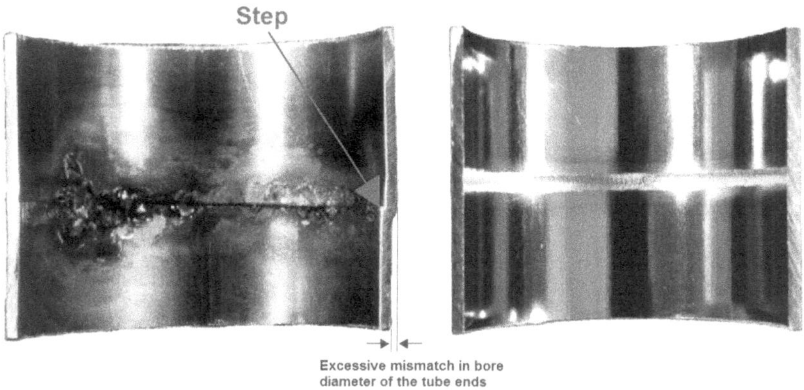

FIGURE 8.27 Step due to excessive mismatch in bore diameter of the tube ends.

FIGURE 8.28 Crevice in weld because both tube ends are not sufficiently pushed together. (Eastwood et al., 1993).

must be sufficiently pushed together to make a weld without gaps (Fig. 8.28). When tubings are welded together, the weld seams finally must be visually inspected on any discoloration, weld and surface-breaking defects, usually by endoscopy and aided by dye penetrant tests that highlight these defects. Inspection personnel should be trained and act with caution to avoid internal surface damages while handling endoscopic tools (Hauser et al., 1993; Kopitzke et al., 2006).

- Pipework also may be designed for rapid regular dismantling to permit cleaning. It is important to avoid crevices and gaps where product residues can accumulate and potentially begin to decompose. Therefore, from a hygienic point of view, the use of threaded piping is not recommended. To make detachable joints, the use of conventional O-ring grooves is also not recommended, because these groove designs leave a considerable free space in the groove. Other hygienic design requirements for detachable joints include coaxial alignment of the two mating bores, axial stop for controlled compression of the seal, room for thermal expansion of the seal, and avoidance of sharp edges such that seals are not damaged. Where there are depressions and steps of more than 0.2 mm in the pipe work, the flow of cleaning fluid may not thoroughly wash the surface and proper drainability of the piping will be hampered. Hence, when making bolted flange fittings, a lot of care should be taken to avoid offsets, gaps, penetrations, and voids.

A further aspect to be considered is that the seal material must be compatible with both the system product and also the cleaning/disinfection fluids which may be at a much higher temperature. A number of specific pipe couplings and also seal arrangements have been developed for hygienic applications. Some types are covered by national, international, or in-company standards. Many of these couplings are in use for a considerable time, but are not considered to be compatible with current hygiene requirements in some areas of the food and drink industry. Where pipes are dismantled, using a new seal or gasket on reassembly is a proof of good maintenance practice. Couplings made must be checked for leaks and retightened as necessary.

8.5.2.8 Other Unsuitable and Suitable Fixing Practices

Making fixings during construction, maintenance and repair is often required. It is obvious that these fixings may not create unhygienic conditions to ensure that safe food is produced once production is started or resumed. Therefore the following review of unsuitable and suitable fixing practices:

- Split pins (Fig. 8.29), self-tapping screws, wire hooks (Fig. 8.30), staples (Fig. 8.31), spring tension pins, bushings, etc. are unsuitable fastenings, because they retain food residues very easily and are difficult to clean and disinfect, which finally may result in the growth of microorganisms. Moreover, they may loosen and cause damage to other equipment or endanger the consumer physically.
- Avoid very small fastenings.
- Stainless-steel or dull-nickel plated fixings should be used as specified in the fixings and fastenings handbook.
- Avoid fixings in plastics which cannot be identified by metal detectors. There exist screws, staples (e.g., plastic staples to join conveyor belt ends), and pins in plastic (e.g., modular belts). However, modular belts with stainless steel pins are not available on the market. The use of plastic strips to fasten electric cables is common practice. A plastic strip of a color that is not omnipresent in the food product and food factory can be used. Nowadays, there are plastic strips on the market with metal content

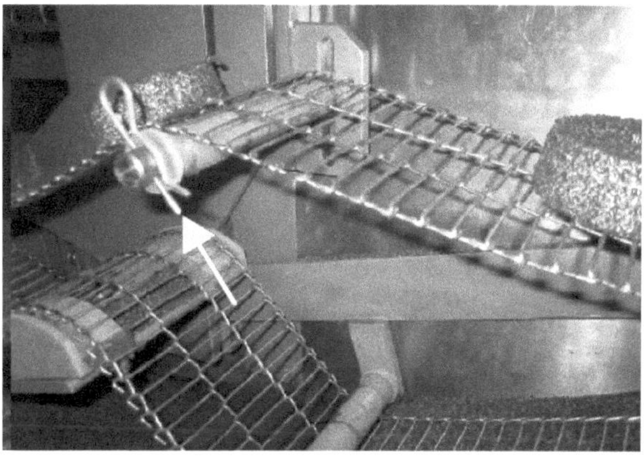

FIGURE 8.29 The split pin (indicated by the arrow) may get lost in the product. *Courtesy of Den Rustfri Stålindustris Kompetencecenter.*

FIGURE 8.30 Wire hooks to make conveyor belts endless are not hygienic. Due to the perforation of the conveyor belt, the reinforcing fabric layer becomes exposed, allowing liquids to penetrate into the interior of the conveyor belt by wicking (capillary action) (Kold et al., 2016).

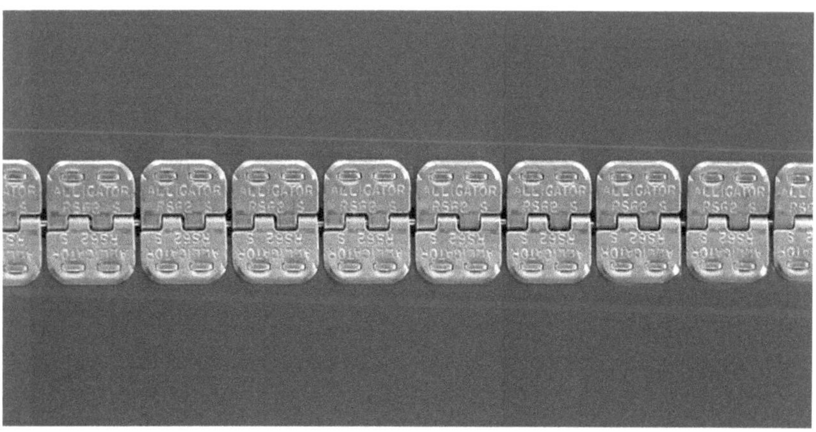

FIGURE 8.31 Staples to joint belt ends together are not hygienic (Kold et al., 2016).

dispersed throughout the head and strap with the objective that even cut-off sections can be detected by means of a metal detector. However, maintenance operators in the food industry must be prudent with such claims, as cases are known where even a lot of one hundred of these plastic strips could not be detected by means of a metal detector. However, it seems that they successfully can be removed from the product stream by means of magnetic separator systems. Stainless-steel cable ties can be used in very aggressive environments and have higher strength and heat resistance, but have the disadvantage that they can't be detected. The strongest neodymium magnetic separation systems, however, can successfully capture very fine stainless-steel particles that cannot be detected with a metal detector. Coated metal ties are available which are easily detectable. When selecting metal cable ties, one must keep in mind that they must not cause galvanic corrosion.

- Temporary repairs may be necessary and are allowed in emergency situations to stop leaks or product spills. However, hastily improvised repairs using rags, string, electric wire, tapes (waterproof, duct, etc.), twist-ties, cardboard, wood, or other porous or nondurable materials must be avoided, especially in product contact equipment. Where temporary repairs may adversely impact the food safety or quality of a product, they must be labeled, dated, and documented, while the maintenance department must be notified immediately to finally replace the temporary repairs with permanent repairs as soon as possible. Temporary repairs in the product area must be replaced by permanent repairs within 24 h and preferably during the next shift. In maintaining and repairing non – product-contact equipment, tapes, or other temporary repair materials must be replaced with appropriate permanent repairs in a timely manner or at a next scheduled down-time.
- All fastenings should be secured firmly.
- Take care for sufficient space around fixings for cleaning (min. 25 mm).

8.5.2.9 Replacement of Insulation

If old insulation containing asbestos has to be removed, all precautions should be taken to avoid the spreading of asbestos fibers in the food processing environment. The food processing equipment and food product must not become contaminated by these asbestos fibers. During the removal of asbestos, maintenance technicians must use the necessary breathing protection, because asbestos fibers may cause long-term health problems such as lung and peritoneum mesothelioma.

US Department of Agriculture (USDA) standards for food, soap, and cosmetic manufacturing plants prohibit use of insulations which sliver or dust, are toxic, or contain glass. Furthermore, non-chloride-releasing insulation should be used. For thermal insulation of vessels, appropriate qualities of rock wool are acceptable. It is highly recommended to install fully welded, vapor-tight, aluminum, or stainless-steel cladding, properly sealed to avoid ingress of dust, liqor, air, and moisture, and with joints facing downwards. The finish must resist washing down with high-pressure water, steam, and detergents without appreciable deterioration and may not support fungus, mildew, or bacteria growth.

To insulate piping, styrofoam (maximum continuous service temperature is 60–65°C), foam glass, or another rigid foam (e.g., foamed nitrile butadiene rubber) is preferred over fibrous materials. Also here vapor-tight aluminum or stainless-steel cladding must be provided at the outside. Damaged or wet insulation should be repaired or immediately replaced.

8.5.2.10 Installation of Cabling

When a new cable has to be installed, it should not be supported from a previously installed cable. Such a practice leads to an uncleanable and hygienically unacceptable entangled cable bundle, where soil can build up. Electrical cables should be routed, and connections made, in such a manner as to create hygienically acceptable installations conforming to the preset hygiene class applicable for that area. The cables should be fastened individually at a distance (no less than 25 mm) from each other to allow for proper cleaning.

8.5.2.11 Reducing the Surface Roughness of Stainless Steel

Note that surface roughness (R_a) is in fact not a suitable method of specifying product contact surfaces. The technique used for achieving the appropriate microsurface finish is of greater importance than the R_a value. Different surface finishing techniques (glass blasting, ceramic beats blasting, electro-polishing, pickling) successfully achieve a surface roughness R_a of 0.8 μm, but the topography/structure of the surface can differ a lot, giving different cleaning results. Electro-polishing (Fig. 8.32) is preferred over mechanical polishing, as physical

(A) (B)

FIGURE 8.32 Remediation of poor "factory" electro-polished equipment: (A) before mechanical and electro-polishing; (B) after mechanical and electro-polishing. *Courtesy of Ultraclean Electropolish Inc.*

and metallurgical damage takes place during mechanical polishing. Electrofinishing of stainless steel is a cost-effective method of obtaining a bright finish, allowing soil to be more readily observed than on dull surfaces. An electropolished surface is also durable and less prone to wear and corrosion. A roughness R_a exceeding 0.8 μm (non-product-contact zone) may be acceptable if test results have demonstrated that the required cleanability can be achieved through other design features or more intensive cleaning methods. Specifically, in the case of polymeric surfaces surface roughness is usually higher, but the hydrophobicity, wettability and reactivity may enhance cleanability

8.5.3 Lubrication According to the Principles of Hygienic Design

Food manufacturers should adopt a lubrication management system, comprising a factory survey to select the correct lubricants based on their technical and potential incidental; lubrication frequency; lubrication monitoring, sampling, and testing; recordkeeping for audit purposes; and operative training in the use of lubricants.

EHEDG guideline N° 23 (Steenaard et al., 2009) suggests that leakage of greases from bearings is a frequently occurring problem. Lubricants also often drip and splash from open lubricating points such as chains and open gears. Oil circulation systems, especially when the oil is under pressure from an oil pump, may allow small leaks to occur, which are difficult to detect. Hydraulic systems and hydraulically operated valves are other examples of potential sources of oil leaks. Leaks from oil-filled heat transfer systems are also difficult to detect. Leaks can also cause materials to corrode or to suffer electrochemical attack over a long period of time. Oil-coated machine surfaces, such as chutes used to transport food, are also a source of contamination risk that has to be managed. Apart from introducing toxic chemicals into the food product, lubricants also may introduce physical and microbial contaminants into the food. If a lubricant does not function properly, this may result in wear and tear, with the associated risk of product contamination by abraded particles. As lubricants may get contaminated with water and food materials, multiplication of microorganisms may occur at appropriate temperatures. Cases are known where *Escherichia coli* and *Listeria monocytogenes* in lubricants have given rise to food product contamination.

As incident contamination of food products with lubricants may occur, food-grade lubricants are required. Some food manufacturers use H1 lubricants only for critical lubricating points, with conventional lubricants being used for lubricating points that could not result in incidental product contact. However, wherever possible, it is recommended that conventional lubricants be replaced by H1 lubricants. This considerably simplifies the management of critical lubricating points, as errors do not lead to the use of potentially toxic lubricants at these points. Moreover, the number of lubricants can be considerably reduced. The EHEDG guideline document N° 23 (Steenaard et al., 2009) describes sequential steps by means of which the maintenance department effectively can replace conventional lubricants by H1 lubricants. When changing the oils in reservoirs, it is recommended to drain the system, change the filters, flush the reservoir with a food-grade product (flushing oil or original lubricant), check or change the filters again, fill from the reservoir with the correct food-grade lubricant, and take a sample for analysis. In the case of a grease application, it is recommended to check if the greases used are compatible. Clean the bearing out with a paraffinic oil or solvent, then fill the bearing(s) one-third to one-half full with food-grade grease and purge any grease lines with food-grade grease. It is recommended to seek assistance from the lubricant supplier as required.

Lubricants may undergo changes during use and may degrade as they become older or are exposed to water and food materials, allowing the growth of food pathogens. It is advisable to carry out regular checks on lubricants to determine whether they are contaminated with microorganisms. Lubrication points where H1 lubricants may become contaminated with beverages are also critical, as such contamination may encourage microbial growth. Therefore, some suppliers incorporate microbial-growth inhibitors in their H1 lubricants. For some applications, such as microbial fermentations, these substances must be effective exclusively against the undesired microbes, so that fermentation is not impaired.

When H1 and non-H1 lubricants are used within the factory, a system must ensure that no errors are made during packaging and labeling of storage and dispensing containers:

- H1 lubricants must be stored separately from toxic substances and dangerous materials. The storage of H1 lubricants in an area where conventional lubricants are also stored can lead to human errors and should be avoided. The use of dedicated transfer and storage containers for H1 lubricants is essential.
- Critical lubricating points must be labeled to reduce the risk of using the wrong lubricant. Text stickers or color codes can be used for this purpose. With the aid of this color code, the maintenance technician knows very well which lubricant he must use. A poster on a wall (Fig. 8.33) could exactly visualize which lubricant fits with which color, allowing prohibition of the use of the wrong lubricants for a given lubrication point.
- To prevent degradation, stored lubricants must not be exposed to extreme temperatures.

The equipment lubrication process should also be undertaken in a hygienic manner with attention to the following:

- Use drip trays where possible in case lubrication points are situated above the product.

ENLUSE
The Lubrication Reliability Specialists

Oil LUBRICANT LABELING CONVENTION

COLOUR	SHAPE	TYPE OF LUBRICANT	NAME OF PRODUCT	VISCOSITY
RED	#1	Gear Box Oil	(Company & product name)	ISO ?
BLUE	#2	Gear Box Oil	(Company & product name)	ISO ?
MID GREEN	#3	Hydraulic Oil	(Company & product name)	ISO ?
BLACK	#4	Hydraulic Oil	(Company & product name)	ISO ?
GREY	#5	Transmission Oil	(Company & product name)	ISO ?
PURPLE	#6	Transmission Oil	(Company & product name)	SAE ?
BEIGE	#7	Compressor Oil	(Company & product name)	ISO ?
DARK GREEN	#8	General Lube Oil	(Company & product name)	ISO ?
YELLOW	#9	Turbine Oil	(Company & product name)	ISO ?
ORANGE	#10	Motor Oil	(Company & product name)	SAE ?

FIGURE 8.33 A poster on a wall could exactly visualize which lubricant fits with which color, prohibiting the use of wrong lubricants for a given lubrication point. *Courtesy of Enluse B.V. (Frank Moerman, © 2016).*

- Use the correct amount of lubricant. Adding too much oil to reservoirs and bearings may cause leaking, which could result in direct food product contamination. The same is true if excessive amounts of grease are applied. Redundant lubricant and grease should be removed.
- To help prevent cross-contamination of different food-grade lubricants, it is recommended to use dedicated lubrication equipment for greases and oils (Figs. 8.34–8.37).
- All containers/implements used for measuring or pouring chemicals are to be restricted from alternative uses (e.g., labeled "for chemical use only") and should be cleaned before use (Fig. 8.37).
- The equipment should be filled carefully with a clean can and a clean funnel (Fig. 8.36), and a suitable cleaned tool should be used to apply grease (Fig. 8.37).
- Any spills must be cleaned up and soiled wipes disposed of correctly. Dirty, greasy, or oily hands should not be placed on any surface with which the product comes into contact.

After the lubrication maintenance has been completed, a paper or electronic job sheet should be completed and records kept for an appropriate period.

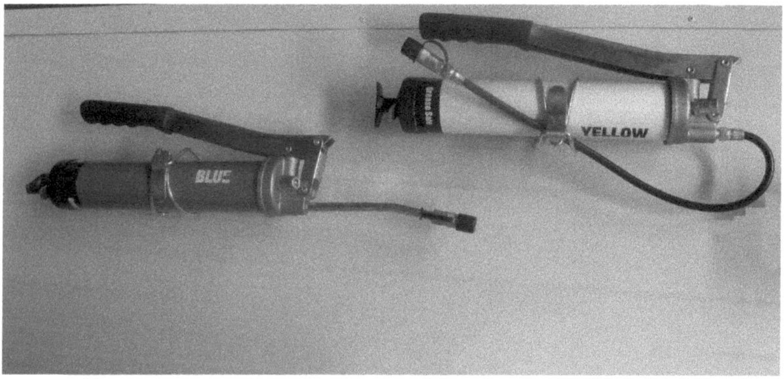

FIGURE 8.34 Color coding of lubrication tools prevents cross-contamination and misapplication of lubricants. *Courtesy of Enluse B.V. (Frank Moerman, © 2016).*

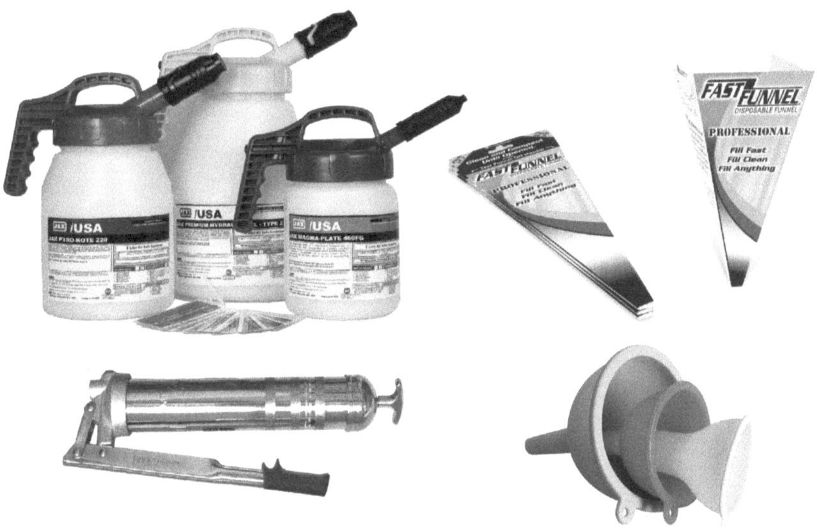

FIGURE 8.35 Dedicated lubrication equipment should be used for lubrication and greasing. In high-hygiene areas, the use of a stainless-steel grease gun and disposable funnels is recommended. Color coding of funnels and dispensing drums prevents cross-contamination and misapplication of lubricants. This color coding also may assign lubrication tools to a specific hygiene area. Additionally, color-coded labeling of all lubrication points will further prevent mixing of lubricants (e.g., food-grade and non – food-grade lubricants). *Courtesy of Justrite Manufacturing Company, L.L.C.; courtesy of S & S Concepts Inc.; courtesy of KitchenWerks.*

Storage of maintenance products should follow these stipulations:

- Maintenance products (oils, greases, lubricants, ammonia, glues, chemical products, etc.) should not be left in the food processing environment when maintenance operations have ceased (e.g., during the night, during weekends, during collective holidays).

FIGURE 8.36 The equipment should be filled carefully with a clean can and a clean funnel (Steenaard et al., 2009). *Courtesy of Van Meeuwen Lubrication B.V.*

FIGURE 8.37 The blue colored brushes used for applying the lubricant are not suitable. They have turned the lubricant from white into blue. Furthermore, the lubricant stored in the "in-use" container has been cross-contaminated with other non – food-grade lubricants, and has been exposed to the environment for too long, which has allowed dirt, pests, water, microorganisms, etc., to contaminate the lubricant. The container has also not been cleaned for a long time (Steenaard et al., 2009). *Courtesy of Van Meeuwen Lubrication B.V.*

FIGURE 8.38 Incorrect storage: contamination and use of incorrect lubricants can easily occur (Steenaard et al., 2009).

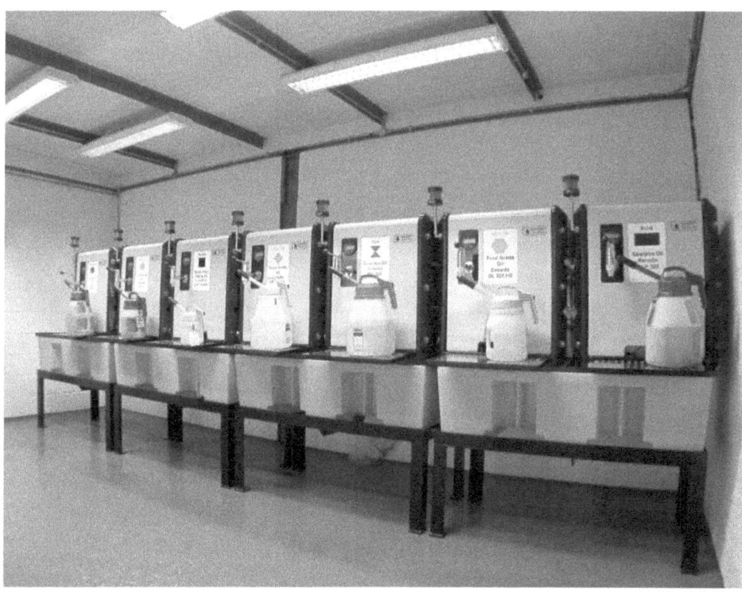

FIGURE 8.39 A correct storage facility helps to prevent contamination. Here, the food-grade lubricants are stored off the floor and in clearly labeled containers (Steenaard et al., 2009) *(Courtesy of Enluse B.V.). Courtesy of Van Meeuwen Lubrication B.V.*

- They shall be stored separately from food products in clearly labeled (identifying the maintenance compound) containers (e.g., bulk supply), that remain closed when not in use. These bulk containers must be stored in dedicated secure storage facilities, that must be kept clean and dry (Figs. 8.38 and 8.39).
- Maintenance compounds that are "in use" or for "immediate use" may be stored in processing and support areas, but only in quantities necessary

for immediate use. When transferred from their original container (e.g., bulk supply) to the new container (e.g., "in use" or for "immediate use"), the latter must be labeled with the name of the maintenance compound.

8.5.4 Recalibration of Measurement Devices

Recalibration is part of every maintenance program in the process industry, including the food processing industry. Instruments and controls used for measuring, regulating, or recording temperatures, pH, acidity, water activity, and other critical factors affecting microbial growth require regular and frequent monitoring for accuracy. Where the accuracy of the measuring instrument is compromised, recalibration is needed. Together with the instrument supplier, the food manufacturer must define the frequency of recalibration of each instrument, as well as the most appropriate method and procedure to do so. A calibration schedule with name, location, frequency and method of calibration should be written for all critical monitoring equipment. Calibrated equipment that is nonconforming (e.g., broken, expired calibration period) must be identified as nonconforming, and not used for critical measurements until it is recalibrated, repaired or replaced. Most operations "tag" their instruments after calibration. The calibration tag may include who did the work, the date the work was done, and the date of the next scheduled calibration. These tags should be made of materials that are water- and oil-resistant so that they will survive the rigors of production, including cleaning.

In no way should the safety of the food manufactured be brought into danger. The following procedures should be observed:

- Preferably, in-place recalibration should be done. The maintenance workers performing the calibration have to follow the hygiene practices applicable in the food factory rigorously (Fig. 8.40).
- The instrument branch as well as the measurement device should be visibly inspected for dirt build-up. They should be cleaned before the start of the calibration process.
- The calibration of certain measurement devices requires the use of calibration liquids. Traditional calibration methods for temperature measurement devices use a bath type where the liquid (e.g., silicone oil for high temperatures, alcohol for low temperatures) is pumped around axially to ensure temperature homogeneity all the way to the surface. In the food industry there is of course a need for "pure" calibration, which means that the sensor may not contaminate the food process after reinstallation. So the silicone oil or anything else that might be located in the bath must be removed from the food contact surface of the sensor before remounting in the process equipment or line. Nowadays, dry calibration of temperature devices is also possible. Calibration of pH measurement devices also requires the use of calibration liquids. Hence, as a general rule, measurement devices should be cleaned and disinfected before reassembly when necessary.

FIGURE 8.40 On-site calibration services have to follow rigorously the hygiene practices applicable in the food factory. *Courtesy of Endress + Hauser.*

Calibration can also be done off-site in an accredited calibration laboratory. Usually measurement instrument suppliers have their own calibration services, which ensures proper calibration according to the highest hygiene standards. Good calibration practices may reduce the risk of food safety hazards.

8.6 PERSONAL HYGIENE PRACTICES DURING MAINTENANCE OPERATIONS IN THE FOOD INDUSTRY

Plant personnel are among the most significant reservoirs and vectors of microorganisms, chemical residues (e.g., allergens) and foreign materials in the food facility, and as such, can be a source of unwanted contamination of products. The transfer of contaminants can occur through a direct route, such as bacteria transferred from the body, skin, mouth, hands or hair to the product, or indirectly via their personal belongings, such as clothing, footwear, utensils, and other tools used in their daily tasks. Because of their activities and movement around the food production site, maintenance operatives should view themselves as more of a hygiene risk than other food operatives and at a very minimum must be obliged to follow the same personal hygiene procedures as all other staff. Therefore, before the onset of maintenance and repair operations, all maintenance workers must comply with the requirements for personal hygiene appropriate to the area where maintenance and repairs will be executed. Many best practices in personal hygiene are well-established in the food industry (Aarnisalo et al., 2006; Smith and Keeler, 2007; NZFSA, 2009; Stier, 2012; Moerman et al., 2014; Margas and Holah, 2014).

- Maintenance technicians suffering from diarrhea, vomiting, uncovered sores or wounds, skin infection, heavy cold, flu, discharges from eyes, ears, nose should not come into the process areas. Particular attention should be paid to contractors, who may be unfamiliar with food hygiene requirements. Like all visitors to the factory they should complete a visitor health check form (a health questionnaire).
- Maintenance workers from contractors should be obliged to follow an introductory training session about personal safety, food safety, and personal hygiene practices before entering the food processing areas. There are many training aids available to food processors in the form of on-line or print manuals, posters and signage, videos, and employee refresher courses. When new hires attend their first orientation, they are typically bombarded with a lot of important information in a relatively short period of time. Providing a follow-up refresher session on personal hygiene policies two weeks after the orientation—even just 30 or 40 minutes to refresh and reiterate policies and protocols on hand washing, outer clothing, sources of cross-contamination, etc.—helps raise employee comprehension. At this point, employees have been in the plant for two weeks and have some experience to better understand how these practices have an impact on food safety.

  Often local community colleges offer simple, short training courses on food safety and hygiene practices for maintenance workers. Contracted maintenance technicians often have this training, allowing them to provide documentation if requested. These types of records are a helpful preparation if the food processing facility is audited for outside contracts or for any regulatory reasons.
- Maintenance workers must wear protective clothing not only to safeguard their clothes during maintenance and repair, but also to protect the product. These items typically include company-provided coats/smocks, plastic aprons or plastic sleeves, when appropriate. Maintenance clothing, like all factory clothing, should be of a food-safe design to prevent foreign bodies from shedding directly (e.g., lint, buttons) or indirectly (e.g., outside pockets from which objects can fall out into product). Whenever possible, smocks should not have outside pockets. Typically many aprons and smocks used in the food industry are constructed and designed to prevent microbial cross-contamination of the product from the employee.

  The protective clothing must be clean. Laundering has to be controlled by the company in order to achieve a greater level of confidence that these items have been cleaned and disinfected adequately before being worn in areas where they may come into contact with finished products. Prior to entering the food processing area where the maintenance and repair work has to be done, the condition of the protective clothing with respect to cleanliness, frayed edges, or loose items such as buttons or snaps must be checked. This is especially the case with maintenance operators of contractors. Preferably, they must be provided with clothing owned by the food manufacturer.

Essentially, protective clothing provided by the company should never be worn outside of the plant premises, should always be worn in the plant production areas, and should be regularly changed. Maintenance and contractor employees who have worked outside the facility, in "raw" or waste areas, must change into clean plant attire prior to entering production areas.

- Nonporous footwear should be worn, especially in the production areas. Footwear should be constructed of material that is cleanable. It should not be made of leather or cloth that will get and stay wet, which is uncomfortable for the wearer and may result in the maintenance technician avoiding the necessary foam sanitizer and foot dips so they don't have to be wet all day. The condition of footwear with respect to cleanliness, frayed edges or loose items must be checked. For access in a dry area, overshoes or shoe covers may be used to cover footwear.

  As footwear can be a vehicle for the transfer of pathogens, footwear must remain at the facility in order to mitigate contaminants carried into the plant from home. It is also good hygienic practice to dedicate footwear to a specific area. The footwear must be cleaned at the facility, and to do so appropriate cleaners and brushes for all maintenance technicians must be available. To decontaminate footwear before entry in the zone where the maintenance or repair work has to be done, foam sanitizers (Fig. 8.41) have proven to be very successful. They also offer the advantage that pallet jacks, forklifts and carts can be decontaminated simultaneously. Foot dips/baths and boot-washers are also quite common in use. Care must be taken that foot dips/baths and boot-washers don't become pools of bacteria, especially because organic material reduces the effectiveness of the disinfectant. Disinfectant solutions in foot dips/baths and boot-washers must be changed regularly. It is also recommended to closely monitor the microbial load and the concentration of the disinfectant, as well as the volume of the disinfectant solution.
- Hair should be kept short and trim, and be covered with a hat, cap, or hairnet. A snood also should be used to cover facial hair.
- In many food factories (especially the smaller ones), as maintenance tasks dictate, maintenance personnel are moving from clean to dirty and/or raw to ready-to-eat areas, as such spreading cross-contamination all over the factory. To prevent cross-contamination between different zones

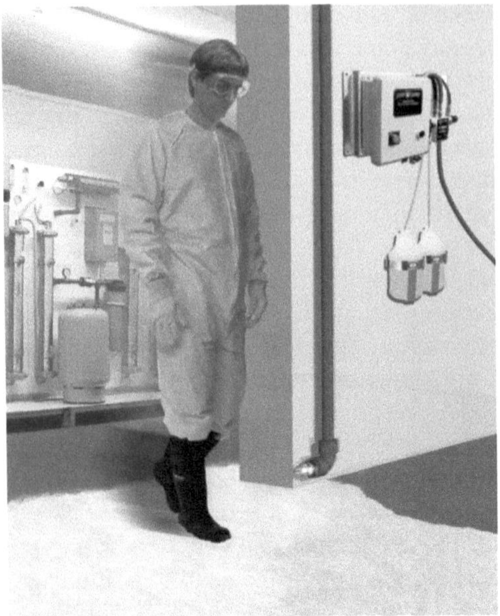

FIGURE 8.41 Foam sanitizers have proven to be very successful in decontaminating footwear before entry in the zone where the maintenance or repair work has to be done. *Courtesy Nelson Jameson Inc.*

from happening, a segregation of maintenance personnel can be implemented. As with their tools and equipment, mechanics can be dedicated by department, and a system of color coding (protective clothing, footwear, boots, hair covering, hats, etc. in a color specific for a given hygiene zone) may prohibit maintenance technicians from operating in a zone where they are not allowed to perform maintenance or repair work. Due to the color coding, their presence in a wrong zone is quickly detected by other staff members. Segregation of maintenance workers is easier to realize in large factories than in smaller ones. In smaller food factories the number of maintenance technicians is usually limited for cost reasons, requiring them to perform maintenance and repair tasks all over the factory.

- Jewelry, including tongue rings and body piercings, in food processing areas is not allowed, not only because jewelry may get lost in the product stream, but also because rings, watches, bangles, etc. can harbor dirt or bacteria which can affect food.
- All personnel entering the food processing area must always adequately wash and disinfect their hands to prevent contamination of foods and food-contact surfaces with microorganisms, allergens, chemical residues (e.g., lubricant residues, adhesive residues). Handwashing policies should require employees to wash after any type of activity that could contaminate the hands. Wearing sterile gloves could be a supplementary option. Training is an important part of instilling good handwashing practices. A handwashing verification method such as swabs or plate counting may act as a means of controlling that proper handwashing practices

are adhered to. They are effective training tools because people often are amazed when they see what is still growing on their hands after a rudimentary wash.

Handwashing stations must be sufficient in number and placed in convenient locations, to avoid employees skipping the handwashing. As maintenance workers need to wash their hands more frequently than process operators (their hands get soiled with residual food debris, dirt, dust, lubricant residues), handwashing stations must be easy to access and well-stocked so as to avoid maintenance operators omitting to wash their hands.

- Fingernails should be kept short and clean, because long nails harbor moisture and dirt, allowing bacteria to grow.
- Visual aids all over the food factory create continual employee awareness of personal hygiene best practices. The best way to accomplish this is to place signs on bulletin boards in hallways, break rooms and other high-traffic areas that reiterate personal hygiene messages, such as showing the sequence steps of good handwashing procedures or images that show how easily jewelry, tape, and pens can get into the finished product.

8.7 HYGIENE PRACTICES DURING MAINTENANCE OPERATIONS IN THE FOOD INDUSTRY

8.7.1 Recommended Hygiene Practices to Be Taken Before the Onset of Maintenance and Repair Operations

The following pre-service practices may create the necessary hygienic conditions allowing maintenance and repair without compromising the safety of the food produced with that equipment once production resumes (Jha, 2006; Smith and Keeler, 2007; NZFSA, 2009, 2010, Moerman et al., 2014):

- A maintenance program must be available that includes procedures that describe how to do the work. Maintenance technicians can only perform their work in a hygienic manner, if they know exactly what is allowed and not allowed during their activities. Mishandling equipment during maintenance and repair, as well as poor lubrication practices, may compromise the safety and quality of the food produced. Therefore the following procedures must be integral parts of a documented maintenance program:
  - maintenance procedures;
  - lubrication procedures;
  - tool reconciliation procedures;
  - procedures for temporary repairs;
  - procedures for emergency repairs;
  - spare parts inventory program;

  - training procedures;
  - "hand-over" procedures;
  - audit procedures to verify that the work is being done properly.

  These procedures should include a title, step-by-step actions to complete the work, who is responsible for the work, how and where records shall be maintained, corrective actions to be carried out, and, finally, procedures for management to verify that the work was not only done, but done properly.
- Some work such as drilling or welding will inevitably produce debris and dust. The area should be examined to assess the potential risk of contamination, and risk areas should be covered. A "Food Safety Maintenance/repair Plan" should be developed and shared with affected employees prior to major construction or renovations.
- Where necessary, traffic inside the food factory should be rerouted.
- Whenever possible, maintenance should be done in a separate room outside the food processing area. As an example, fabrication and repair could occur in the maintenance workshop, but then care must be taken that there is sufficient access for machinery/equipment to be brought into the shop. Weld, thread or cut operations also can be done in any other area screened (isolated) from ingredient or packaging material storage and product handling.
- Alternatively, when possible, production operators should remove food processing equipment from the processing room before repairs will be made, or the maintenance area must be segregated from production by use of tarps. No maintenance work shall be allowed during production if the necessary protection is not ensured.
- Adequate screening is of paramount importance in exposed food areas, to prevent exposed food from becoming contaminated with metal shavings, filings and other airborne particles generated during maintenance and repair, and to protect it from grease removers, lubricants, paints, and paint odors.
- An appropriate number of shields and plastic sheets must be available to contain possible contamination in the work area during on-line repairs.
- The tarps or plastic sheeting (polyethylene or equivalent film) draped over adjacent equipment must be clean and free from dirt and water.
- If it is necessary to stand on machinery, the equipment must be covered to prevent dirt and debris from contaminating the surface. If entry into process equipment is required, a plastic cover film must be laid down on the bottom of the process equipment.
- Maintenance workers must necessarily use many tools in the production area. The maintenance tools of contractors should be company-owned. Where practical, maintenance tools should be dedicated to a specific area, because tools may be a source of *L. monocytogenes*, and other harmful microorganisms, as well as physical hazards (foreign bodies).

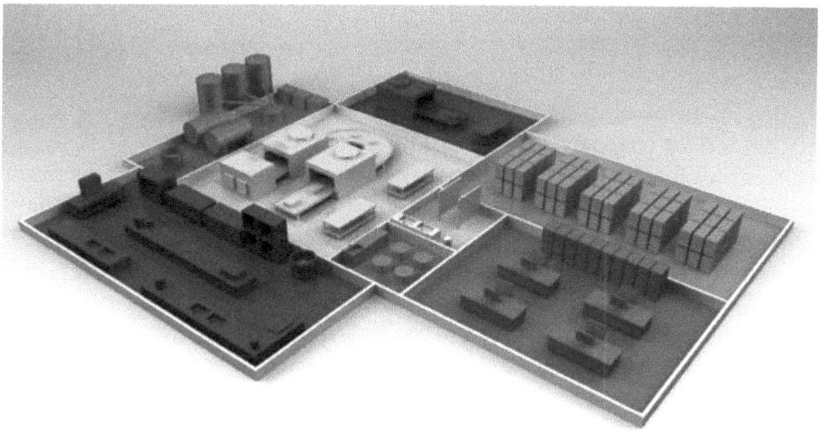

FIGURE 8.42 To improve hygiene and prevent cross-contamination, the production premises may be divided into visually separated zones, where each zone is assigned a specific color. In analogy with the color coding of cleaning tools, every zone can have its own colored set of maintenance tools and its own color-coded service equipment. Color coding is a tremendous tool because it works in all languages and makes these tools easily distinguishable from other production tools for better accountability. *Courtesy of Carlisle Sanitary Maintenance Products.*

No transfer of these hazards may occur from a less to a more hygienic area. By dividing the production premises into visually separated zones (Fig. 8.42) where each zone has its own colored set of maintenance tools and equipment, hygiene can be improved and cross-contamination can be eliminated. Also tool bags, tool boxes, receptacles, and trolleys may be provided with a color code. It is recommended that the color coding be the same as that implemented for cleaning equipment and cleaning tools.

- Segregation of maintenance tools and equipment may result in the following categories: food contact versus nonfood contact, allergenic versus nonallergenic contact, wet versus dry, areas with basic/medium/high hygiene requirements, equipment versus nonequipment (e.g., repair of drains), halal versus nonhalal, kosher versus nonkosher, etc. Tools used in areas where foods are exposed should be exclusively used in these areas.
- To avoid cross-contamination, workshops and storage areas for maintenance tools can also be designated to a specific zone. When each zone has its own workshop, the service crew must not cross less-hygienic areas. It is important to minimize the crossing of flows of people, goods, process equipment, cleaning tools, maintenance equipment, and maintenance tools. Floors as well as air currents may act as carriers of food pathogens, spoilage microorganisms, food debris (e.g., allergens), aerosols, etc. The fewer doors between different zones are opened, the better.

- Color coding also may promote hygienic behavior among employees and help to ensure that they understand the maintenance plans of the company.
- Broken parts of colored maintenance tools are visually more easy to detect if they become a foreign body in the product.
- Tools used for repair and maintenance must not compromise the hygienic status of any product or packaging material. The maintenance tools must be free of rust, peeling paint and niches for bacteria (scratches, cracks, pockets, threads, etc.); and without wooden handles or knurling soft rubber grips. They should be noncorrosive, easy to inspect, clean, and disinfect, with smooth finish and hard plastic grips, and with fitted heads for equipment longevity. They must be designed in such a way that they can't damage the process equipment (Fig. 8.43).
- Maintenance tools and aids preferably should be designed with a minimum of parts. As an example, it is recommended that employees use one-piece ink pens made of metal, rather than plastic cap-and-pen types. This reduces the chance of a pen cap falling into product and ensures that the metal detector will find the entire pen.
- Carbon steel tools with electroplated surface coating are available but have their limitations. The surface coating temporarily prevents the interaction of the free iron in the carbon steel tool with the oxygen in the environment, hence reducing the level of oxidation. However, because of intensive use and frequent cleaning/disinfection, the plating deteriorates

(A) (B)

FIGURE 8.43 (A) The maintenance tools suffer from corrosion, are painted (paint may peel off), contain niches and threads; have wooden handles or knurling soft rubber grips; or the tool heads may damage the stainless-steel surfaces of equipment and components. (B) These maintenance tools are made of stainless steel (noncorrosive), easy to inspect, clean, and disinfect, with smooth finish and hard plastic grips, and with fitted heads for equipment longevity. A system of color coding allows maintenance tools to be dedicated for use in specific zones, such as low, medium, or high hygiene areas. *Courtesy of John Butts, Land O'Frost; courtesy of Steritool (Frank Moerman, © 2016).*

with time, causing it to chip, flake, and peel. The tiny flakes and chips from the deteriorating chrome plating finally may become foreign matter in the food product. Furthermore, the exposed ferrous surface of the tool will transfer ferrous contaminants to the surface of a stainless-steel fastener or equipment, which subsequently will result in the formation of iron oxide, even at low levels of humidity. Moreover, bimetallic corrosion may occur. Therefore, stainless-steel maintenance tools suitable for intensive use and easy to clean and disinfect must be used. Stainless steel 316L, however, is not the ideal material of construction for maintenance tools, because it is not capable of achieving the hardness level required for high torque applications. As an example, the small serrations in the jaws of common slip joint pliers would likely wear too easily under normal use if they were manufactured from stainless steel 316L. Precipitation-hardened low carbon stainless steel 465 has proven to be excellent as a material of construction for maintenance tools due to its greater hardness, excellent tensile strength, and sufficient corrosion resistance over a long period of time. Maintenance tools with a long life reduce the need for replacement (Pekarsky, 2007).

- In certain conditions the use of nonmetal tools is preferred over metal tools, especially if the latter can damage process equipment parts.
- When maintenance or repair is performed at height, technicians need to use tools that are secured or tethered to prevent a hazard. Accidental dropping of tools while working at height can have devastating consequences. Smaller objects such as wrenches or sockets can cause severe injury to employees working beneath, while heavier tools such as a hammer can kill them. An injured employee may cost a food producer or maintenance contractor a lot of money: medical bills, workmen's compensation claims, increasing safety insurance policies, potential lawsuits, bad publicity and reputation damage, workforce and labor unions discontent, strikes, etc. A wrench slipping out of a utility worker's hand or a socket slipping out of a tool pouch also could cause damage to sensitive infrastructure (e.g., sensors) and equipment (e.g., damage to insulation, inside damage in a tall tank). A falling tool isn't necessarily going to drop straight down, but often ricochets when hitting several objects on its way down to the ground. It thus can damage a multitude of components.

Reliable retention of tools to prevent them from falling when working at height not only may prevent employees from injury and death and protect equipment against damage, it also may reduce the possibility that tools get into product, and it provides accountability to prevent tool loss. Moreover, a built-in drop-prevention system reduces the chance that tools may be laid on food-contact surfaces or places prone to dirt build up, microbial contamination, etc.

Tools with built-in coils or other fastening devices are available allowing lanyard attachment. These lanyards can be attached to the operator's wrist, belt, harness or other suitable tether site location, while ensuring that the original functionality and quality is maintained (Fig. 8.44). Tools weighing more than 2.5 kg should be attached by lanyard to the operator, more specifically by means of one lanyard hook attached and locked to an approved attachment point on the technician's safety harness, belt, or wrist strap, and another lanyard hook attached and locked to the selected tool (Fig. 8.45). Tools weighing more than 5 kg should be attached by lanyard to a suitable fixed point.

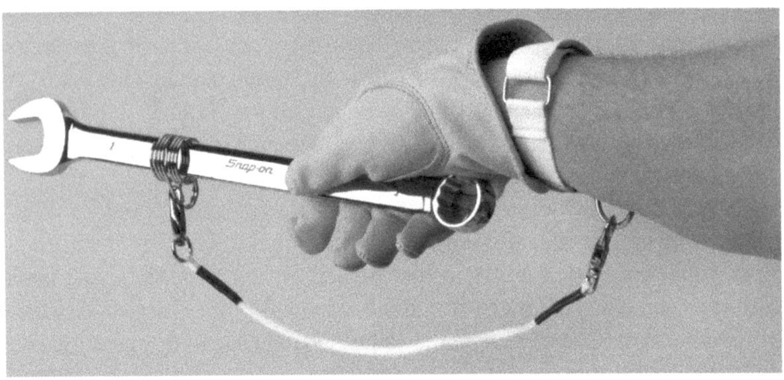

FIGURE 8.44 Although the tool is secured, the lanyard gives technicians unrestricted tool use and doesn't impede mobility or compromise personal safety equipment. *Courtesy of Snap-on Industrial.*

FIGURE 8.45 Means of attachment such as coils or other fastening devices are built into tools, and are not added as an accessory later. *Courtesy of Snap-on Industrial.*

- Ordinary steel wool or steel brushes and scrapers should never be used on stainless steel surfaces, as particles of steel may get embedded in stainless-steel surfaces and rust.
- Utensils such as knives, spoons, scoops, and ladles used for handling or measuring toxic chemicals and other nonfood materials should not be used for food contact.
- Maintenance tools must be used with care so that they cannot be left in the production equipment. Use only the maintenance tools that are required for the job. The fewer tools that are taken in the area where servicing of the equipment must be done, the less chance that tools will get lost in the product stream or will be forgotten once the maintenance/repair job is finished.
- Maintenance tools must be stored off the floor, and in no way should service technicians pick up items from an uncovered floor. Furthermore, replacement equipment and parts must be kept at least 0.5 m away from walls to avoid pest infestation and breeding.
- Maintenance tools must be inspected for broken parts. As soon as the slightest sign of deterioration is observed on a tool, the tool needs to be discarded and replaced with a new one.
- Maintenance tools must be inspected for cleanliness. Where required, they must be recleaned. In all circumstances, it is essential to disinfect maintenance tools before entry in the food processing area. Use cleaners and disinfectants that are less harsh on these tools (less corrosive), while still being effective in the removal of food residues and dirt, and in the inactivation of food pathogens and spoilage microorganisms. Note that the maintenance tools must be thoroughly dried after cleaning.
- Because maintenance staff are a foreign body risk, all unsecured objects (e.g., pens, pocket notebooks, small screwdrivers, pencils behind the ear, nonattached ear plugs, nuts and bolts in shirt pocket) which could fall into the product must be stored in the toolbox or the carrier used to bring parts to the work site.
- Receptacles for maintenance tools should be marked in a clearly visible fashion, to show that they are "only used for maintenance operations."
- Tool bags and boxes, receptacles, and trolleys used by the maintenance staff and contractors should be assessed for their suitability for application in the intended environment (Fig. 8.46). Where necessary they should be cleaned and disinfected before being brought into processing and/or support areas. To do so, the tool bags and boxes should be made of a material that can be easily cleaned, e.g., not fabric. A documented cleaning and disinfection procedure must be implemented for that purpose.
- Doing maintenance work and reparations over exposed product while standing on ladders and platforms is not allowed, because maintenance debris, nuts, bolts, screws, tape, lubricant, or any other dirt, as well as

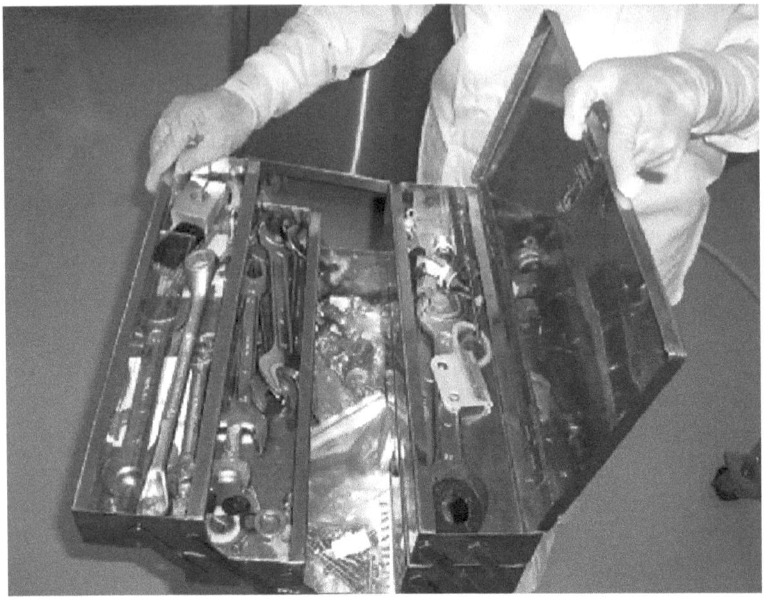

FIGURE 8.46 Toolbox is suitable for use in this high hygienic area. *Frank Moerman, © 2016.*

maintenance tools, may fall onto production lines and into the food products beneath. During maintenance activities at height, production of exposed food products must be stopped. If open process equipment and lines used in the processing of exposed food can't be moved aside, they must be protected by means of covers, and thoroughly cleaned and disinfected once maintenance and repairs are finished.

- Stairs, ladders, scaffolds, platforms, pallet jacks (to move heavy equipment components), cherry pickers, etc. (for work at height) must be free of damage and corrosion, and clean. Where possible, they should be cleaned and disinfected.
- Stairs, ladders, scaffolds, platforms, etc. must be made of impervious, noncorrodable, easy to clean and impact-resistant materials of construction. They must not have sharp corners and niches (scratches, cracks, pockets, threads, etc.) for bacteria to hide. Wooden planking as scaffold platform is not allowed. Work platforms must consist of steel plates (provided with a coating as protection against corrosion) containing a raised antislip material and preferably also kickplates (toeboard) over the whole perimeter of the walkway or platform. Open grating is not allowed. Stairs and ladders are often made of aluminum which has the advantage of being low in weight. Detergent and disinfectant solutions used to clean and disinfect stairs, ladders, etc. may not adversely affect the aluminum (must not induce corrosion).

- Anything that moves within the plant (stairs, ladders, scaffolds, platforms, pallet jacks, cherry pickers, etc.) has to be controlled: use, location, etc.
- Debris from engineering workshops (such as swarf and other unwanted materials) must be prevented from entering processing or support areas. This is especially important where engineering workshops have access ways (e.g., doorways) that lead into processing or support areas. This may be achieved by keeping doors closed, the use of swarf mats, boot washes, etc.

8.7.2 Recommended Hygiene Practices During Maintenance and Repair

The following hygiene practices should be followed during maintenance and repair (Smith and Keeler, 2007; NZFSA, 2009, 2010; Moerman et al., 2014):

- Before maintenance and repair can start, residual product often must be removed from the equipment, line, etc. It is essential to avoid spills and splashing onto ingredients, products, packaging materials or adjacent process equipment. Contaminated ingredients and products, as well as spills, have to be removed as soon as possible. In some cases of product spills, leaking water, etc., enhanced microbiological monitoring for food pathogens may be required.
- As opening of certain systems may endanger maintenance operators (e.g., utilities such as steam piping, compressed air lines, food gas lines), the necessary actions must be taken to avoid any injury.
- Ensure adequate lighting, particularly where detailed or intricate work is required. Lamps with higher light output may permit the factory staff to perform inspections of the food processing equipment and the process environment more easily and profoundly, enhancing the detection of grease, leaking oil, failures, maintenance residues, etc. However, light sources should not be placed above open process equipment, or the lamp should be housed in a shatter-resistant fixture to avoid shattering of glass that may lead to broken fragments falling into open processing equipment. By using a protective PTFE or acrylic coating, one may also maintain the integrity of the lamp in the event of breakage. Light sources used during maintenance operations should not contain mercury. Torches to light dark places within process equipment should be resistant to breakage.
- Glass is not allowed in food processing areas. However, in older facilities, glass still can be found. The only glass (other than jars and bottles for packaging of food and beverages) permitted in food factories should be windows and fluorescent light tubes, light bulbs and lamps. Care should be taken to ensure that glass does not enter food products. As an example, when changing light tubes, light bulbs and lamps, glass should

be prevented from breaking and showering the production line. As a means of protection, while still working at height, put removed lamps, light tubes, and bulbs immediately in a fully secured and lockable damage-resistant container or bag. Where possible, the production line should be moved aside.

- Replace light sources in glass by lamps, light tubes, and bulbs covered with a PTFE or acrylic sheath, or use shatter-proof plastic light sources as substitute. Additionally, shield the lighting with a plastic cover.
- Opening the distribution system will expose the system to particles from the outside environment. To keep the interior of the process equipment and components as free as possible from any exterior contamination, the food producer can protect the equipment openings. Cap open ends on lines while performing the maintenance and repair work.
- Any risk of contamination during maintenance/repair or when altering the system can be minimized by performing maintenance/repair during shutdown periods (e.g., collective holidays) and by using strict specifications on how to conduct activities, such as cutting pipework, and handling pipes and components before the actual installation.
- Avoid placing dirty hands on any surface with which the product comes into contact.
- Precautions should be taken to prevent the distribution of any contamination residues or mechanical damage residues in the surroundings. Doors and windows should remain closed during maintenance operations, to prevent high-velocity air currents from entraining maintenance debris. These high-velocity air currents also may occur in the neighborhood of exhaust openings and the air supply. However, natural or mechanical ventilation (at low air velocity) should be provided to minimize the likelihood of airborne contamination of food and to provide a safe working environment by effectively removing smoke, fumes, combustion gases, toxic gases, oil vapor, metal vapor, obnoxious odors, dust, etc.
- Facilities for good housekeeping in the maintenance area must be present. Maintenance spills must be removed to avoid cross-contamination. It is recommended to collect most of the maintenance debris at its source. Vacuum cleaners (Fig. 8.47) should be applied to extract maintenance debris at the place where the maintenance takes place, drip pans should be used to collect oil, etc.
- Equipment components subjected to maintenance, spare parts and tools should not be placed on the ground or walking surface (e.g., platform deck). Storage should occur (i) in a receptacle, box, carrier or trolley provided with a plastic cover, (ii) on a hygienic rubber mat or plastic pallet provided with a plastic cover, or (iii) hanging on a parts rack. Also they eventually should be designated by color for their intended use. In the

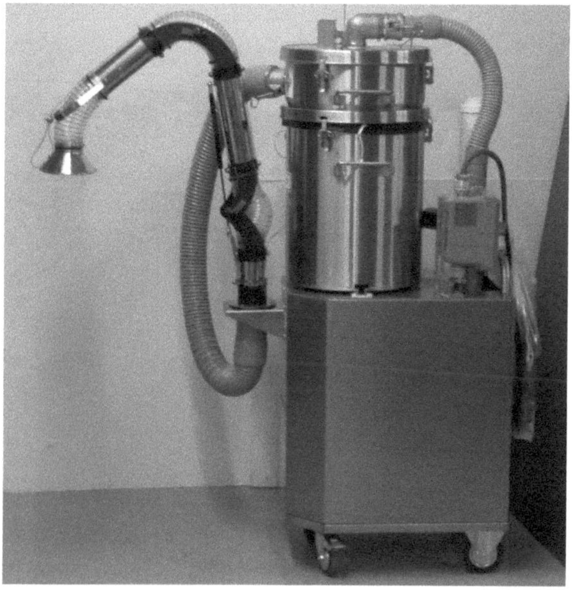

FIGURE 8.47 Vacuum cleaners should be applied to extract maintenance debris at the place where the maintenance takes place.

food processing area, no wooden pallets should be used to store new or replaced equipment components.

- Dirty parts (e.g., caps and gaskets) must not be stored in baskets intended for clean parts.
- Whenever parts and tools are stored in the production area, they should preferably be kept in rooms or lockers reserved for that use.
- Equipment components in service should be clearly indicated and placed in quarantine.
- To facilitate correct reassembly, disassembled equipment parts should be positioned in chronological order of disassembly.
- Take care not to lose nuts, bolts, etc. when removing them from machinery. Because small parts easily can be misplaced, loose bolts, nuts, screws, rivets, washers, etc. should be stored in maintenance receptacles.
- Bolts, nuts, screws, etc. of a lower alloy composition may not be used with stainless steel, because they may induce corrosion.
- Proper access for maintenance should be ensured, e.g., by stepladders or mobile platforms (cherry pickers) to prevent, e.g., maintenance personnel from stepping on the cladding of insulated piping which could result in it becoming damaged. When the insulation is torn, it could become a sanitation problem (harborage of dust, insects and rodents; absorption of moisture from air or spills may allow the growth of molds). Damaged or wet insulation should be repaired or immediately replaced.

- Personnel must be trained and suitably skilled in the correct access, handling and use of approved maintenance compounds, or have access to documented directions. Documented directions must be followed, either at the point where the compound is used (e.g., on the container label), or on information data sheets available to the person using the compound. Processing areas or equipment contaminated by a maintenance compound must be corrected according to the maintenance compound's properties and its effect on the product's fitness for its intended purpose.

8.7.3 Recommended Hygiene Practices After Maintenance and Repair

After maintenance and repair operations, the following practices should be followed (Smith and Keeler, 2007; NZFSA, 2009, 2010; Moerman et al., 2014):

- If emergency repairs were required during production, any product that may have been left sitting for long periods of time or may have become contaminated during repairs should be disposed of or scheduled for reprocessing to prevent any potential for poor quality or contaminated finished product.
- Maintenance tools or machinery must be removed or returned to storage without delay once maintenance or repair work is completed. Therefore, maintenance technicians must verify that all maintenance tools and components are removed after maintenance and repair to ensure nothing is left, as it may enter the product or damage the equipment. An inventory can be made of all tools prior to maintenance. It is good practice to "count-in" all tools and replacement parts taken to a job and "count-out" all tools and removed parts to ensure that everything is removed from the food processing area. It is possible to make someone responsible for tool reconciliation, and that only he receives access to the tool storage place.

  Tool containers with foam cutouts may help technicians in that task. When tools are returned at the end of a project, any open space alerts the technician of a potentially missing tool (Fig. 8.48). Nowadays, electronic tool storage and control systems to automatically track individual tools by user with bar codes, scanners, RFID tags, or other add-ons are available. They are provided with a warning alert system, which is activated when one or more tools are missing, or incorrectly positioned. As a further advancement, these toolboxes can be networked, either wirelessly or by Ethernet and managed by powerful, easy-to-use software. The administrator can view all the boxes on the network, whether they are on- or off-line, and the status of each box—including the number of tools issued, the active users, and all history.

FIGURE 8.48 Tool containers with foam cutouts may support technicians in tool accountability. When tools are returned at the end of a project, any open space alerts the technician of a potentially missing tool. *Courtesy of Snap-on Industrial.*

- To aid in tool identification, tools can be etched with company name, logo, serial number or any other ID marker. The identification number also can be etched into the specific storage space for the tool. The engraving must be designed for minimum dirt accumulation and maximum cleanability. Engraving may not be too deep and too small, and should not have sharp corners.
- Before storage, tools must be cleaned, disinfected, and maintained in a hygienic manner. If tools are not cleaned and disinfected before storage, the storage room can get contaminated with food pathogens, spoilage microorganisms, allergens, maintenance debris, lubricant residues, etc. The maintenance staff must be provided with the necessary facilities to clean, disinfect and maintain their tools, toolboxes, trolleys, etc.
- Broken maintenance and repair tools must be discarded. Making a service technician responsible for control and maintenance of service equipment and maintenance tools guarantees that all service equipment and tools are well-maintained and timely replaced when they are worn or in poor condition.
- If required for continuous use, tools must have a designated storage place (e.g., racks, trolleys, tool boxes, etc.) adequately segregated from ingredients, products and packaging materials, and protected from pest access and harborage.
- Tools dedicated to specific uses must be stored separately.
- The maintenance workshop and storage room for replacement parts, maintenance tools and maintenance aids should be neat and clean, and

FIGURE 8.49 Storage room for replacement parts, maintenance tools and maintenance aids should be neat and clean, and allow storage in dry conditions. *Courtesy of Mondelēz International, © 2016.*

allow storage in dry conditions. No poor maintenance housekeeping is allowed (Fig. 8.49).

- Also maintenance aids such as safety goggles, gloves, sample bags, ink pens, thermometers, tape rolls, etc. must be removed and returned to storage. A "count-in" inventory made prior to maintenance will facilitate their "count-out" and return to storage at the end of the maintenance and/or repair job.
- Handling and storage conditions should be monitored and recorded to assure that the specifications are met and that the controls are effective.
- Any part removed from equipment that is suspected of being microbiologically contaminated must immediately be sealed into a container or plastic bag to ensure that it does not "drip" contamination around the food processing area.
- Maintenance debris (e.g., abraded particles, swarf) may act as an abrasive that grinds off more particles from the pipe or equipment wall. Therefore, it is necessary to flush the system after maintenance and repairs.
- Appropriate waste removal and disposal procedures should be in place. All nuts, bolts, screws, nails should be accounted for and removed. Any other refuse (e.g., packaging materials, broken components, failed parts, drill bits, metal shavings, dirt, dust, spilled oil) must not be allowed to accumulate in production areas, but should be regularly removed to a suitable storage area and without delay. Maintenance waste and refuse should not be collected in ingredient or product containers but disposed of in covered garbage containers labeled with "MAINTENANCE WASTE."
- Damaged, decommissioned, or idle equipment must be removed as soon as it has served its purpose. It must be stored in an appropriate way to ensure that it does not become a source of contaminants or harbor pests. Equipment that could be a source of contamination must be physically isolated from processing lines and product, or removed from processing

FIGURE 8.50 Decommissioned or idle equipment must be stored in an appropriate way to ensure it does not become a source of contaminants or harbor pests. When stored outdoors, it should be placed on a hard standing (e.g., concrete, sealed, or paved area) and covered, which is not the case in this example. Moreover, the equipment is not appropriately covered.

areas. Nothing should be left in a food production facility that is not part of the production process. Damaged or decommissioned equipment that remains in processing areas must be clearly identified as such, to ensure that it is not used. Decommissioned equipment may be stored outdoors, but should be placed on a hard standing (e.g., concrete, sealed, or paved area) and covered (Fig. 8.50).

- When it is necessary to "break in" to the system for maintenance or inspection, documented "Clean Before Use" procedures should be in place to ensure that equipment is clean and will not compromise product integrity when returned to service. Therefore, equipment should be thoroughly cleaned any time maintenance or repairs of any type (e.g., drilling, cutting, polishing and welding) are performed in a food processing facility. The equipment and area should be cleaned with hot solutions of detergent (s) and disinfectant(s) in the right concentration, then rinsed, and finally dried prior to resuming production. Cleanliness and microbiological condition of the equipment should be confirmed by taking indicator and/or food pathogen swabs. The equipment may need to be recleaned, disinfected, and rechecked before being placed into service.
- Cleaning the equipment and area after maintenance and repair also provides the opportunity to check a last time for misplaced tools or parts.
- After servicing, all covers and guards must be put back in place. Missing guards may cause injury to process operators, while missing covers will allow debris, dirt, or water to collect in the equipment.

8.8 EVALUATION OF THE QUALITY OF MAINTENANCE WORK DONE AND RECORDKEEPING

Before production resumes, the food manufacturer must evaluate if finished maintenance operations and repairs meet the expectations with respect to the quality of the maintenance and repairs. From this perspective, the following practices should be followed (Moerman et al., 2014; BRC, 2015):

- Equipment must be subjected to a preoperational check before processing recommences. Have all technical problems been solved and is the equipment operating correctly? Have maintenance and repairs been done in a way that the process equipment allows the production of safe food products once production resumes? Maintenance technicians should read and sign off on Good Maintenance Practices.
- Equipment operating under validated conditions must be revalidated if the repairs and maintenance activity may affect its validated status (e.g., replacing temperature probes/sensors in ovens/freezers).
- Both routine (preventive) and emergency maintenance work should be documented. Maintenance records or job sheets (including when and how the defect/breakdown was repaired, who conducted the work, who has signed off that it was completed, and that appropriate equipment return-to-use procedures were followed) should be completed. Comprehensive maintenance records will assist the food manufacturer to verify that the repairs and maintenance program is working correctly. Where required, maintenance and repair procedures should be adapted. Maintenance records also may provide management with a tool for more intelligent purchasing of new equipment. If the records indicate that a piece of equipment was down often or very expensive to maintain, maybe it is not a brand that one should buy again.
- Regardless of whether maintenance has been carried out in a workshop or within the food production environment, the equipment must be cleaned by the cleaning crew. The cleaning crew must also keep records to document the cleaning and disinfection operations undertaken, as well as any visual, ATP, or microbiological sampling to verify cleanliness.
- Finally, the production department should sign off (e.g., on their daily record sheets) to indicate that they are content to accept the equipment back into production. For that purpose, processors must establish procedures for hand-over.
- Electronic maintenance management systems exist that can support the maintenance staff in documenting their maintenance and repair activities.

8.9 EVALUATION OF THE MAINTENANCE PRACTICES

Maintenance practices should be consistent with Good Maintenance and Manufacturing Practices. It is especially the task of maintenance managers and supervisors to implement and guarantee "Maintenance Best Practice" so as to eliminate the sources of contamination that cause downtime, quality holds and lost profits. The maintenance and quality assurance department must regularly conduct audits to verify if the maintenance staff or contractors have adopted the correct hygiene practices during the preoperational activities and the maintenance operations. Preoperational activities include personal hygiene practices (e.g., garment, footwear, hair protection, washing of hands, etc.), selection of tools, placing barriers, draping of tarps, etc., while maintenance practices include hygiene practices during the maintenance and repair work (e.g., break-in to the equipment while reducing contamination to a minimum, removal of maintenance debris, accountability for equipment parts and maintenance tools).

8.10 CONCLUSION

Food manufacturers who resort to breakdown maintenance instead of preventive maintenance are often not aware of the real cost of an unplanned failure and line shutdown. Not only can less food be produced, but also a workforce of well-paid process operators may be unemployed for many hours. Moreover, contamination of the food produced may already have taken place for a long period, far before the failure occurs. The result is food of inferior quality and risky to consume, which may force the food manufacturer to do costly and painstaking product recalls. The company's reputation may be compromised, with a final result being that the company may be bankrupt. Moreover, breakdown maintenance usually proceeds under pressure in order to reduce costly downtime, and therefore may result in (i) the use of unsuitable materials, (ii) lack of care, or (iii) once-made temporary repairs that tend to be forgotten. In contrary to emergency maintenance, preventive maintenance can provide a food manufacturer with significant cost savings, as it may maintain the high throughput, reduce the amount of low-quality and/or contaminated food, improve the energy efficiency, achieve savings in spare parts and maintenance aids, extend the equipment's life, etc.

However, both during emergency and preventive maintenance, maintenance technicians by the nature of their work run a high risk of contaminating the product, and therefore must show a high level of diligence in the workplace. Correct personal hygiene and maintenance attitudes are required to ensure that the process equipment and utilities, the product area and products are kept free from contamination by undesirable microorganisms, filth, maintenance debris, or machine parts and to comply with the requirements outlined in national and international legislation and regulations, as well as standards. Furthermore, besides legislative bodies, also operators involved in the certification of food operations (e.g., BRC-2015 issue 7, SQF, FSSC 22000) have detected the value of personal hygiene and good maintenance practices during service operations, the result being that preventive maintenance is acknowledged as one of the prerequisite programs in a Hazard

Analysis and Critical Control Points (HACCP) protocol. Therefore, with this book chapter, we aimed to provide food manufacturers and maintenance operators with guidance in the implementation of appropriate hygiene procedures during the maintenance of food processing equipment and utilities.

REFERENCES

Aarnisalo, K., Tallavaara, K., Wirtanen, G., Maijala, R., Raaska, L., 2006. The hygienic working practices of maintenance personnel and equipment hygiene in the Finnish food industry. Food Control 17 (12), 1001–1011.

BRC (2015), Global Standard Food Safety - Issue 7, BRC Global Standards, British Retail Consortium, London, United Kingdom, 118 p.

Campden Food Preservation Research Association (CFPRA) 1983. Hygienic Design of Food Processing Equipment, in Dudley, K. (Ed.), Report prepared by the Working Party on Hygienic Design of the Heat Preserved Foods Panel in conjunction with Campden Research Association, Technical Manual No. 7, Chipping Campden, Gloucestershire, United Kingdom, 93 p.

Eastwood, C.A., Woodall, D.L., Timperley, D.A., Curiel, G.J., Peschel, P., Hauser, G., 1993. Welding stainless steel to meet hygienic requirements, EHEDG Guideline N° 9, 1st ed. EHEDG working group "Design Principles", EHEDG, Frankfurt, Germany, pp. 1–21.

Hauser, G., Curiel, G.J., Bellin, H.-W., Cnossen, H.J., Hofmann, J., Kastelein, J., et al., 2004a. Hygienic Equipment Design Criteria, EHEDG Guideline N° 8, 2nd ed. EHEDG working group "Design Principles", EHEDG, Frankfurt, Germany, pp. 1–16.

Hauser, G., Curiel, G.J., Bellin, H.-W., Cnossen, H.J., Hofmann, J., Kastelein, J., et al., 2004b. Hygienic Design of Open Equipment for Processing of Food, EHEDG Guideline N° 13, 2nd ed. EHEDG working group "Design Principles", EHEDG, Frankfurt, Germany, pp. 1–24.

Higgins, K.T., 2013. "Food Plant 2023", Food Engineering, January magazine, pp. 89–100.

Jha, S.N., 2006. Dairy and Food Processing Plant Maintenance: Theory & Practice, 1st ed International Book Distribution Co., Lucknow, India, p. 140.

Kopitzke, T., Barnickel, M., Gasparetti, M., Merhof, P., Wahlers, J., 2006. Hygienic Welding of Stainless Steel Tubing in the Food Processing Industry, EHEDG Guideline N° 35, 1st ed. EHEDG working group "Welding", EHEDG, Frankfurt, Germany, pp. 1–29.

Lelieveld, H.L.M., Mostert, M.A., Curiel, G.J., 2003. Hygienic Equipment Design. In: Lelieveld, H. L.M., Mostert, M.A., Holah, J., White, B. (Eds.), Hygiene in Food Processing, 1st ed. Woodhead Publishing, Cambridge, United Kingdom, pp. 122–166. Book N°88. (Chapter 8).

Margas, E., Holah, J., 2014. Personal Hygiene in the Food Industry, Ch. 12. In: Lelieveld, H.L.M., Holah, J., Napper, D. (Eds.), Hygiene in Food Processing: principles and practice, 2nd ed. Woodhead Publishing, Cambridge, United Kingdom, pp. 408–440. Book N° 258.

Moerman, F., Partington, E., 2014. Materials of Construction for Food Processing Equipment and Services: Requirements, Strengths and Weaknesses. J. Hyg. Eng. Des. 6, 1–37.

Moerman, F., Holah, J., Steenaard, P., 2014. Hygiene Practice during Maintenance Operations in the Food Industry. In: Lelieveld, H.L.M., Holah, J., Napper, D. (Eds.), Hygiene in Food Processing: principles and practice, 2nd ed. Woodhead Publishing, Cambridge, United Kingdom, pp. 384–407. Book N° 258. (Chapter 11).

Moerman, F., Fikiin, 2015. Guiding Principles for Hygienic Design of Evaporators to Mitigate Contamination-Related Risks in Air Blast Freezing Systems. In: Gaspar, P.D., da Silva, P.D. (Eds.), Handbook of Research on Advances and Applications in Refrigeration Systems and Technologies, 1st ed. IGI Global, Hershey, Pennsylvania, United States, pp. 490–542. (Chapter 14).

Moerman, F., Fikiin, K., 2016. Hygienic Design of Air-Blast Freezing Systems, Ch. 20. In: Lelieveld, H.L.M., Holah, J., Gabrić, D. (Eds.), Handbook of Hygiene Control in the food industry, 2nd ed. Woodhead Publishing/Elsevier, Duxford-Cambridge, United Kingdom, pp. 271–316.

NZFSA (2009), Code of Practice - Processing of Poultry, Part 2: Good Manufacturing Practice, Chapter 3: Hygiene and Sanitation, New Zealand Food Safety Authority, Wellington, New Zealand, 39 p.

NZFSA (2010), Code of Practice - Processing of Poultry, Part 2: Good Manufacturing Practice, Chapter 2: Repairs and Maintenance, New Zealand Food Safety Authority, Wellington, New Zealand, 22 p.

Pekarsky, B., 2007. Eliminating Ferrous Contamination in Critical and Sterile Equipment Maintenance, in-Pharma Technologist.com, William Reed Business Media, Montpellier, France, 8 p.

Smith, D.A., Keeler, L.J. 2007. Maintenance in a Food Manufacturing Facility – Keeping a Sanitary Process Environment during Repairs, Food Processing for Entrepreneurs Series, NebGuide G1815, University of Nebraska – Lincoln Extension, Institute of Agriculture and Natural Resources, 2 p.

Steenaard, P., Bowmer, I., Mets, T., Joosten, F., Mistry, A., van der Vlugt, R., 2009. Part 1: Use of H1 Registered Lubricants, Part 2: Production of H1 Registered Lubricants, EHEDG Guideline N° 23, 1st ed. EHEDG working group "Lubricants", EHEDG, Frankfurt, Germany, pp. 1–23 (part 1), pp. 1–10 (part 2).

Stier, R.F., 2012. Personal Hygiene: A Basic Prerequisite Program for Ensuring Food Safety. Food Saf. Mag. 18 (5), 20–26.

Index

Note: Page numbers followed by "*f*" and "*t*" refer to figures and tables, respectively

G

H

R

FSC
www.fsc.org
MIX
Papier | Fördert gute Waldnutzung
FSC® C083411

Kolibri 360 GmbH
Ferdinand-Jühlke-Straße 7
99095 Erfurt, Deutschland
produktsicherheit@kolibri360.de